Smart Nanomaterials Technology

Series Editors

Azamal Husen, Wolaita Sodo University, Wolaita, Ethiopia

Mohammad Jawaid, Department of Chemical and Petroleum Engineering, United Arab Emirates University, Al Ain, United Arab Emirates

Nanotechnology is a rapidly growing scientific field and has attracted a great interest over the last few years because of its abundant applications in different fields like biology, physics and chemistry. This science deals with the production of minute particles called nanomaterials having dimensions between 1 and 100 nm which may serve as building blocks for various physical and biological systems. On the other hand, there is the class of smart materials where the material that can stimuli by external factors and results a new kind of functional properties. The combination of these two classes forms a new class of smart nanomaterials, which produces unique functional material properties and a great opportunity to larger span of application. Smart nanomaterials have been employed by researchers to use it effectively in agricultural production, soil improvement, disease management, energy and environment, medical science, pharmaceuticals, engineering, food, animal husbandry and forestry sectors.

This book series in Smart Nanomaterials Technology aims to comprehensively cover topics in the fabrication, synthesis and application of these materials for applications in the following fields:

- Energy Systems—Renewable energy, energy storage (supercapacitors and electrochemical cells), hydrogen storage, photocatalytic water splitting for hydrogen production
- Biomedical—controlled release of drugs, treatment of various diseases, biosensors,
- Agricultural—agricultural production, soil improvement, disease management, animal feed, egg, milk and meat production/processing,
- Forestry—wood preservation, protection, disease management
- Environment—wastewater treatment, separation of hazardous contaminants from wastewater, indoor air filters.

Devendra Sillu · Goh Bey Hing · Nadeem Akhtar
Editors

Nanobiosensors for the Food Industry

Springer

Editors
Devendra Sillu
Department of Environmental Science and Engineering
Guangdong Technion-Israel Institute of Technology
Shantou, Guangdong, China

Goh Bey Hing
Department of Biological Sciences
Sunway Biofunctional Molecules Discovery Centre (SBMDC)
Subang Jaya, Selangor, Malaysia

Nadeem Akhtar
CBS Bio Platforms Inc.
Calgary, AB, Canada

ISSN 3004-8273 ISSN 3004-8281 (electronic)
Smart Nanomaterials Technology
ISBN 978-981-95-0135-9 ISBN 978-981-95-0136-6 (eBook)
https://doi.org/10.1007/978-981-95-0136-6

This Springer imprint is published by the registered company Springer Nature Singapore Pte Ltd.
The registered company address is: 152 Beach Road, #21-01/04 Gateway East, Singapore 189721, Singapore

Contents

Nanotechnology in Food Industry: Current Status and Future Trends

Manisha Mohapatra

Abstract Food production, influenced by geographical and topographical factors, faces challenges in meeting demand due to social, technical, and manpower issues. To address this, advancements in food safety, preservation, processing, and packaging are crucial, with nanotechnology playing a transformative role. Nanotechnology has significantly improved food quality, texture, taste, flavor, nutrient bioavailability, and shelf life through nanoencapsulation and nanoprecipitation. These methods enhance nutrient bioavailability, offering better thermal stability, solubility, and long-term preservation while strengthening barrier properties against toxins and pathogens. Innovative nanomaterials in active, improved, and intelligent packaging possess antimicrobial properties and facilitate food pathogen detection. The integration of artificial intelligence with nanosensors for advanced food processing and preservation represents future trends. Nanodelivery mediated nutraceutical and synergistic actions of nanomaterials are anticipated to further protect food products, nanosensors in smart packaging will enable real-time monitoring of food quality. Over the recent decades, nanotechnology has gained attention for its significant impact in food industry, enhancing material properties for agriculture and food. Applications include targeted material delivery, flavor improvement, antibacterial nanoparticles, and consumer awareness. Labeling with nanobarcodes ensures food security and traceability. This review will discuss current advances in nanotechnology and its applications in food-tech, packaging, and security, along with the future opportunities in the field.

Keywords Nanotechnology · Nanobiosensor · Food processing · Smart packaging · Nutraceutical

M. Mohapatra (✉)
Department of Biotechnology, Trident Academy of Creative Technology (TACT), Infocity, Chandrasekharpur, Odisha 751024, India
e-mail: manishamohapatra7@gmail.com

D. Sillu et al. (eds.), *Nanobiosensors for the Food Industry*, Smart Nanomaterials Technology, https://doi.org/10.1007/978-981-95-0136-6_1

1 Introduction

Nanotechnology is relatively recent innovation that have begun to significantly impact the food sector, offering transformative potential in areas ranging from global food production to improving nutritional value, quantity, quality, and safety [1]. It is poised to revolutionize food manufacturing, including aspects such as food protection and the molecular synthesis of new products and ingredients including enhancing the diverse mechanism of food packaging [2]. It is anticipated that nanotechnology will advance genetically modified crops, animal, and fisheries production with precision farming, which is crucial for enhancing crop yields by monitoring climatic factors and applying targeted interventions [3]. The growing need for efficient, sensitive, and cost effective methods to detect food contaminants has driven the development of nanotechnology enabled sensors and systems, enabling rapid and noninvasive contamination detection methods over the past decade.

2 Food Industries: Recent Trend

The global food market is experiencing rapid growth, with revenues projected to reach $9.43 trillion in 2023, and an annual growth rate of 6.21%. The Confectionery and Snacks sector is expected to be the largest, generating $1.64 trillion in 2023. As the food industry evolves, there is an increasing demand for innovations that enhance the convenience, authenticity, and taste of food products, thereby maintaining the sector's competitive edge in processing and distribution [4]. Nanotechnology is emerging as a key innovation with significant applications in the food market. Nanoparticles, for instance, are being utilized to extend the shelf life of food products by preventing spoilage, and are found in various food related applications such as preservatives, antimicrobial sensors, flavoring agents, packaging materials, and food components. These nanomaterials and nanoscale additives help improve the nutritional profile, product longevity, quality, and appeal of food items [1].

Nanotechnology is advancing in production of novel food products thereby expanding in sectors of functional foods, nutraceutical, and bioactive. For instance, nanoencapsulation techniques are used to encase nanosized food components, additives like flavors and colors, and nutritional supplements such as proteins and antioxidants, enhancing their solubility, masking undesirable tastes, and providing protective barriers with controlled release. Moreover, nanotechnology holds promise in addressing food related health issues, including obesity and diabetes, by enabling the design of precise nutritional diets tailored to specific populations and lifestyles. The development of smart systems for nanoencapsulated, regulated nutrient release, along with nanoscale enzymatic reactors, could lead to the creation of innovative food products through nutrient fortification [5]. Additionally, nanomaterials are being explored for use in food packaging, offering benefits such as improved mechanical strength, microbial contamination detection, and enhanced nutrient bioavailability.

This represents one of the most prevalent applications of nanotechnology in the food industry, offering a promising avenue for enhancing food safety and quality through advanced packaging solutions [6].

Nanotechnology, often referred to as the new industrial revolution, is witnessing substantial investments from both developed and developing nations as they vie for a competitive edge in this emerging field. The United States currently leads with a significant investment of 3.7 billion USD over four years through its National Nanotechnology Initiative, followed by Japan and the European Union, which allocate 750 million and 1.2 billion USD per year, respectively, including contributions from individual countries [7]. Other countries including India are also making strides, focusing on applications tailored to their specific economic needs. In food processing, nanotechnology is being utilized in various ways, including the integration of nutraceuticals, viscosifying agents, nutrient propagation, mineral and vitamin fortification, and the nanoencapsulation of flavors [4]. This technology's influence extends to aspects like food safety and molecular synthesis, with nanosensors being actively researched for their potential to enhance pathogen detection in food systems. Nanofoods products grown, processed, or packaged using nanotechnology—represent a significant development in the food industry, promising improvements in both quality and safety [5].

3 Nanotechnology as a Novel Approach for Food Security

Nanotechnology has introduced innovative techniques in food processing, enhancing physicochemical qualities and increasing nutrient stability and bioavailability [8]. Due to their unique properties, such as larger surface area, high reactivity, and tiny particle size, nanoparticles are widely used across various industries, including food and agriculture [9]. The modernization of these industries parallels advancements in drug delivery and pharmaceuticals [10]. Nanomaterials, with their distinctive physicochemical and biological characteristics, have driven scientific innovation, leading to improvements in texture, taste, stability, and other food quality attributes [11]. Applications of nanotechnology span the entire food sector, from agriculture and processing to safety, packaging, and nutrient delivery [5]. Research has explored the feasibility of nanoparticles in food quality assurance, packaging, and the development of functional and nutritious food products [2, 12, 13]. Additionally, nanotechnology aids in addressing food related diseases, designing nutrition regimens, and ensuring sustainable food production through nanoencapsulation and smart nutrient delivery systems and many more as depicted in Fig. 1.

Nanotechnology offers promising solutions for enhancing global food security by increasing production and reducing waste. This technology can improve agricultural practices by minimizing nutrient losses in fertilizers, enhancing pest control, and boosting crop yields. Additionally, nanotechnology aids in food processing, enabling the creation of novel food products with improved taste and flavor profiles.

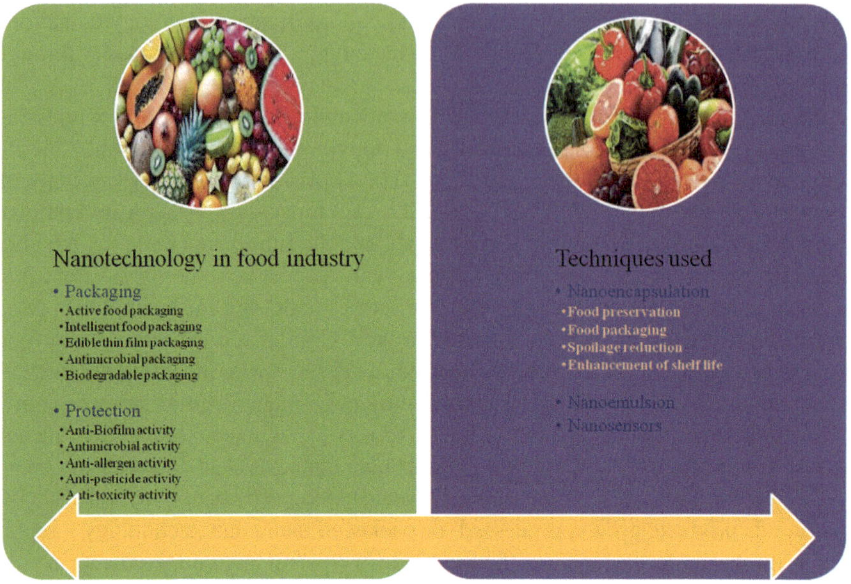

Fig. 1 Nanotechnology as a novel approach for food security

It also plays a crucial role in extending the shelf life of food and developing nano-based techniques for detecting and preventing spoilage, further reducing food waste [14]. It is revolutionizing the entire food chain, from production to consumption, by enhancing the thermal stability, water solubility, and oral bioavailability of nutrients [15]. Nanoparticles, ranging in size up to 100 nm, are the key in delivering edibles due to their unique properties, such as high surface-to-volume ratio, solubility, and strength [16]. These characteristics make nanoparticles valuable in fields like drug delivery, tissue engineering, and food safety [17, 18]. In food technology, nanoparticles improve the solubility of micronutrients and address challenges associated with incorporating nutraceuticals into foods, such as chemical instability and poor solubility [19]. Encapsulation in nanoparticle-based delivery systems offers a promising solution for enhancing bioavailability, though it requires precise methods tailored to each micronutrient's properties [20]. As global demand for nutraceuticals rises, driven by increasing health awareness and chronic disease rates, nanotechnology is seen as a key tool in overcoming these challenges [21, 22].

4 Nanoparticles in Food Packaging

The evolution of food packaging, from natural materials like leaves and animal skins to ceramic and glass containers, has been essential in protecting food from spoilage and contamination [23]. Traditional packaging has primarily served to contain, transport, and protect food, but it is inert, offering no interaction with the contents. However, with the growth of the global food industry and the demand for fresher, more nutritious foods, traditional packaging has become inadequate [6]. The main challenge with current food packaging materials is their permeability, meaning they do not fully block atmospheric gases or water vapor. Organic polymers like polypropylene, polyethylene, and others are widely used due to their low cost, ease of processing, and lightweight nature. However, their primary drawback is their natural permeability, which allows gases and small molecules to pass through, potentially compromising food quality and shelf life. This limitation remains a significant concern in the food packaging industry [24]. This has led to the development of active and intelligent packaging, which enhances food quality, safety, and traceability, improving both consumer experience and supply chain efficiency. In addition, nanotechnology is playing a crucial role in addressing global food security by increasing production, reducing waste, and enhancing food processing and preservation through advanced methods like nutrient management and nano-based spoilage detection [25, 26].

Nanotechnology plays a crucial role in food preservation, additives, and packaging, utilizing nanoparticles and nanoscale additives to enhance nutritional content, extends shelf life, and improves product attributes such as aroma and texture [2, 27]. Nanoparticles like Cu/CuO, Ag, MgO, TiO_2, ZnO, carbon dots, and graphene are widely employed for their antimicrobial properties in food packaging (Table 1), offering advantages over traditional methods, including improved mechanical strength, heat resistance, and biodegradability [9]. Nanomaterials, particularly in antimicrobial packaging, inhibit the growth of microorganisms on food surfaces by releasing active substances, thus maintaining quality and extending shelf life [28]. Additionally, nanosensors in packaging can detect food spoilage, further ensuring food safety and freshness by leveraging the enhanced antibacterial properties of these materials [29, 30]. Incorporating antimicrobial nanoparticles into active packaging also strengthens it, reduces oxygen permeability, and makes it lighter, contributing to better food protection [9, 31].

4.1 Smart Packaging

Smart packaging systems primarily use nanosensors, often developed with nanoparticles, to detect food contaminants and monitor changes during food processing and preservation [32]. These nanoparticle-based nanosensors are effective in tracing chemical, physical, and biological modifications, helping to detect chemicals, toxins,

Table 1 Types of nanostructures for food packaging

Sl no.	Nanostructure name	Nanostructure type	Applications
1	Chitosan	Nanofibers	Antimicrobial activity
2	Cellulose and lignin	Cellulose and lignin nanostructures	Antibacterial activity
3	Starch	Polymer-based nanofilms	Antioxidant activity
4	Chitosan with polycaprolactone	Nanofibers	Intelligent packaging
5	Titanium dioxide and copper oxide	Nanocomposite films	Antibacterial activity
6	Lignocellulose and wheat gluten	Nanofibers	Antibacterial activity water resistant activity

and pathogens in food products [33]. In addition to ensuring food safety, smart packaging with nanosensors and indicators tracks the quality of packaged foods during transport and storage, recording responses to environmental stimuli [34]. Commonly used indicators measure packaging integrity, freshness, and time or temperature dependent changes, thereby helping to maintain food quality and extend shelf life throughout the supply chain. Moreover, nanoparticle-based nano-barcodes can serve as identity tags in smart packaging [35]. Nano-based packaging can be developed by incorporating nanoparticles into traditional packaging materials or by fabricating nanocomposite multilayer packaging and coatings through methods like immersion and spraying [36]. This technology facilitates the creation of improved packaging with enhanced physical and mechanical properties, active packaging with antibacterial and antioxidative features, and smart packaging that monitors food conditions [37].

4.2 Active Packaging

Nanotechnology enhances food packaging by improving the properties of polymer matrices, such as elasticity, gas barrier efficiency, and stability against temperature and moisture fluctuations. Active packaging incorporates nanotechnology to interact with environmental factors, helping to preserve food quality and extend shelf life through mechanisms triggered by intrinsic or extrinsic elements [24, 38]. It differs from conventional packaging by intentionally incorporating components that either release substances, like antimicrobial or antioxidant agents, or absorb materials, such as oxygen or water vapor, to protect the food. By combining active compounds like antimicrobial agents, preservatives, and absorbers with polymers, this packaging effectively enhances the shelf life and quality of food products [39]. Active packaging systems interact directly with food to enhance safety and extend shelf life by incorporating antibacterial properties that protect against microbial contamination.

These packaging films also serve as UV blockers, preventing oxidation caused by UV exposure. This combination of features helps to reduce spoilage and maintain the quality of food products [40, 41].

4.3 Intelligent Packaging

Intelligent packaging is defined by the European Commission (2004) as materials and articles that monitor the condition of packaged food or the environment surrounding the food [42]. This packaging system integrates advanced devices such as barcodes, Radio Frequency Identification Tags, sensors, and indicators to communicate, monitor, sense, record, track, and provide information on food safety, quality, and history throughout the supply chain. Among the innovations in this field, nanosensors have emerged as one of the most significant outputs of nanotechnology. These sensors are particularly effective in monitoring the external and internal conditions of both fresh and processed food products [43]. Intelligent packaging uses nanotechnology-based indicators or sensors to provide real time feedback on the food's quality, offering additional protection against fraud and enhancing communication and marketing efforts [44]. These innovations significantly improve the functionality, safety, and effectiveness of food packaging.

A colorimetric hydrogen sulfide sensor made of gellan gum-capped silver nanoparticles was developed, which detects spoilage in chicken breast and silver carp by changing color from yellow to colorless upon exposure to hydrogen sulfide, with a detection limit of 0.18 μM at pH 7 [45]. Other advancements include a soluble soybean polysaccharide film containing curcumin and SiO_2 was created to indicate shrimp spoilage through distinct color changes due to pH fluctuations during spoilage. This film also exhibits strong antibacterial activity, highlighting its potential for active and intelligent packaging applications [46]. These developments in intelligent packaging are instrumental in enhancing food safety, reducing waste, and improving overall consumer confidence in food products. By providing dynamic feedback on the condition of the food, intelligent packaging not only enhances the consumer experience but also supports a more transparent and trustworthy food supply chain.

4.4 Thin Film Packaging

Edible thin film packaging, made from materials such as chitosan, carrageenan, polylactic acid, gelatin, and various starch blends, extends the freshness and quality of perishable foods by preventing oxidation and acting as a barrier against gases like ethene and oxygen. These films are versatile, protecting a range of products from fruits and vegetables to meats and baked goods [34]. Edible nanolaminates, which can be created using polysaccharides, proteins, or lipids, offer enhanced protective properties by blocking moisture, gases, and off-flavors, though they vary in effectiveness based on their composition. Polysaccharide and protein films are good at blocking gases but less effective against moisture, while lipid films excel in moisture control but lack mechanical strength and gas resistance. Combining functional compounds such as antimicrobials and antioxidants with these films can further enhance their protective qualities [32, 40].

4.5 Nanoencapsulation Packaging

Nanoencapsules, with their core-shell architecture, offer numerous benefits for food storage and supplementation through nanoencapsulation. These vesicular systems enhance bioavailability, stability, and controlled release of active ingredients while protecting them from oxidation and improving sensory attributes [47]. Nanocapsules, which can enclose agents in liquid, solid, or molecular dispersion forms, facilitate the delivery of essential nutrients and the trapping of undesirable substances in food. Various methods for manufacturing nanocapsules include emulsion-diffusion, nanoprecipitation, layer by layer, double emulsification, and polymer coating [48]. Nanoliposomes, a type of nano-based carrier, enable targeted delivery of bioactive compounds like minerals and vitamins, demonstrating superior solubility and specificity compared to traditional systems. This technology shows promise in enhancing the nutritional value and safety of food products through innovative encapsulation approaches [6].

5 Nanoemulsions in Food Industries

A nanoemulsion is a type of nanomaterial composed of tiny droplets, typically with a diameter of 500 nanometers or less, dispersed within another immiscible liquid, like oil in water or water in oil. Unlike regular emulsions, nanoemulsions have much smaller droplet sizes, which give them unique properties, such as improved stability and increased surface area. Due to their small size and high stability, nanoemulsions are used in various applications, including food, pharmaceuticals, and cosmetics, where they help deliver active ingredients more effectively [49]. Depending on the dispersion, they are classified as oil in water or water in oil emulsions. In the food industry, nanoemulsions are utilized for delivering hydrophobic substances, including nutraceuticals, colorants, flavor agents, and antimicrobials. They also play a role in creating biodegradable coatings and packaging films to enhance food quality and extend shelf life. Nanoclays, composed of layered mineral silicates, are categorized into several types, such as montmorillonite and bentonite, based on their shape and composition [50]. These cost effective and eco-friendly nanomaterials are particularly valued in food packaging due to their natural nanolayer structure, which provides effective barriers against gas and moisture [51]. In one study, curcumin dispersed with colloidal nanoparticles exhibited higher absorption efficacy than natural curcumin powder [52]. Another study showed that the color degradation rate of beta-carotene in nanoemulsions slowed in the presence of a chemical inhibitor, likely due to its ability to chelate transition metals, which typically promote oxidation [53]. Furthermore, the oral bioavailability of curcumin can be increased by nine times when combined with piperine, an absorption enhancer [54].

Nanoemulsions offer several advantages over traditional emulsions, particularly in food and beverage applications, due to their small droplet size which ensures high optical clarity, improved physical stability, and better bioavailability of encapsulated ingredients. They can be produced through various low energy or high energy methods, and their encapsulation capabilities enhance the delivery of active components, such as vitamins and omega-3 fatty acids, while also facilitating controlled release and taste masking [4]. Companies like Nestle and Unilever have utilized nanoemulsions to create low fat ice creams, showcasing their potential in reducing fat content while maintaining product quality. Despite these benefits, challenges such as the potential adverse effects of high lipid content, the need for cosolvents or cosurfactants in production, and public concerns about nanoparticle safety need to be addressed to fully capitalize on the advantages of nanoemulsions in food systems [49, 52].

6 Nanosenors in Food Industries

Nanosensors and nanobiosensors, including electrochemical, optical, nanoparticle based, and quantum dot sensors, have emerged as powerful tools for detecting a wide range of contaminants and adulterants in food. These advanced sensors offer superior sensing capabilities compared to traditional methods, enabling the precise detection of chemical contaminants such as pesticides and heavy metals, biological threats like toxins and pathogens, and food adulterants. They also enhance food packaging by improving thermal and mechanical properties, monitoring freshness, and ensuring traceability. Current research highlights the diverse applications of these nanosensors in improving food safety and quality, underscoring their role in the ongoing evolution of food safety technologies [55].

Nanosensors have facilitated the development of various indicators, such as time temperature indicators, pH indicators, and leak indicators, which provide real-time feedback on food quality and storage conditions. For example, freshness indicators visually represent color changes to inform consumers whether a product is still safe to consume, while time temperature sensors track the duration and exposure of food to certain conditions after packaging is opened [56]. The freshness of packed food is compromised by oxygen exposure, which accelerates oxidation and bacterial growth. To address this, colorimetric oxygen indicators have been developed using methylene blue/titanium dioxide hybrid nanocomposites and luminescent metal ligand complex functionalized poly-(styrene-block-vinylpyrrolidone) nanobeads. Additionally, trimethylamine, a byproduct of protein metabolism indicating loss of freshness, has been detected using tin dioxide–zinc oxide nanocomposite sensors in fish samples, as well as polyvinylpyrrolidone-capped zinc oxide nanoparticles and branched iron oxide–titanium dioxide heteronanostructures [57]. For meat products, degradation of xanthine and hypoxanthine releases hydrogen peroxide, which can be detected using gold nanoparticles [58]. Types of nanosensors employed include gas detectors, time–temperature indicators, and array biosensors, among others. Additionally, nanosensors have been developed for detecting organophosphates in agriculture due to their high sensitivity and selectivity, crucial for managing pesticide residues. Recent advancements include magnetic immune sensors with gold nanoparticles for broad food applications and SPR based biosensors for rapid pathogen identification. These technologies are also applied to enhance agricultural productivity and safety by improving pathogen detection and optimizing agrochemical use [59].

Nanosensors integrated with polymers are revolutionizing food safety and quality monitoring through smart packaging solutions by detecting pathogens, chemicals, and environmental conditions during storage and transit. These sensors, including nanogas sensors, nanosmart dust, and nanobarcodes, offer enhanced capabilities such as monitoring temperature, humidity, and gas composition, as well as identifying metabolites and byproducts from microbial growth and food degradation [2]. Electrochemical glucose biosensors, nanofabricated using layer by layer self-assembly, exemplify the precision and efficiency of nanosensors in food analysis. Nanosensors, including carbon nanotube-based sensors, electronic tongues, and microfluidic devices, provide rapid, sensitive, and specific detection, improving on traditional methods by offering speed, reduced cost, and simplicity in both online and offline applications [13, 59]. These advancements make nanosensors a powerful tool in ensuring food safety and verifying product authenticity. Nanosensors are increasingly utilized to assess the quality of various foods, such as wine, coffee, juice, and milk, through advanced technologies like layer by layer macromolecule films, which offer greater sensitivity than the human tongue. These sensors, which can be integrated into food packaging, allow for real-time detection of contaminants and quality indicators without requiring laboratory analysis; they often utilize color changes for easy consumer interpretation [4].

7 Nanotechnology in Food Protection

Nanotechnology enhances food packaging by incorporating organic, inorganic, and hybrid nanoparticles into polymer matrices to improve vapor and gas barrier properties, antimicrobial activity, and smart packaging capabilities. These advancements contribute to better food quality and safety. However, nanoparticles exhibit unique properties compared to bulk materials, such as a larger surface area and potential migration into food, which could pose health risks. Consequently, while nanotechnology offers significant benefits, its potential hazards must be carefully managed [60].

7.1 *Stress Tolerance*

Salinity is a significant abiotic stress that negatively impacts plant growth, development, and yield, including in rapeseed. Studies have shown that selenium and ZnO nanoparticles, synthesized using plant hormones, can mitigate salt stress by reducing abscisic acid levels and regulating gibberellic acid, thereby promoting early seedling growth and germination in rapeseed plants [61]. Additionally, carbon quantum dots functionalized with proline have been found to improve various physiological parameters in grape vine under salt stress, such as increasing leaf weight, enhancing antioxidant enzyme activities, and reducing oxidative damage [62]. In *Brassica napus*, yttrium-doped γ Fe_2O_3 NPs in nutrient solutions helped manage drought stress by reducing hydrogen peroxide and lipid peroxidation, increasing chlorophyll content, and improving overall plant growth, suggesting their potential as effective fertilizers [63]. Moreover, nanotechnology has been applied to support arbuscular mycorrhizal fungi activity under salinity stress, with TiO_2 nanoparticles showing positive effects on arbuscular mycorrhizal fungi in bean roots, though the benefits were limited to arbuscular mycorrhizal fungi rather than the plants themselves [64].

7.2 *Enhancing Shelf Life*

Food packaging is essential for maintaining the nutritional and sensory qualities of food while protecting it from spoilage and adverse environmental conditions. Nanotechnology has significantly enhanced packaging by improving barrier properties, thermal stability, strength, and durability, thus extending the shelf life of food products [61]. Nanotechnology has significantly enhanced the taste, absorption, and bioavailability of nutraceuticals and additives through the use of nanosized components such as antimicrobials, nutrients, antioxidants, and preservatives [36]. For instance, vegetables and fruits coated with nanosilver maintain their activity during transit and storage due to altered respiration mechanisms. Nanoencapsulation, which involves packaging materials at the nanoscale using nanocapsules, provides benefits like extended shelf life, improved stability, and controlled release of active components [60]. This technique also enables the inclusion of various functional components, including antioxidants, vitamins, and probiotics, enhancing their stability and effectiveness [65]. Silver nanoparticles are used in food packaging for their antimicrobial properties, which extend food shelf life by preventing microbial growth. They are effective against both gram positive and gram negative bacteria and also possess antioxidant properties that enhance food safety. To optimize their use, there is a focus on developing cost effective and eco-friendly synthesis methods. These methods ensure that silver nanoparticles are both effective and sustainable in food packaging applications [66]. Functional nanoparticles enhance food packaging by improving barrier properties, thermal stability, and strength, and durability, thus extending shelf life. They can be integrated into traditional packaging materials like films and containers or used in nanocomposite multilayer packaging and coatings. Nanotechnology enables the creation of advanced packaging with improved physical properties, active features such as antibacterial and UV protection, and smart capabilities for monitoring food conditions [37].

7.3 Detection of Pathogens

Nanoparticle-assisted methods for detecting pathogenic bacteria in food are more efficient and sensitive compared to traditional techniques. Magnetic iron oxide nanoparticles, for instance, facilitate the isolation of DNA from bacteria like *Listeria monocytogenes* in milk, with subsequent quantification via polymerase chain reaction [67]. While PCR-based assays can be expensive and the 16S ribosomal RNA-based microarray methods often lack sensitivity, nanoparticle-based detection of 16S rRNA offers an easier and more sensitive alternative. Various nanoparticles are now employed for detecting pathogenic bacteria in both standard and complex food samples [68]. For example, Chitosan, derived from chitin, is an environmentally friendly and nontoxic polysaccharide with strong antimicrobial properties, making it ideal for use in packaging. When incorporated into biodegradable polymers like polylactic acid films, chitosan enhances gas and moisture barrier properties. Chitosan nanoparticles improve the physical, mechanical, and barrier qualities of biocomposite films, while their antimicrobial effects stem from chelation and ionic interactions with bacterial cell surfaces, causing cell death. Additionally, their hydrophobic nature and bonding with biopolymers reduce moisture permeability [60].

8 Nanotechnology in Food Industries: Current Challenges and Future Scope

Despite significant advances in applying nanotechnology to the food sector, progress in understanding nanostructures and their implications remains limited. Nanotechnology holds promise for developing novel food products and processes, but there are substantial barriers, particularly concerning the safety of nanotechnology based food items for human consumption [69]. The primary concern is ensuring that these products are safe before they reach the market, as there are no standardized guidelines for assessing the safety of nanomaterials in foodstuffs. Issues include the migration of nanoparticles from packaging into food and their potential health risks, which are complicated by the unique behaviors of materials at the nanoscale [70]. Further research is needed to understand these risks better and to develop cost effective, safe edible delivery systems. The migration and toxicity of nanoparticles are complex due to their varied physicochemical properties, including morphology, composition, and surface characteristics [20]. Nanoparticles can cross biological barriers, penetrate cells, and potentially cause oxidative stress, protein denaturation, and DNA damage [70]. Additionally, the environmental impact of nanoparticle synthesis and the potential for harmful by products complicate safety evaluations. Comprehensive risk assessments, biosafety evaluations, and regulatory frameworks are essential to address these concerns [65].

9 Conclusion

Nanotechnology has made significant strides in the food industry, revolutionizing areas such as packaging, safety, and nutritional enhancement. The integration of nanoparticles into food packaging materials has led to improved barrier properties, enhanced mechanical strength, and extended shelf life of food products. Active and intelligent packaging systems, empowered by nanotechnology, offer advanced functionalities such as antimicrobial properties, real-time freshness monitoring, and precise detection of spoilage indicators. These innovations not only preserve food quality but also ensure safety by detecting contaminants and providing critical information to consumers.

Looking ahead, the future of nanotechnology in the food industry holds tremendous potential but also faces challenges. Continued research is needed to address safety concerns, particularly regarding the migration of nanoparticles into food and their potential health impacts. Developing standardized guidelines and regulations for the use of nanotechnology in food products will be crucial in ensuring consumer safety and gaining public trust. As technological advancements progress, the focus will likely to shift toward creating more sustainable and eco-friendly nanomaterials, while also exploring novel applications that enhance both the quality and safety of food.

References

1. Chung IM, Rajakumar G, Gomathi T, Park SK, Kim SH, Thiruvengadam M (2017) Nanotechnology for human food: advances and perspective. Front Life Sci 10(1):63–72. https://doi.org/10.1080/21553769.2017.1365775
2. Pathakoti K, Manubolu M, Hwang H (2017) Nanostructures: current uses and future applications in food science. J Food Drug Anal 25(2):245–253. https://doi.org/10.1016/j.jfda.2017.02.004
3. Chen H, Yada R (2011) Nanotechnologies in agriculture: new tools for sustainable development. Trends Food Sci Technol 22:585–594. https://doi.org/10.1016/j.tifs.2011.09.004

4. Thiruvengadam M, Rajakumar G, Chung IM (2018) Nanotechnology: current uses and future applications in the food industry. Biotech 8(1):74. https://doi.org/10.1007/s13205-018-1104-7
5. Sahoo M, Vishwakarma S, Panigrahi C, Kumar J (2021) Nanotechnology: current applications and future scope in food. Food Front 2(1):3–22
6. Singh R, Dutt S, Sharma P, Sundramoorthy AK, Dubey A, Singh A, Arya S (2023) Future of nanotechnology in food industry: challenges in processing, packaging, and food safety. Global Challen 7(4):2200209. https://doi.org/10.1002/gch2.202200209
7. Kour H, Malik AA, Ahmad N, (2015) Nanotechnology-new lifeline for the food industry. Crit Rev Food Sci Nutr. 5. https://doi.org/10.1080/10408398.2013.802662
8. Patra JK, Shin HS, Paramithiotis S (2018) Application of nanotechnology in food science and food microbiology. Front Microbiol 9:714
9. Omerovic N, Djisalov M, Živojevic K, Mladenovic M, Vunduk J, Milenkovic I, Knezevic NZ, Gadjanski I, Vidic J (2021) Antimicrobial nanoparticles and biodegradable polymer composites for active food packaging applications. Compr Rev Food Sci Food Saf 20(3):2428–2454
10. López-Rubio A, Fabra MJ, Martínez-Sanz M (2019) Multidisciplinary digital publishing institute, Food packaging based on nanomaterials
11. Anandharamakrishnan C, Parthasarathi S (2019) CRC Press; food nanotechnology: principles and applications
12. Ariyarathna IR, Rajakaruna R, Karunaratne DN (2017) The rise of inorganic nanomaterial implementation in food applications. Food Control 77:251–259
13. Singh T, Shukla S, Kumar P, Wahla V, Bajpai VK, Rather IA (2017) Application of nanotechnology in food science: perception and overview. Front Microbiol 8:1501
14. Rizvi SS, Moraru CI, Bouwmeester H, Kampers FW, Cheng Y (2022) Ensuring global food safety. Academic Press; Cambridge, MA, USA, Nanotechnology and food safety, pp 325–340
15. Ndlovu N, Mayaya T, Muitire C, Munyengwa N (2020) Nanotechnology applications in crop production and food systems. Int J Plant Breeding Crop Sci 7(1):624–634
16. Priyanka S (2018) Nanotechnology in food preservation. Food Sci Res J 9(2):441–447. https://doi.org/10.15740/HAS/FSRJ/9.2/441-447
17. Jain KK (2020) Role of nanobiotechnology in drug delivery. In: Jain K (eds) Drug delivery systems, methods in molecular biology, pp 55–73. https://doi.org/10.1007/978-1-4939-9798-5-2
18. Sajid A, Manzoor Q, Sajid A, Imran M, Khalid S, Arshad Z, Saleem A, Haq IU, Aslam F (2020) Characterization and antibacterial properties of Eriobotrya japonica extract loaded silver-nanoparticles. Curr Bioact Compd. https://doi.org/10.2174/15734072169992012082 03418
19. McClements DJ (2015) Nanoscale nutrient delivery systems for food applications: improving bioactive dispersibility, stability, and bioavailability. J Food Sci 80(7):N1602–N1611. https://doi.org/10.1111/1750-3841.12919
20. McClements DJ, Xiao H (2017) Is nano safe in foods? establishing the factors impacting the gastrointestinal fate and toxicity of organic and inorganic food-grade nanoparticles. NPJ Sci Food 1(1):1–13
21. Prasad E (2016) Nutraceuticals market-global opportunity analysis and industry forecast, 2014–2022. https://www.alliedmarketresearch.com/nutraceuticals-market
22. Arshad R, Gulshad L, Haq IU, Farooq MA, Al-Farga A, Siddique R, Manzoor MF, Karrar E (2021) Nanotechnology: a novel tool to enhance the bioavailability of micronutrients. Food sci Nutr 9(6):3354–3361. https://doi.org/10.1002/fsn3.2311
23. Saha NC (2022) Food packaging: concepts and its significance. In: Food packaging. Springer, Singapore, pp 1–45
24. Sharma C, Dhiman R, Rokana N, Panwar H (2017) Nanotechnology: an untapped resource for food packaging. Front Microbiol 8:1735. https://doi.org/10.3389/fmicb.2017.01735
25. Aswathi VP, Meera S, Maria CGA, Nidhin M (2022) Green synthesis of nanoparticles from biodegradable waste extracts and their applications: a critical review. Nanotechnol Environ Eng 8:377–397

26. de Sousa MS, Schlogl AE, Estanislau FR, Souza VGL, dos Reis Coimbra JS, Santos IJB (2023) Nanotechnology in packaging for food industry: past, present, and future. Coatings 13:1411. https://doi.org/10.3390/coatings13081411
27. Stoma M, Dudziak A (2022) Eastern Poland consumer awareness of innovative active and intelligent packaging in the food industry: exploratory studies. Sustainability 14:13691
28. Drago E, Campardelli R, Pettinato M, Perego P (2020) Innovations in smart packaging concepts for food: an extensive review. Foods 9(11):1628
29. Hoseinnejad M, Jafari SM, Katouzian I (2018) Inorganic and metal nanoparticles and their antimicrobial activity in food packaging applications. Crit Rev Microbiol 44(2):161–181
30. Fadiji AE, Mthiyane DMN, Onwudiwe DC, Babalola OO (2022) Harnessing the known and unknown impact of nanotechnology on enhancing food security and reducing postharvest losses: constraints and future prospects. Agronomy 12:1657
31. Sharma R, Jafari SM, Sharma S (2020) Antimicrobial bio-nanocomposites and their potential applications in food packaging. Food Control, 112
32. Nile SH, Baskar V, Selvaraj D, Nile A, Xiao J, Kai G (2020) Nanotechnologies in food science: applications, recent trends, and future perspectives. Nano-Micro Lett 12:1–34
33. Berekaa MM (2015) Nanotechnology in food industry; advances in food processing, packaging and food safety. Int J Current Microbiol Applied Sci 4(5):345–357
34. Dias MV, Nilda de Fátima FS, Borges SV, de Sousa MM, Nunes CA, de Oliveira IRN, Medeiros EAA (2013) Use of allyl isothiocyanate and carbon nanotubes in an antimicrobial film to package shredded, cooked chicken meat. Food Chem 141(3):3160–3166
35. Mohammad ZH, Ahmad F, Ibrahim SA (2022) Application of nanotechnology in different aspects of the food industry. Discov Food 2:12. https://doi.org/10.1007/s44187-022-00013-9
36. Shukla P, Chaurasia P, Younis K, Qadri OS, Faridi SA, Srivastava G (2019) Nanotechnology in sustainable agriculture: studies from seed priming to postharvest management, Springer Science and Business Media Deutschland GmbH. Nanotechnol Environ Eng 4(1)
37. Primoˇziˇc M, Knez Z, Leitgeb M (2021) (Bio) nanotechnology in food science—food packaging. MDPI AG Nanomater 11(2):1–31. https://doi.org/10.3390/nano11020292
38. Lim LT (2011) Active and intelligent packaging materials. comprehensive biotechnology. Elsevier, Amsterdam
39. Majid I, Nayik GA, Dar MS, Nanda V (2016) Novel food packaging technologies: innovations and future prospective. J Saudi Society Agric Sci. https://doi.org/10.1016/j.jssas.2016.11.003
40. Roy S, Rhim JW (2020) Carboxymethyl cellulose-based antioxidant and antimicrobial active packaging film incorporated with curcumin and zinc oxide. Int J Biol Macromol 148:666–676
41. Alves J, Gaspar PD, Lima TM, Silva PD (2023) What is the role of active packaging in the future of food sustainability? a systematic review. J Sci Food Agric 103:1004–1020
42. Ghaani M, Cozzolino CA, Castelli G, Farris S (2016) An overview of the intelligent packaging technologies in the food sector. Trends in Food Sci Technol 51:1–11. https://doi.org/10.1016/j.tifs.2016.02.008.May01
43. Kalpana S, Priyadarshini SR, Leena MM, Moses JA, Anandharamakrishnan C (2019) Intelligent packaging: trends and applications in food systems. Trends Food Sci Technol 93:145–157. https://doi.org/10.1016/j.tifs.2019.09.008
44. Mihindukulasuriya SDF, Lim LT (2014) Nanotechnology development in food packaging: a review. Trends Food Sci Tech 40:149–167. https://doi.org/10.1016/j.tifs.2014.09.009
45. Zhai X et al (2019) A colorimetric hydrogen sulfide sensor based on gellan gum-silver nanoparticles bionanocomposite for monitoring of meat spoilage in intelligent packaging. Food Chem 290:135–143. https://doi.org/10.1016/j.foodchem.2019.03.138
46. Salarbashi D, Tafaghodi M, Bazzaz BSF, Aboutorabzade SM, Fathi M (2021) PH-sensitive soluble soybean polysaccharide/SiO_2 incorporated with curcumin for intelligent packaging applications. Food Sci Nutr 9(4):2169–2179. https://doi.org/10.1002/fsn3.2187
47. Ansari MA (2023) Nanotechnology in food and plant science: challenges and future prospects. Plants 12:2565. https://doi.org/10.3390/plants12132565
48. Godwin HA, Chopra K, Bradley KA, Cohen Y, Harthorn BH, Hoek EM, Nel AE (2009) The University of California center for the environmental implications of nanotechnology

49. Grumezescu AM, Holban AM (2018) Impact of nanoscience in the food industry, vol 12, Academic Press
50. Aswathanarayan JB, Vittal RR (2019) Nanoemulsions and their potential applications in food industry. Front Sustain Food Syst 3:95
51. Guo F, Aryana S, Han Y, Jiao Y (2018) A review of the synthesis and applications of polymer–nanoclay composites. Appl Sci 8(9):1696
52. Sasaki H, Sunagawa Y, Takahashi K, Imaizumi A, Fukudu H, Hashimoto T, Wada H, Katanasaka Y, Kakeya H, Fujita M, Hasegawa K, Morimoto T (2011) Innovative preparation of curcumin for improved oral bioavailability. Biolog Pharmacal Bull 34(5):660–665. https://doi.org/10.1248/bpb.34.660
53. Qian C, Decker EA, Xiao H, McClements DJ (2012) Inhibition of beta-carotene degradation in oil-in-water nanoemulsions: influence of oil-soluble and water-soluble antioxidants. Food Chem 135(3):1036–1043. https://doi.org/10.1080/10408398.2016.1223599
54. Haq IU, Imran M, Nadeem M, Tufail T, Tanweer AG, Mubarak MS (2020) Piperine: a review of its biological effects. Phytother Res 35(2):680–700. https://doi.org/10.1002/ptr.6855
55. Toydemir G et al (2020) Nanosensors for Foods. In: Hebbar U, Ranjan S, Dasgupta N, Mishra RK (eds) Nano-food engineering. food engineering series. Springer, Cham. https://doi.org/10.1007/978-3-030-44552-2_12
56. Caon T, Martelli SM, Fakhouri FM (2017) New trends in the food industry: application of nanosensors in food packaging. In: Nano biosensors, Elsevier, pp 773–804
57. Lou Z, Li F, Deng J, Wang L, Zhang T (2013) Branch-like hierarchical heterostructure (α-Fe_2O_3/TiO_2): a novel sensing material for trimethylamine gas sensor. ACS Appl Mater Interfaces 5(23):12310–12316
58. Kumar V, Guleria P, Mehta SK (2017) Nanosensors for food quality and safety assessment. Environ Chem Lett 15:65–177. https://doi.org/10.1007/s10311-017-0616-4
59. Vidotti M, Carvalhal RF, Mendes RK (2011) Biosensors based on gold nanostructures. J Braz Chem Soc 22:3–20. https://doi.org/10.1590/S0103-50532011000100002
60. Alweera A, Nazia K, Samra F, Zayeema A, Kaiser Y (2022) Application of nanotechnology in food packaging: pros and cons. J Agricult Food Res 7:100270
61. El-Badri AM, Batool M, Wang C, Hashem AM, Tabl KM, Nishawy E, Kuai J, Zhou G, Wang B (2021) Selenium and zinc oxide nanoparticles modulate the molecular and morpho-physiological processes during seed germination of Brassica napus under salt stress. Ecotoxicol Environ Saf 225:112695. https://doi.org/10.1016/j.ecoenv.2021.112695
62. Gohari G, Panahirad S, Sepehri N, Akbari A, Zahedi SM, Jafari H, Dadpour MR, Fotopoulos V (2021) Enhanced tolerance to salinity stress in grapevine plants through application of carbon quantum dots functionalized by proline. Environ Sci Pollut Res 28:42877–42890. https://doi.org/10.1007/s11356-021-13794-w
63. Palmqvist NM, Seisenbaeva GA, Svedlindh P, Kessler VG (2017) Maghemite nanoparticles acts as nanozymes, improving growth and abiotic stress tolerance in *Brassica napus*. Nanoscale Res Lett 12:631. https://doi.org/10.1186/s11671-017-2404-2
64. El-Gazzar N, Almaary K, Ismail A, Polizzi G (2020) Influence of Funneliformis mosseae enhanced with titanium dioxide nanoparticles (TiO_2NPs) on *Phaseolus vulgaris* L. under salinity stress. PLoS ONE 15:e0235355. https://doi.org/10.1371/journal.pone.0235355
65. Bajpai VK, Kamle M, Shukla S, Mahato DK, Chandra P, Hwang SK, Kumar P, Huh YS, Han YK (2018) Prospects of using nanotechnology for food preservation, safety, and security. J Food Drug Anal 26(4):1201–1214
66. Biswal AK, Misra PK (2020) Biosynthesis and characterization of silver nanoparticles for prospective application in food packaging and biomedical fields. Mater Chem Phys 250. https://doi.org/10.1016/j.matchemphys.2020.123014
67. Yang H, Qu L, Wimbrow AN, Jiang X, Sun Y (2007) Rapid detection of Listeria monocytogenes by nanoparticle-based immunomagnetic separation and real-time PCR. Int J Food Microbiol 118(2):132–138
68. Joung HA, Lee NR, Lee SK, Ahn J, Shin YB, Choi HS, Kim MG (2008) High sensitivity detection of 16s rRNA using peptide nucleic acid probes and a surface plasmon resonance biosensor. Anal Chim Acta 630(2):168–173

69. Biswas R, Alam M, Sarkar A, Haque MI, Hasan MM, Hoque M (2022) Application of nanotechnology in food: processing, preservation, packaging and safety assessment. Heliyon 8(11):e11795. https://doi.org/10.1016/j.heliyon.2022.e11795
70. Jokar M, Pedersen GA, Loeschner K (2017) Six open questions about the migration of engineered nano-objects from polymer-based food-contact materials: a review. Food Addit Contam 34(3):434–450

Emerging Technologies in Nanobiosensors Design and Principles

Anushka Garg, Sachin Jaidka, and Vikasdeep Singh Mann

Abstract This chapter explores the cutting-edge developments in the field of nano-nanobiosensors, focusing on the innovative technologies driving their design and underlying principles. This chapter provides a comprehensive overview of the latest materials and fabrication techniques that enhance the sensitivity, specificity, and overall performance of nano-nanobiosensors. The topics covered here are the integration of nanomaterials such as graphene, carbon nanotubes, and quantum dots, as well as the application of novel biorecognition elements and signal transduction mechanisms. The importance of the interdisciplinary nature of nanobiosensor development, highlighting the conjunction of nanotechnology, biology, and electronics, is also highlighted. The insights into the practical applications of these sensors in medical diagnostics, environmental monitoring, industrial processes, etc., are also discussed in detail. The current challenges in the field of nanobiosensors and future directions have also been explored in the present chapter, offering a further perspective on the potential impact of emerging technologies on nanobiosensor design.

Keywords Nanobiosensors · Nanomaterials · Biorecognition elements · Signal transduction · Interdisciplinary technologies

A. Garg
Department of Chemistry and Biochemistry, TIET-VT Center of Excellence in Emerging Materials, Thapar Institute of Engineering and Technology, Patiala 147004, India

S. Jaidka (✉)
Division of Research and Innovation, Department of Applied Sciences, Chandigarh Engineering College, Chandigarh Group of Colleges, Jhanjeri, Mohali, Punjab 140307, India
e-mail: sachinjaidka@gmail.com

V. S. Mann
Division of Research and Innovation, Department of Mechanical Engineering, Chandigarh Engineering College, Chandigarh Group of Colleges, Jhanjeri, Mohali, Punjab 140307, India

D. Sillu et al. (eds.), *Nanobiosensors for the Food Industry*, Smart Nanomaterials Technology, https://doi.org/10.1007/978-981-95-0136-6_2

1 Introduction

Nanobiosensors convert biological signals into measurable electrical or optical outputs, which have played an important role in the scientific and medical advancements since their discovery in the early twentieth century. Early nanobiosensors, like the ones used for glucose monitoring in diabetes management, relied on enzymes like glucose oxidase to catalyze reactions with blood glucose, producing a signal relative to glucose concentration. This laid the groundwork for modern biosensing technologies [1–3]. However, limitations such as bulkiness, slow response times, and inadequate sensitivity limited their pertinency to a narrow range of biomedical tasks [4, 5]. The modern-day nanobiosensors are defined as a subset of nanobiosensors that incorporate nanomaterials into their design, having enhanced sensitivity, selectivity, and potential for miniaturization [6–9]. Nanomaterials have some very peculiar properties, such as small size, high surface area-to-volume ratio, and tunable characteristics (physical as well as chemical), making them crucial in the development of next-gen nanobiosensors [7, 10, 11]. These nanomaterials include nanoparticles, nanowires, nanotubes, and quantum dots, which can interact with the biological systems at a molecular level, leading to noteworthy progress in the detection and estimation of biological processes [12–16].

A nanobiosensor generally contains two primary components: a bioreceptor and a transducer. A bioreceptor could be an antibody, a nucleic acid, an enzyme, or other biomolecule used to recognize and bind to the target material. For example, in glucose nanobiosensors, glucose oxidase acts as the bioreceptor, reacting only with glucose molecules. On the other hand, the transducer converts the biological interface between the bioreceptor and the material into a digital signal (in the form of an electrical current or optical change) [1, 3]. The incorporation of nanomaterials into both the bioreceptor and transducer leads to the enhancement in the performance of the biosensor by amplifying the signal and increasing the sensitivity of the device [16, 17].

2 Nanomaterials in Nanobiosensor Design

2.1 Graphene and Its Derivatives

Graphene is a single layer of carbon atoms arranged in a 2D honeycomb lattice. It has emerged as a versatile material in the field of nano-nanobiosensors due to its excellent electrical, thermal, and mechanical properties. The large surface area, high electron mobility, and tunable surface chemistry of graphene make it a perfect candidate for biosensing applications [18, 19]. Graphene-based nanobiosensors have been widely explored for the detection of a variety of biological analytes such as glucose, DNA, proteins, pathogens, etc. In glucose sensing, graphene is generally used to boost the electron transfer between the glucose oxidase enzyme and the electrode, leading to a significant improvement in the sensitivity and response time of the nanobiosensor. Studies have shown that graphene-based glucose sensors can achieve a resolution of as low as nanomolar concentrations, which is critical for accurate diabetes management [20–23].

Graphene derivatives, such as graphene oxide (GO) and reduced graphene oxide (rGO), have also been widely studied throughout the globe. GO, generally contains oxygenated functional groups, offering easy functionalization with biomolecules, making it highly useful in biosensing applications. On the other hand, rGO provides high conductivity while maintaining some functional groups for biomolecule attachment [24–28]. A recent study demonstrated the use of rGO in an electrochemical DNA biosensor, where the electrode was modified using rGO, which further led to a highly sensitive platform for the recognition of some definite DNA sequences linked with cancer biomarkers [29]. Figure 1 shows different types of nanomaterials that can be used in the fabrication of nanobiosensors.

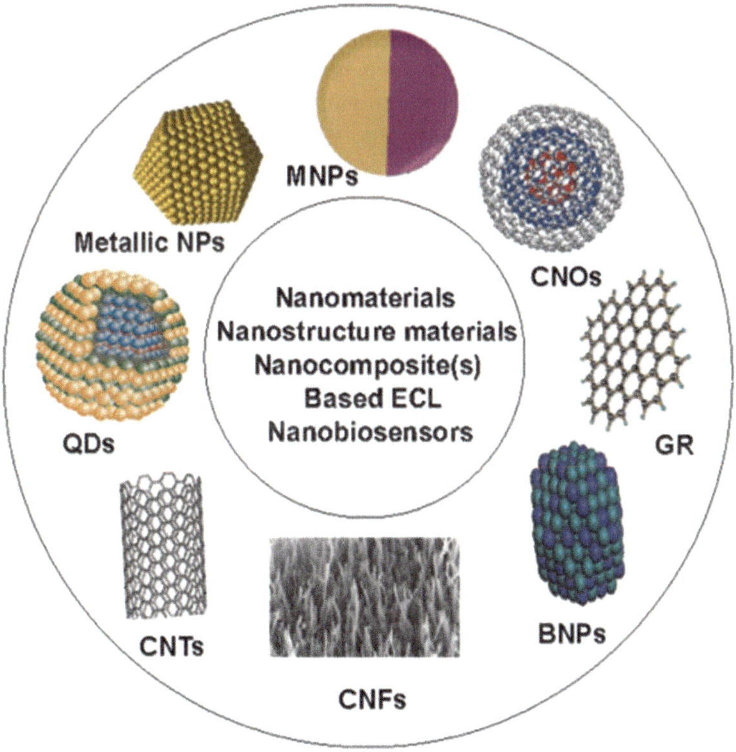

Fig. 1 Different types of nanomaterials used for nanobiosensors [37]

2.2 Carbon Nanotubes

Carbon nanotubes (CNTs), cylindrical nanostructures composed of rolled graphene sheets, have garnered significant attention in nanobiosensor design due to their exceptional electrical conductivity, high surface area, and chemical stability [30, 31]. CNTs come in two main forms: single-walled carbon nanotubes (SWCNTs) and multi-walled carbon nanotubes (MWCNTs) [32]. Both forms have been employed in various biosensing platforms, offering unique advantages in terms of sensitivity and selectivity. CNTs have been used to detect a wide range of analytes, including glucose, cholesterol, lactate, and pathogens. One notable application is in the development of electrochemical nanobiosensors, where CNTs serve as the electrode material, facilitating rapid electron transfer and enhancing the overall sensitivity of the sensor [3, 33–35]. For example, a SWCNT-based biosensor for the detection of *Escherichia coli* had a detection limit in the range of picomolar, highlighting the potential of CNTs in pathogen detection [36].

Furthermore, the functionalization ability of CNTs (with various biomolecules, such as enzymes, antibodies, or aptamers) enables the creation of highly specific nanobiosensors. Functionalized CNTs have been used to develop sensors for detecting specific cancer biomarkers, providing a powerful tool for early cancer diagnosis [3, 33, 35]. A recent study reported the use of antibody-functionalized SWCNTs in a biosensor for detecting prostate-specific antigen (PSA) with high sensitivity and selectivity [38].

2.3 Quantum Dots

Quantum dots (QDs) are semiconductor nanocrystals that exhibit unique optical properties, such as size-tunable fluorescence and high photostability, making them highly attractive for use in nanobiosensors [14]. The optical properties of QDs can be precisely controlled by adjusting their size, which allows for the development of sensors with tailored sensitivity and selectivity. QDs have been widely used in optical nanobiosensors for the biomolecule detection, such as proteins, nucleic acids, and small molecules. For instance, QDs have been employed in fluorescence-based nanobiosensors to detect cancer biomarkers, where the binding of the target analyte to a QD-conjugated biorecognition element results in a detectable change in fluorescence [14, 15]. A recent study highlighted the use of QDs in a multiplexed biosensor capable of simultaneously detecting multiple cancer biomarkers with high sensitivity, demonstrating the potential of QDs in personalized medicine [39]. The combination of QDs with other nanomaterials, such as graphene or CNTs, has also been explored to enhance the performance of nanobiosensors further.

2.4 *Other Nanomaterials*

Gold nanoparticles (AuNPs) have unique optical properties, due to which they are frequently used in nanobiosensors, particularly due to their strong surface plasmon resonance (SPR). When AuNPs are conjugated with antibodies or other bioreceptors, they can produce a significant color change in the presence of the target analyte, making them highly effective for colorimetric biosensing applications [40–43]. This principle has been widely applied in the development of lateral flow assays, such as those used for rapid diagnostic tests, including COVID-19 antigen tests [42]. Additionally, AuNPs have been utilized in electrochemical sensors, where they enhance electron transfer between the bioreceptor and the electrode, thereby increasing the sensitivity of the sensor [44].

Silver nanoparticles (AgNPs), known for their antimicrobial properties, have been integrated into nanobiosensors for detecting bacterial pathogens in food and water. These sensors rely on the ability of AgNPs to interact with bacterial cells and measure changes in the sensor's optical or electrical properties [45–47]. AgNPs have been used in colorimetric sensors where the aggregation of nanoparticles causes a visible color change, indicating the presence of bacterial contamination [48, 49]. Despite these advancements, challenges remain in the widespread adoption of nanobiosensors, particularly in terms of scalability, cost-effectiveness, and regulatory approval. The reproducible synthesis of biorecognition elements and nanomaterials with consistent properties is critical for the reliable performance of nanobiosensors, yet it remains a significant challenge [50, 51]. Additionally, the complexity of biological interactions and the potential for interference from nontarget substances can affect the accuracy and reliability of nanobiosensors, necessitating ongoing research to improve their robustness and specificity. Regulatory hurdles also pose challenges, as nanobiosensors, particularly those intended for medical applications, must undergo rigorous testing and validation to ensure their safety and efficacy.

3 Fundamental Principles of Nanobiosensors

The fundamental principles of nanobiosensors are grounded in the integration of biological and technological systems to change biological exchanges into measurable signals. The choice of biorecognition elements, transducers, and amplification strategies plays a vital role in deciding the performance of the biosensor [9, 17]. While significant progress has been made in the development of nanobiosensors with high sensitivity, specificity, and stability, ongoing research and innovation are necessary to overcome the remaining challenges and fully realize the potential of nanobiosensors in diverse applications. The fundamentals of nanobiosensors are rooted in the integration of biology with advanced technological systems designed to translate biological interactions into quantifiable signals that can be processed and interpreted. A biosensor is a complex device that typically consists of three main components: a biorecognition element, a transducer, and a signal processor [52]. The biorecognition element is central to the biosensor's function, as it is responsible for the specific interaction with the target analyte. This element can be composed of various biological entities such as enzymes, antibodies, nucleic acids, or cell receptors, each chosen based on the nature of the analyte and the desired specificity and sensitivity of the sensor [17, 50, 52].

The transducer in a biosensor is the component that translates the biological interaction into a measurable signal, which can be electrical, optical, piezoelectric, or thermal in nature [52, 53]. The choice of transduction mechanism significantly influences the sensitivity, response time, and potential applications of the biosensor. Electrochemical transducers are famous for their simplicity, high sensitivity, and miniaturization ease. In an electrochemical glucose sensor, for instance, the transducer measures the current generated by the oxidation of glucose, providing a quantitative measure of glucose concentration [3, 54, 55]. Optical transducers, on the other hand, are favored in applications requiring high sensitivity and specificity, such as DNA hybridization assays [29]. These transducers detect changes in light absorption, fluorescence, or refractive index upon the requirement of the target analyte, enabling label-free detection and real-time monitoring [56–58]. Piezoelectric transducers detect changes on the sensor surface and are particularly useful in applications like immunosensors, where the binding of an antibody to an antigen induces a measurable frequency shift in a piezoelectric crystal [59–61].

Signal amplification and processing are critical steps in the operation of a biosensor, as they ensure that the signal generated by the transducer is strong enough to be detected and accurately interpreted. Various strategies are employed to amplify the signal, often involving the use of nanomaterials with high surface areas, which can enhance the communication between the biorecognition element and the analyte. Gold nanoparticles (AuNPs) are commonly used in nanobiosensors for their ability to increase the surface area and enhance the electrochemical signal through catalytic effects [29, 42, 62]. Enzymatic amplification is another powerful technique, where enzymes like horseradish peroxidase are used to produce a large amount of product from a small amount of analyte, thereby amplifying the signal. In nucleic acid detection, external amplification systems such as polymerase chain reaction (PCR) are employed to exponentially increase the amount of target DNA or RNA, making it detectable even at very low concentrations [29, 45, 60].

The signal generated by the biosensor, after amplification, is processed through various electronic and computational means to filter noise, enhance signal clarity, and convert the raw data into a readable output. This output can be displayed on electronic devices like digital meters or computers, providing real-time monitoring and analysis capabilities. Signal processing is particularly important in complex biological environments where the presence of multiple interfering substances can affect the accuracy of the sensor. Advanced algorithms and machine learning techniques are increasingly being integrated into biosensor systems to improve the interpretation of signals and to enable the sensor to adapt to varying conditions. The performance of a biosensor is evaluated based on several key metrics, which include sensitivity, specificity, response time, stability, and reproducibility [4, 5, 17, 50, 52]. Sensitivity refers to the sensor's ability to detect low concentrations of the target analyte, which is crucial in applications like early disease diagnosis, where biomarkers may be present at very low levels. Specificity is the ability of the sensor to distinguish the target analyte from other similar molecules, reducing the likelihood of false positives. This is particularly important in clinical diagnostics, where the accurate identification of a specific pathogen or biomarker can significantly impact patient outcomes [52, 63]. Response time, or the time it takes for the biosensor to produce a stable signal after exposure to the analyte, is critical in applications requiring rapid detection, such as point-of-care testing or environmental monitoring. Stability refers to the sensor's ability to maintain its performance over time, under varying environmental conditions, which is essential for long-term monitoring applications. Reproducibility, or the consistency of the sensor's performance across multiple measurements, ensures that the data generated by the sensor is reliable and can be trusted for decision-making. The development and optimization of nanobiosensors, with numerous studies demonstrating the effectiveness of different biorecognition elements, transduction mechanisms, and amplification strategies, has been done by the research community across the globe [10, 15, 17, 42, 45, 50, 57, 63].

4 Emerging Technologies in Designing the Nanobiosensors

Nanobiosensors have evolved remarkably since their launch, driven by advancements in materials science, nanotechnology, biotechnology, and information technology. The continuous push toward enhanced sensitivity, specificity, portability, and real-time monitoring has spurred the development of novel technologies and innovative designs in biosensor engineering [4, 5, 17, 50, 52]. This comprehensive exploration investigates the cutting-edge technologies shaping the future of nanobiosensors, highlighting the materials, methodologies, and scientific breakthroughs that are redefining the landscape of biosensing.

4.1 Nanomaterials and Their Multifaceted Roles

As discussed above, one of the most significant advancements in biosensor design is the integration of nanomaterials, which offer unique physical, chemical, and biological properties that enhance sensor performance. Beyond traditional nanomaterials like gold nanoparticles (AuNPs) and carbon nanotubes (CNTs), emerging nanomaterials such as graphene derivatives, metal–organic frameworks (MOFs), and 2D materials are playing pivotal roles [20, 23, 24, 29, 31, 36]. Graphene is one such material with exceptional electrical conductivity, mechanical strength, and large surface area [19]. These properties make graphene an ideal candidate for transducers in electrochemical and field-effect transistor (FET) nanobiosensors. Recent studies have shown that the graphene-based FET nanobiosensors are capable of detecting biomarkers at femtomolar concentrations, highlighting their potential in early disease diagnostics [64]. Additionally, graphene oxide (GO) and reduced graphene oxide (rGO) offer functional groups that ease the attachment of biorecognition elements, leading to the enhancement of the sensor specificity and sensitivity, as discussed above [23, 24]. In contrast, metal–organic frameworks (MOFs) are porous crystalline materials with metal ions coordinated to organic ligands, providing them with a highly tunable structure having high surface areas and selective adsorption properties [65–67]. With remarkable sensitivity and selectivity, many MOF-based electrochemical sensors have also been developed for the detection of glucose and lactate [68]. Other 2D materials, such as molybdenum disulfide (MoS_2) and transition metal dichalcogenides (TMDs), are also emerging as promising components in biosensor design. Highly sensitive photonic and electronic nanobiosensors can be synthesized using these materials due to their unique electronic and optical properties [69, 70]. MoS_2-based nanobiosensors have already been tested for the detection of cancer biomarkers through enhanced fluorescence quenching and signal amplification mechanisms [71].

4.2 Advancements in Biorecognition Elements

The specificity and sensitivity of nanobiosensors are largely determined by the biorecognition elements employed. Aptamers are synthetic oligonucleotides that can bind to a wide range of targets with high affinity and are emerging as an alternative to traditional antibodies. They can be selected through the systematic evolution of ligands by exponential enrichment (SELEX) processes to bind specific bodies, including small molecules, proteins, whole cells, etc. [72, 73]. Their synthetic nature reduces the risk of immunogenicity, making them attractive for medical and environmental applications. Another breakthrough in biorecognition elements is the use of molecularly imprinted polymers (MIPs). MIPs are synthetic polymers that have specific recognition sites for target molecules due to the polymerization of monomers in the presence of a template analyte. Once the template is removed, the polymer retains corresponding binding sites, enabling selective recognition of the target, offering high stability, reusability, and cost-effectiveness [74–78]. Figure 2 shows different nanobiosensors from starting to testing.

Furthermore, the introduction of CRISPR-Cas systems has introduced a new standard in biosensor design. CRISPR-Cas-based nanobiosensors can control the programmable nature of CRISPR technology, achieving highly specific nucleic acid detection. These nanobiosensors utilize the recognizing ability of Cas protein and cut specific DNA or RNA sequences, generating detectable signals upon target binding [79, 80]. This technology has been successfully applied in the rapid detection of viral pathogens, including SARS-CoV-2, demonstrating its potential for point-of-care diagnostics and pandemic response [81].

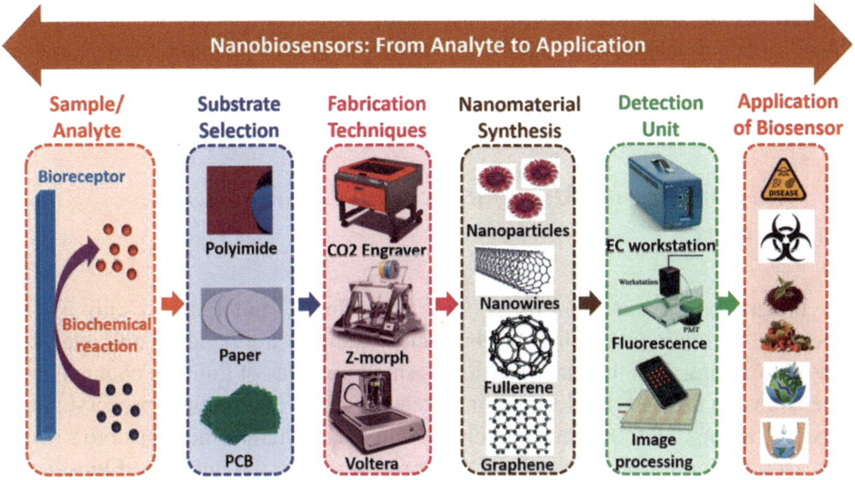

Fig. 2 Different nanobiosensors from starting to testing [6]

4.3 *Innovative Transduction Mechanisms*

The transduction mechanism is a critical component of nanobiosensors, determining how biological interactions are converted into measurable signals. Emerging transduction technologies are pushing the boundaries of biosensor performance by enhancing sensitivity, enabling label-free detection, and facilitating real-time monitoring. Surface plasmon resonance (SPR) and localized surface plasmon resonance (LSPR) are optical transduction techniques that have seen significant advancements through the use of nanostructured materials. SPR-based nanobiosensors detect changes in the refractive index near the sensor surface upon analyte binding, allowing for label-free and real-time monitoring of biomolecular interactions. The integration of nanostructures such as gold nanorods and nanostars has improved the sensitivity and resolution of SPR nanobiosensors, enabling the detection of low-abundance biomarkers with high precision [40–43]. Another innovative transduction mechanism is electrochemical impedance spectroscopy (EIS), which measures the impedance changes at the sensor-electrolyte interface caused by analyte binding. EIS-based nanobiosensors offer high sensitivity and the ability to perform label-free detection, making them suitable for applications in medical diagnostics and environmental monitoring. Recent developments in nanostructured electrodes, such as nanoporous gold and carbon nanofibers, have enhanced the surface area and electrical properties of EIS nanobiosensors, leading to improved detection limits and faster response times [82, 83]. Photonic crystal nanobiosensors represent a cutting-edge optical transduction technology that manipulates the flow of light through periodic nanostructures to achieve high sensitivity in biomolecule detection. These nanobiosensors utilize the unique optical properties of photonic crystals, such as photonic bandgaps and defect modes, to create highly sensitive platforms for label-free biosensing. Recent research has demonstrated the use of photonic crystal nanobiosensors for the detection of cancer biomarkers and pathogens, highlighting their potential for multiplexed and high-throughput applications [84–86].

4.4 Integration with Microfluidics and Lab-on-a-Chip Systems

The integration of nanobiosensors with microfluidic and lab-on-a-chip (LOC) systems is revolutionizing the way biological samples are handled and analyzed. Microfluidics involves the manipulation of small volumes of fluids in channels with dimensions on the microscale, enabling precise control over sample processing and reagent handling. LOC systems combine multiple laboratory functions onto a single chip, facilitating the development of compact, portable, and automated biosensing devices [87–89]. Emerging technologies in microfluidics are enhancing biosensor performance by enabling rapid sample processing, reducing reagent consumption, and allowing for high-throughput analysis [90]. The droplet-based microfluidics enables the compartmentalization of individual analyte molecules in picoliter droplets, allowing for single-molecule detection and digital biosensing. This approach has been applied in applications such as single-cell analysis and digital PCR, where high sensitivity and quantification accuracy are essential. The integration of nanobiosensors with LOC systems is also paving the way for point-of-care diagnostics and personalized medicine [87, 89, 90].

4.5 Wearable and Flexible Nanobiosensors

The demand for continuous health monitoring and real-time diagnostics has driven the development of wearable and flexible nanobiosensors. These devices are designed to conform to the contours of the body, providing unobtrusive and seamless integration with daily activities [91]. Emerging materials and fabrication techniques are enabling the creation of flexible, stretchable, and biocompatible nanobiosensors that can monitor a wide range of physiological parameters. Flexible electronics, such as organic thin-film transistors (OTFTs) and stretchable interconnects, are at the forefront of wearable biosensor technology. These materials can bend and stretch without compromising their electrical performance, making them ideal for integration into wearable devices like smartwatches, fitness trackers, and skin patches [92–94]. Recent innovations include graphene-based flexible electrodes and polymer nanocomposites, which offer high conductivity and mechanical resilience for reliable biosensor performance under dynamic conditions [91, 92].

In addition to electronic flexibility, the development of soft and biocompatible substrates is crucial for wearable nanobiosensors. Materials such as polydimethylsiloxane (PDMS), hydrogels, and bioresorbable polymers provide a comfortable interface with the skin or other biological tissues, minimizing irritation and enhancing sensor adhesion. These substrates can be functionalized with nanomaterials and biorecognition elements to create highly sensitive and selective wearable nanobiosensors capable of monitoring biomarkers in sweat, interstitial fluid, and other bodily fluids [95–97]. Recent advancements in printed electronics and inkjet printing techniques have further facilitated the mass production and customization of wearable nanobiosensors [47]. These additive manufacturing processes allow for the deposition of conductive inks and functional nanomaterials onto flexible substrates, enabling the rapid prototyping and scaling of wearable sensor arrays. The integration of energy harvesting technologies, such as flexible solar cells and triboelectric generators, is also enhancing the autonomy and sustainability of wearable nanobiosensors, reducing the need for external power sources [98, 99].

4.6 Advanced Analytics and Artificial Intelligence Integration

The proliferation of biosensor data necessitates sophisticated data analytics and artificial intelligence (AI) techniques to extract meaningful insights and drive decision-making. Emerging technologies in machine learning, deep learning, and edge computing are being integrated with nanobiosensors to enhance data processing, pattern recognition, and real-time analysis. Machine learning algorithms can be employed to analyze complex biosensor data, identifying patterns and correlations that may not be discernible through traditional statistical methods [100, 101]. AI-driven algorithms have been utilized to interpret multi-analyte biosensor data for disease diagnosis, enabling the identification of biomarker signatures associated with specific conditions. Deep learning models, with their ability to process large datasets and capture intricate relationships, are particularly effective in handling high-dimensional biosensor data from multiplexed and integrated sensor arrays [102].

Edge computing, which involves processing data locally on the biosensor device rather than relying on cloud-based servers, is another emerging technology enhancing biosensor functionality. By integrating AI algorithms directly onto biosensor platforms, edge computing enables real-time data analysis, reducing latency and improving the responsiveness of biosensor systems [100, 101]. This is particularly advantageous for applications requiring immediate feedback, such as glucose monitoring for diabetes management or real-time pathogen detection in environmental monitoring. The convergence of nanobiosensors with Internet of Things (IoT) technologies is also transforming biosensor applications by enabling connectivity, data sharing, and remote monitoring. IoT-enabled nanobiosensors can transmit data wirelessly to centralized databases or cloud platforms, facilitating large-scale health monitoring, environmental surveillance, and personalized healthcare [51]. The integration of secure communication protocols and data encryption techniques ensures the privacy and integrity of biosensor data, addressing critical concerns in data-sensitive applications.

4.7 Biocompatible and Implantable Nanobiosensors

The development of biocompatible and implantable nanobiosensors is expanding the horizons of continuous health monitoring and therapeutic applications. These nanobiosensors are designed to operate within the body, providing real-time monitoring of physiological parameters and facilitating closed-loop therapeutic interventions. Biocompatibility is a paramount consideration in the design of implantable nanobiosensors, as these devices must operate without eliciting adverse immune responses or causing tissue damage [103, 104]. Emerging materials such as biocompatible polymers, hydrogels, and bioresorbable metals are being employed to create sensors that integrate seamlessly with biological tissues [105, 106]. Implantable nanobiosensors are being developed for a variety of applications, including continuous glucose monitoring, neurotransmitter detection, and implantable drug delivery systems [106]. Advanced fabrication techniques, such as microelectromechanical systems (MEMS) and flexible electronics, are enabling the creation of miniature and minimally invasive nanobiosensors that can be implanted with minimal discomfort and risk. Recent innovations include wireless power transfer and data transmission systems, which eliminate the need for battery replacements and facilitate remote monitoring of implantable devices [107, 108]. The integration of nanobiosensors with therapeutic systems is leading to the development of closed-loop systems that provide automated and responsive treatments. Neuroprosthetic devices equipped with neurotransmitter sensors can modulate neural activity in response to detected changes, offering promising applications in the treatment of neurological disorders [109, 110].

4.8 3D Printing and Additive Manufacturing in Biosensor Fabrication

3D printing is revolutionizing the fabrication of nanobiosensors by enabling the creation of complex, customized, and multifunctional sensor architectures. 3D printing allows for precise control over the geometry and composition of nanobiosensors, facilitating the integration of multiple components and the incorporation of intricate nanostructures. One of the key advantages of 3D printing in biosensor design is the ability to create customized sensor platforms tailored to specific applications [111]. This flexibility is particularly valuable in the development of personalized medical devices, where nanobiosensors can be designed to fit individual anatomical features or to monitor specific biomarkers relevant to a patient's condition. 3D printing also enables the incorporation of multiple sensing elements and functionalities within a single device [112]. Multimaterial 3D printing allows for the simultaneous deposition of conductive, insulating, and biocompatible materials, facilitating the integration of electronic components, microfluidic channels, and biorecognition elements into a unified sensor platform. This capability is essential for the development of multianalyte nanobiosensors that can simultaneously detect and quantify multiple biomarkers, enhancing the diagnostic power and versatility of biosensor systems. Advanced 3D printing techniques, such as two-photon polymerization (2PP) and direct ink writing (DIW), offer high-resolution and high-precision fabrication capabilities, enabling the creation of nanoscale features and intricate nanostructures within nanobiosensors [113, 114]. These techniques have been utilized to fabricate nanoelectrodes, nanowires, and nanogap structures that enhance the sensitivity and selectivity of nanobiosensors by increasing the surface area and facilitating efficient electron transfer.

5 Practical Applications of Nano Nanobiosensors

5.1 Medical Diagnostics

Nano-nanobiosensors have had a profound impact on medical diagnostics, providing highly sensitive and specific tools for detecting biomarkers associated with various diseases. One of the most significant applications is in the detection of cancer biomarkers, where early diagnosis can greatly improve patient outcomes. As discussed above, graphene-based nanobiosensors can be used for the detection of circulating tumor DNA (ctDNA) in the blood of cancer patients. These sensors demonstrate high sensitivity and specificity, with a detection limit in the femtomolar range, making them suitable for early cancer diagnosis and monitoring [29, 115]. Another example is the use of QD-based nanobiosensors for detecting proteins associated with neurodegenerative diseases, such as Alzheimer's disease. Several studies have demonstrated the use of QD-based fluorescence nanobiosensors for the detection of beta-amyloid, a protein associated with Alzheimer's disease, with a detection limit in the picomolar range [54].

5.2 Environmental Monitoring

Nano-nanobiosensors have also been widely used in environmental monitoring, providing highly sensitive tools for detecting pollutants and toxins in water, air, and soil. CNT-based nanobiosensors can detect heavy metals, such as lead and mercury, in water samples. These sensors demonstrate high sensitivity and specificity, with detection limits in the nanomolar range, making them suitable for real-time environmental monitoring [31, 116]. Another example is the use of MIP-based nanobiosensors for detecting pesticides in agricultural runoff. Recent studies have demonstrated that the use of a MIP-based electrochemical sensor for the detection of atrazine, a commonly used herbicide, in water samples [74, 117].

5.3 Industrial Applications

Nano-nanobiosensors have also found applications in industrial processes, particularly in the monitoring of chemical and biological processes. Some graphene-based nanobiosensors can help in monitoring the fermentation processes in the production of biofuels. The sensor demonstrated high sensitivity and specificity, with the ability to detect key metabolites in real time, enabling more efficient and controlled fermentation processes [27]. Another example is the use of CNT-based nanobiosensors in the monitoring of wastewater treatment processes. A recent study demonstrated the use of a CNT-based electrochemical sensor for detecting pathogens in wastewater, with high sensitivity and specificity, making it suitable for ensuring the safety and efficiency of wastewater treatment processes [31].

6 Limitations and Challenges in Broad Usage of Nanobiosensors

6.1 Technical Challenges

Despite the significant advancements in nanobiosensor technology, several technical challenges remain that limit their widespread adoption. One of the primary challenges is the reproducibility and stability of nanobiosensors. The performance of these sensors can be affected by factors such as environmental conditions, sensor fabrication inconsistencies, and the degradation of biorecognition elements over time. Ensuring consistent performance across different batches of sensors is critical for their commercialization.

6.2 Cost and Scalability

The cost of manufacturing nanobiosensors is another major limitation. The fabrication of nanoscale materials and devices often requires sophisticated equipment and processes, which can be expensive and time-consuming. Scaling up production while maintaining quality and performance is a significant challenge that needs to be addressed to make nanobiosensors commercially viable for large-scale applications.

6.3 Regulatory and Standardization Issues

Governments around the world are recognizing the potential of nanobiosensors and are supporting research and development in this field. In the United States, the National Institutes of Health (NIH) and the National Science Foundation (NSF) have funded numerous projects aimed at developing advanced nanobiosensors for healthcare and environmental applications. In the European Union, the Horizon 2020 program has provided significant funding for nanotechnology research, including the development of nanobiosensors for various applications. These initiatives are aimed at accelerating the translation of nanobiosensor technology from the laboratory to the market, ensuring that these innovative devices can be widely adopted for the benefit of society.

6.4 Integration with Existing Systems

Integrating nanobiosensors with existing diagnostic and analytical systems can be challenging, particularly in terms of data management and interpretation. The large volumes of data generated by nanobiosensors, especially when multiple analytes are detected simultaneously, require advanced data processing and storage solutions. Ensuring compatibility with existing infrastructure and training personnel to operate and interpret results from these advanced sensors are essential steps for their successful deployment. Figure 3 indicates the contemplative progression of nanobiosneosrs into the future.

6.5 Challenges in Nanobiosensor Development

Despite the significant advancements in nanobiosensor technology, several challenges remain. One of the key challenges is the reproducibility and scalability of sensor fabrication. The fabrication of nano-nanobiosensors often involves complex processes, and achieving consistent performance across different batches of sensors

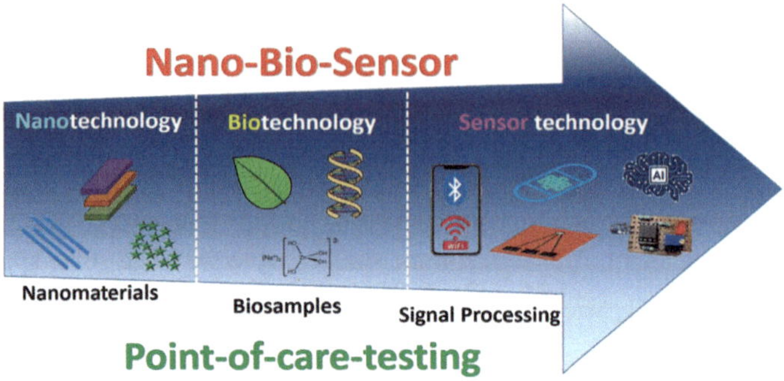

Fig. 3 Future perspectives of nanobiosensors [6]

can be difficult. Furthermore, scaling up the production of nano-nanobiosensors for commercial applications presents additional challenges, particularly in terms of cost and manufacturing efficiency. Another challenge is the stability and durability of nano-nanobiosensors, particularly in harsh environmental conditions. While nanomaterials offer unique properties that enhance sensor performance, they can also be sensitive to environmental factors such as temperature, humidity, and pH. Ensuring the long-term stability and reliability of nano-nanobiosensors in real-world applications remains a critical challenge.

7 Conclusion

Nano nanobiosensors represent a rapidly evolving field at the intersection of nanotechnology, biology, and electronics. The integration of nanomaterials such as graphene, carbon nanotubes, and quantum dots has significantly enhanced the performance of nanobiosensors, enabling the detection of biomolecules with unprecedented sensitivity and specificity. Novel biorecognition elements, such as aptamers and molecularly imprinted polymers, have further expanded the range of detectable analytes. At the same time, advances in signal transduction mechanisms have improved the sensitivity and versatility of these sensors. While several challenges remain, such as the reproducibility and stability of nano nanobiosensors, emerging technologies such as artificial intelligence, flexible electronics, and microfluidics offer promising solutions for overcoming these challenges and further advancing the field. As nano-nanobiosensors continue to evolve, they hold the potential to revolutionize numerous fields, providing powerful tools for disease diagnosis, environmental monitoring, and industrial applications. The interdisciplinary nature of nanobiosensor development, driven by the convergence of nanotechnology, biology, and electronics, ensures that this field will continue to be a vibrant and rapidly advancing area of research for years to come.

References

1. Wang J (2008) Electrochemical glucose biosensors. Chem Rev 108:814–825
2. Sabu C, Henna TK, Raphey VR, Nivitha KP, Pramod K (2019) Advanced biosensors for glucose and insulin. Biosens Bioelectron 141:111201
3. Juska VB, Pemble ME (2020) A critical review of electrochemical glucose sensing: evolution of biosensor platforms based on advanced nanosystems. Sensors 20:6013
4. Tu J, Torrente-Rodríguez RM, Wang M, Gao W (2020) The era of digital health: a review of portable and wearable affinity biosensors. Adv Funct Mater 30:1906713
5. Estevez M-C, Otte MA, Sepulveda B, Lechuga LM (2014) Trends and challenges of refractometric nanoplasmonic biosensors: a review. Anal Chim Acta 806:55–73
6. Kulkarni MB, Ayachit NH, Aminabhavi TM (2022) Recent advancements in nanobiosensors: current trends, challenges, applications, and future scope. Biosensors (Basel) 12:892
7. Welch EC, Powell JM, Clevinger TB, Fairman AE, Shukla A (2021) Advances in biosensors and diagnostic technologies using nanostructures and nanomaterials. Adv Funct Mater 31:2104126
8. Vestergaard MC, Tamiya E (2015) Nanobiosensors and nanobioanalyses: a review. Nanobiosens Nanobioanal 3–20
9. Bhattarai P, Hameed S (2020) Basics of biosensors and nanobiosensors. Nanobiosens From Design Appl 1–22
10. Abdel-Karim R (2024) Nanotechnology-enabled biosensors: a review of fundamentals, materials, applications, challenges, and future scope. Biomed Mater Dev 2:759–777
11. Tran TB, Son SJ, Min J (2016) Nanomaterials in label-free impedimetric biosensor: current process and future perspectives. Biochip J 10:318–330
12. Jain KK (2005) Nanotechnology in clinical laboratory diagnostics. Clin Chim Acta 358:37–54
13. Yue Z, Lisdat F, Parak WJ, Hickey SG, Tu L, Sabir N, Dorfs D, Bigall NC (2013) Quantum-dot-based photoelectrochemical sensors for chemical and biological detection. ACS Appl Mater Interfaces 5:2800–2814
14. Soldado A, Barrio LC, Díaz-Gonzalez M, de la Escosura-Muñiz A, Costa-Fernandez JM (2022) Advances in quantum dots as diagnostic tools. Adv Clin Chem 107:1–40
15. Farzin MA, Abdoos H (2021) A critical review on quantum dots: from synthesis toward applications in electrochemical biosensors for determination of disease-related biomolecules. Talanta 224:121828
16. Holzinger M, Le Goff A, Cosnier S (2014) Nanomaterials for biosensing applications: a review. Front Chem 2:63
17. Naresh V, Lee N (2021) A review on biosensors and recent development of nanostructured materials-enabled biosensors. Sensors 21:1109
18. Bin Wu J, Lin ML, Cong X, Liu HN, Tan PH (2018) Raman spectroscopy of graphene-based materials and its applications in related devices. Chem Soc Rev 47:1822–1873. https://doi.org/10.1039/c6cs00915h
19. Akashah MHN, Sahudin MR, Rani RA, Scully PJ, Makhsin SR (2022) Chemical properties of graphene. In: Graphene, nanotubes and quantum dots-based nanotechnology, Elsevier, pp 43–67
20. Abdelbasset WK, Jasim SA, Bokov DO, Oleneva MS, Islamov A, Hammid AT, Mustafa YF, Yasin G, Alguno AC, Kianfar E (2022) Comparison and evaluation of the performance of graphene-based biosensors. Carbon Lett 32
21. Jiang Z, Feng B, Xu J, Qing T, Zhang P, Qing Z (2020) Graphene biosensors for bacterial and viral pathogens. Biosens Bioelectron 166:112471
22. Terse-Thakoor T, Badhulika S, Mulchandani A (2017) Graphene based biosensors for healthcare. J Mater Res 32:2905–2929
23. Peña-Bahamonde J, Nguyen HN, Fanourakis SK, Rodrigues DF (2018) Recent advances in graphene-based biosensor technology with applications in life sciences. J Nanobiotechnology 16:75

24. Shahriari S, Sastry M, Panjikar S, Singh Raman RK (2021) Graphene and graphene oxide as a support for biomolecules in the development of biosensors, Nanotechnol Sci Appl 197–220
25. Li D, Zhang W, Yu X, Wang Z, Su Z, Wei G (2016) When biomolecules meet graphene: from molecular level interactions to material design and applications. Nanoscale 8:19491–19509
26. Singh DP, Herrera CE, Singh B, Singh S, Singh RK, Kumar R (2018) Graphene oxide: an efficient material and recent approach for biotechnological and biomedical applications. Mater Sci Eng C 86:173–197
27. Wang Y, Li Z, Wang J, Li J, Lin Y (2011) Graphene and graphene oxide: biofunctionalization and applications in biotechnology. Trends Biotechnol 29:205–212
28. Nanda SS, Papaefthymiou GC, Yi DK (2015) Functionalization of graphene oxide and its biomedical applications. Crit Rev Solid State Mater Sci 40:291–315
29. Rahman M, Niu J, Cui X, Zhou C, Tang N, Jin H, Cui D (2022) Electrochemical biosensor based on l-arginine and rGO-AuNSs deposited on the electrode combined with DNA probes for ultrasensitive detection of the gastric cancer-related PIK3CA gene of ctDNA. ACS Appl Bio Mater 5:5094–5103
30. Islam MA, Hasan M, Rahman M, Mobarak MH, Mimona MA, Hossain N (2024) Advances and significances of carbon nanotube applications: a comprehensive review. Eur Polym J 113443
31. Gergeroglu H, Yildirim S, Ebeoglugil MF (2020) Nano-carbons in biosensor applications: an overview of carbon nanotubes (CNTs) and fullerenes (C60). SN Appl Sci 2:603
32. Yumura M (1999) Synthesis and purification of multi-walled and single-walled carbon nanotubes. In: The science and technology of carbon nanotubes, Elsevier, pp 2–13
33. Kumar S, Ahlawat W, Kumar R, Dilbaghi N (2015) Graphene, carbon nanotubes, zinc oxide and gold as elite nanomaterials for fabrication of biosensors for healthcare. Biosens Bioelectron 70:498–503
34. Mondal J, An JM, Surwase SS, Chakraborty K, Sutradhar SC, Hwang J, Lee J, Lee Y-K (2022) Carbon nanotube and its derived nanomaterials based high performance biosensing platform. Biosensors (Basel) 12:731
35. Monošík R, Stred'anský M, Šturdík E (2012) Application of electrochemical biosensors in clinical diagnosis. J Clin Lab Anal 26:22–34
36. Das A, Wazeer A (2022) Graphene and carbon nanotubes (CNTs)-based biosensor for life sciences applications. In: Next generation smart nano-bio-devices, Springer, pp 61–79
37. Rizwan M, Mohd-Naim NF, Ahmed MU (2018) Trends and advances in electrochemiluminescence nanobiosensors. Sensors 18:166
38. Fattah ARA, Abdalla AM, Mishriki S, Meleca E, Geng F, Ghosh S, Puri IK (2017) Magnetic printing of a biosensor: inexpensive rapid sensing to detect picomolar amounts of antigen with antibody-functionalized carbon nanotubes. ACS Appl Mater Interfaces 9:11790–11797
39. Smith AM, Dave S, Nie S, True L, Gao X (2006) Multicolor quantum dots for molecular diagnostics of cancer. Expert Rev Mol Diagn 6:231–244
40. Karnwal A, Kumar Sachan RS, Devgon I, Devgon J, Pant G, Panchpuri M, Ahmad A, Alshammari MB, Hossain K, Kumar G (2024) Gold nanoparticles in nanobiotechnology: from synthesis to biosensing applications. ACS Omega 9:29966–29982
41. Tai J, Fan S, Ding S, Ren L (2022) Gold nanoparticles based optical biosensors for cancer biomarker proteins: a review of the current practices. Front Bioeng Biotechnol 10:877193
42. Tessaro L, Aquino A, de Carvalho APA, Conte-Junior CA (2021) A systematic review on gold nanoparticles based-optical biosensors for Influenza virus detection. Sens Actuat Rep 3:100060
43. Sharifi M, Hosseinali SH, Alizadeh RH, Hasan A, Attar F, Salihi A, Shekha MS, Amen KM, Aziz FM, Saboury AA (2020) Plasmonic and chiroplasmonic nanobiosensors based on gold nanoparticles. Talanta 212:120782
44. Zhu C, Yang G, Li H, Du D, Lin Y (2015) Electrochemical sensors and biosensors based on nanomaterials and nanostructures. Anal Chem 87:230–249
45. Ibrahim N, Jamaluddin ND, Tan LL, Mohd Yusof NY (2021) A review on the development of gold and silver nanoparticles-based biosensor as a detection strategy of emerging and pathogenic RNA virus. Sensors 21:5114

46. Jasni AH, Ali AA, Sagadevan S, Wahid Z (2021) Silver nanoparticles in various new applications. In: Silver micro-nanoparticles—properties, synthesis, characterization, and applications
47. Rosati G, Ravarotto M, Scaramuzza M, De Toni A, Paccagnella A (2019) Silver nanoparticles inkjet-printed flexible biosensor for rapid label-free antibiotic detection in milk. Sens Actuat B Chem 280:280–289
48. Pomal NC, Bhatt KD, Modi KM, Desai AL, Patel NP, Kongor A, Kolivoška V (2021) Functionalized silver nanoparticles as colorimetric and fluorimetric sensor for environmentally toxic mercury ions: an overview. J Fluoresc 31:635–649
49. Kim D-M, Yoo S-M (2022) Colorimetric systems for the detection of bacterial contamination: strategy and applications. Biosensors (Basel) 12:532
50. Ramesh M, Janani R, Deepa C, Rajeshkumar L (2022) Nanotechnology-enabled biosensors: a review of fundamentals, design principles, materials, and applications. Biosensors (Basel) 13:40
51. Pateraki M, Fysarakis K, Sakkalis V, Spanoudakis G, Varlamis I, Maniadakis M, Lourakis M, Ioannidis S, Cummins N, Schuller B (2020) Biosensors and internet of things in smart healthcare applications: challenges and opportunities. Wear Implant Med Dev 25–53
52. Malhotra BD (2017) Biosensors: fundamentals and applications. Smithers rapra
53. Bhatia D, Paul S, Acharjee T, Ramachairy SS (2024) Biosensors and their widespread impact on human health. Sens Int 5:100257
54. Shah NS, Thotathil V, Zaidi SA, Sheikh H, Mohamed M, Qureshi A, Sadasivuni KK (2022) Picomolar or beyond limit of detection using molecularly imprinted polymer-based electrochemical sensors: a review. Biosensors (Basel) 12:1107
55. Li D, Xu K, Wang J, Ye Z, Ying Y, Li Y (2008) Quartz crystal au electrode-based electrochemical immunosensor for rapid detection of *Escherichia Coli* O157:H7, Trans ASABE 51:1847–1852. https://doi.org/10.13031/2013.25291.
56. Rho D, Breaux C, Kim S (2020) Label-free optical resonator-based biosensors. Sensors 20:5901
57. Li B, Zhang R, Bi R, Olivo M (2022) Applications of optical fiber in label-free biosensors and bioimaging: a review. Biosensors (Basel) 13:64
58. Peltomaa R, Glahn-Martínez B, Benito-Peña E, Moreno-Bondi MC (2018) Optical biosensors for label-free detection of small molecules. Sensors 18:4126
59. Montoya A, Ocampo A, March C (2009), Fundamentals of piezoelectric immunosensors. In: Piezoelectric transducers and applications, Springer, pp 289–306
60. Pohanka M (2018) Overview of piezoelectric biosensors, immunosensors and DNA sensors and their applications. Materials 11:448
61. Tombelli S (2012) Piezoelectric biosensors for medical applications. In: Biosensors for medical applications, Elsevier, pp 41–64
62. Li L, Zhao H, Chen Z, Mu X, Guo L (2010) Aptamer-based electrochemical approach to the detection of thrombin by modification of gold nanoparticles. Anal Bioanal Chem 398:563–570
63. Lino C, Barrias S, Chaves R, Adega F, Martins-Lopes P, Fernandes JR (2022) Biosensors as diagnostic tools in clinical applications. Biochimica et Biophysica Acta (BBA)-Rev Cancer 1877:188726
64. Sengupta J, Hussain CM (2023) CNT and graphene-based transistor biosensors for cancer detection: a review. Biomolecules 13:1024
65. Giménez-Marqués M, Hidalgo T, Serre C, Horcajada P (2016) Nanostructured metal–organic frameworks and their bio-related applications. Coord Chem Rev 307:342–360
66. Devaraj M, Sasikumar Y, Rajendran S, Ponce LC (2021) Metal organic framework based nanomaterials for electrochemical sensing of toxic heavy metal ions: progress and their prospects. J Electrochem Soc 168:037513
67. Lustig WP, Mukherjee S, Rudd ND, Desai AV, Li J, Ghosh SK (2017) Metal–organic frameworks: functional luminescent and photonic materials for sensing applications. Chem Soc Rev 46:3242–3285

68. Li M, Zhang G, Boakye A, Chai H, Qu L, Zhang X (2021) Recent advances in metal-organic framework-based electrochemical biosensing applications. Front Bioeng Biotechnol 9:797067
69. Wang Y-H, Huang K-J, Wu X (2017) Recent advances in transition-metal dichalcogenides based electrochemical biosensors: a review. Biosens Bioelectron 97:305–316
70. Sengupta J, Hussain CM (2024) Molybdenum disulfide-based field effect transistor biosensors for medical diagnostics: exploring a decade of advancements (2014–2024). TrAC Trends Anal Chem 117742
71. Khaleghi M, Chaji M, Pishbin F, Sillanpää M, Sheibani S (2024) A review of molybdenum disulfide-based 3d printed structures for biomedical applications. J Mater Res Technol
72. Yüce M, Ullah N, Budak H (2015) Trends in aptamer selection methods and applications. Analyst 140:5379–5399
73. DeRosa MC, Lin A, Mallikaratchy P, McConnell EM, McKeague M, Patel R, Shigdar S (2023) In vitro selection of aptamers and their applications. Nat Rev Meth Primers 3:54
74. Lowdon JW, Diliën H, Singla P, Peeters M, Cleij TJ, van Grinsven B, Eersels K (2020) MIPs for commercial application in low-cost sensors and assays–an overview of the current status quo. Sens Actuat B Chem 325:128973
75. Rahman S, Bozal-Palabiyik B, Unal DN, Erkmen C, Siddiq M, Shah A, Uslu B (2022) Molecularly imprinted polymers (MIPs) combined with nanomaterials as electrochemical sensing applications for environmental pollutants. Trends Environ Anal Chem 36:e00176
76. Azizi A, Bottaro CS (2020) A critical review of molecularly imprinted polymers for the analysis of organic pollutants in environmental water samples. J Chromatogr A 1614:460603
77. Si B, Song E (2018) Recent advances in the detection of neurotransmitters. Chemosensors 6:1
78. Zaidi SA (2018) Development of molecular imprinted polymers based strategies for the determination of Dopamine. Sens Actuat B Chem 265:488–497
79. Wang S-Y, Du Y-C, Wang D-X, Ma J-Y, Tang A-N, Kong D-M (2021) Signal amplification and output of CRISPR/Cas-based biosensing systems: a review. Anal Chim Acta 1185:338882
80. Masi A, Antonacci A, Moccia M, Frisulli V, De Felice M, De Falco M, Scognamiglio V (2023) CRISPR-Cas assisted diagnostics: a broad application biosensing approach. TrAC, Trends Anal Chem 162:117028
81. Mohammad N, Katkam SS, Wei Q (2022) Recent advances in CRISPR-based biosensors for point-of-care pathogen detection. CRISPR J 5:500–516
82. Eveness J, Cao L, Kiely J, Luxton R (2022) Equivalent circuit model of a non-faradaic impedimetric ZnO nano-crystal biosensor. J Electroanal Chem 906:116003
83. Kokab T, Shah A, Khan MA, Nisar J, Ashiq MN (2021) Electrochemical sensing platform for the simultaneous femtomolar detection of amlodipine and atorvastatin drugs. RSC Adv 11:27135–27151
84. Shrikrishna NS, Sharma R, Sahoo J, Kaushik A, Gandhi S (2024) Navigating the landscape of optical biosensors. Chem Eng J 151661
85. Saini TS, Kumar A, Sinha RK (2022) Photonic crystal cavity-based sensors and their potential applications. In: Nanomaterials-based sensing platforms, Apple Academic Press, pp 75–116
86. Ramirez JC, Grajales García D, Maldonado J, Fernández-Gavela A (2022) Current trends in photonic biosensors: advances towards multiplexed integration. Chemosensors 10:398
87. Nikoleli GP, Siontorou CG, Nikolelis DP, Bratakou S, Karapetis S, Tzamtzis N (2018) Biosensors based on microfluidic devices lab-on-a-chip and microfluidic technology. Nanotechnol Biosens 375–394
88. Lopez GA, Estevez M-C, Soler M, Lechuga LM (2017) Recent advances in nanoplasmonic biosensors: applications and lab-on-a-chip integration. Nanophotonics 6:123–136
89. Luka G, Ahmadi A, Najjaran H, Alocilja E, DeRosa M, Wolthers K, Malki A, Aziz H, Althani A, Hoorfar M (2015) Microfluidics integrated biosensors: A leading technology towards lab-on-a-chip and sensing applications. Sensors 15:30011–30031
90. Sharma B, Sharma A (2022) Microfluidics: recent advances toward lab-on-chip applications in bioanalysis. Adv Eng Mater 24:2100738

91. Liu K, Ouyang B, Guo X, Guo Y, Liu Y (2022) Advances in flexible organic field-effect transistors and their applications for flexible electronics. Flexible Electronics 6:1
92. Liu F, Lorenzelli L (2024) Toward all flexible sensing systems for next-generation wearables. Wearable Electron
93. Trung TQ, Lee N (2016) Flexible and stretchable physical sensor integrated platforms for wearable human-activity monitoringand personal healthcare. Adv Mater 28:4338–4372
94. Ling H, Liu S, Zheng Z, Yan F (2018) Organic flexible electronics. Small Meth 2:1800070
95. Rajitha G, Dash RK (2018) Optically transparent and high dielectric constant reduced graphene oxide (RGO)-PDMS based flexible composite for wearable and flexible sensors. Sens Actuat A Phys 277:26–34. https://doi.org/10.1016/j.sna.2018.04.040
96. Veeralingam S, Bharti DK, Badhulika S (2022) Lead-free PDMS/PPy based low-cost wearable piezoelectric nanogenerator for self-powered pulse pressure sensor application. Mater Res Bull 151. https://doi.org/10.1016/j.materresbull.2022.111815
97. Chen G, Chen G, Pan L, Chen D (2022) Electrospun flexible PVDF/GO piezoelectric pressure sensor for human joint monitoring, Diam Relat Mater 129. https://doi.org/10.1016/j.diamond.2022.109358
98. Khan S, Ali S, Bermak A (2019) Recent developments in printing flexible and wearable sensing electronics for healthcare applications. Sensors 19:1230
99. Xu S, Wu W (2020) Ink-based additive nanomanufacturing of functional materials for human-integrated smart wearables. Adv Intell Syst 2:2000117
100. Volk MJ, Lourentzou I, Mishra S, Vo LT, Zhai C, Zhao H (2020) Biosystems design by machine learning. ACS Synth Biol 9:1514–1533
101. Youssef AE, Alfarraj O, Alkhalaf M, Hassanein AS (2022) A supervised biosensor-based non-variant structuring approach for analyzing infectious disease data. Measurement 193:110903
102. Amethiya Y, Pipariya P, Patel S, Shah M (2022) Comparative analysis of breast cancer detection using machine learning and biosensors. Intell Med 2:69–81
103. Liu G, Lv Z, Batool S, Li M, Zhao P, Guo L, Wang Y, Zhou Y, Han S (2023) Biocompatible material-based flexible biosensors: from materials design to wearable/implantable devices and integrated sensing systems. Small 19:2207879
104. Bullock CJ, Bussy C (2019) Biocompatibility considerations in the design of graphene biomedical materials. Adv Mater Interfaces 6:1900229
105. Alam F, Ashfaq Ahmed M, Jalal AH, Siddiquee I, Adury RZ, Hossain GMM, Pala N (2024) Recent progress and challenges of implantable biodegradable biosensors. Micromachines (Basel) 15:475
106. Khan AA, Kim JH (2024) Recent advances in materials and manufacturing of implantable devices for continuous health monitoring. Biosens Bioelectron 116461
107. Rodrigues D, Barbosa AI, Rebelo R, Kwon IK, Reis RL, Correlo VM (2020) Skin-integrated wearable systems and implantable biosensors: a comprehensive review. Biosensors (Basel) 10:79
108. Chircov C, Grumezescu AM (2022) Microelectromechanical systems (MEMS) for biomedical applications. Micromachines (Basel) 13:164
109. Tiwari S, Sharma V, Mujawar M, Mishra YK, Kaushik A, Ghosal A (2019) Biosensors for epilepsy management: state-of-art and future aspects. Sensors 19:1525
110. Cheran L-E, Benvenuto P, Thompson M (2008) Coupling of neurons with biosensor devices for detection of the properties of neuronal populations. Chem Soc Rev 37:1229–1242
111. Remaggi G, Zaccarelli A, Elviri L (2022) 3D printing technologies in biosensors production: recent developments. Chemosensors 10:65
112. Muñoz J, Pumera M (2020) 3D-printed biosensors for electrochemical and optical applications. TrAC, Trends Anal Chem 128:115933
113. Tay RY, Song Y, Yao DR, Gao W (2023) Direct-ink-writing 3D printed bioelectronics. Mater Today
114. Nguyen AK, Narayan RJ (2017) Two-photon polymerization for biological applications. Mater Today 20:314–322

115. Dejous C, Krishnan UM (2021) Sensors for diagnosis of prostate cancer: looking beyond the prostate specific antigen. Biosens Bioelectron 173:112790
116. Yu ES, Lee SH, Bae YG, Choi J, Lee D, Kim C, Lee T, Lee SY, Lee SD, Ryu YS (2018) Highly sensitive color-tunablility by scalable nanomorphology of dielectric layer in liquid permeable metal-insulator-metal structure. ACS Appl Mater Interf. https://doi.org/10.1021/acsami.8b12553
117. Xie M, Zhao F, Zhang Y, Xiong Y, Han S (2022) Recent advances in aptamer-based optical and electrochemical biosensors for detection of pesticides and veterinary drugs. Food Control 131:108399

Advancements in Nanobiosensors and Their Prospects in the Agriculture and Food Sectors

Balaji Govindaswamy

Abstract Nanotechnology applications in the agriculture and food sectors are becoming increasingly widespread. The food and agriculture industries are deeply ingrained in people's daily lives worldwide, and the success of any country's economy is heavily reliant on the performance of these two critical businesses. The cultivation of healthy food is inextricably linked to good health, and nanosensors have the potential to play an important role in protecting foods from harmful conditions. Owing to their high sensitivity, specificity, and quick reaction times, nanobiosensors are critical in tackling current difficulties in the agricultural and food sectors. These technologies, which use nanomaterials such as nanoparticles, nanotubes, and quantum dots, allow for unparalleled precision in detecting infections, pollutants and nutritional levels. Nanobiosensors are used in agriculture to monitor soil health, diagnose crop diseases and practice precision farming, all of which improve resource efficiency and crop yields. This chapter focuses on current advances in nanobiosensor technology, their applications in the agricultural and food sectors with future prospects, and breakthroughs in nanotechnology development.

Keywords Nanotechnology · Quantum dots · Biosensors · Precision farming · Agriculture

1 Introduction

Numerous issues facing the food and agricultural industries jeopardize the sustainability and security of the world's food supply. Crop yield is hampered in agriculture by problems such as deteriorating soil health, the waste of water and fertilizers, and pests' increasing resistance to traditional insecticides [1]. Climate change exacerbates these issues by increasing the frequency of extreme weather events, which further disrupts food production. Since forborne illnesses, contamination, and spoilage pose

B. Govindaswamy (✉)
Department of Biotechnology, Bhupat and Jyoti Mehta School of Biosciences, Indian Institute of Technology Madras, Chennai, Tamil Nadu 600036, India
e-mail: balajishilou.08@gmail.com

D. Sillu et al. (eds.), *Nanobiosensors for the Food Industry*, Smart Nanomaterials Technology, https://doi.org/10.1007/978-981-95-0136-6_3

serious dangers to public health, the food industry faces ongoing challenges in maintaining the safety and quality of its goods [2]. Conventional approaches for monitoring and controlling these issues are frequently inefficient, expensive and imprecise, which makes timely responses difficult and results in resource waste, environmental harm, and financial losses. Innovative approaches that can improve the productivity, security, and sustainability of farming methods and food production are needed to address these problems [3, 4].

Precise monitoring of crop conditions, insect populations, and soil health is essential in agriculture to maximize resource use and increase crop yields. Conventional monitoring techniques frequently lack the speed and sensitivity needed for decision making, which results in inefficiencies and an excessive reliance on chemical inputs such as pesticides and fertilizers [5]. Nanobiosensors, with their increased detection capabilities, provide a possible solution by providing real time, in situ monitoring of different parameters, including soil nutrient levels, moisture content, and the presence of diseases or pests [6]. This enables more focused treatments, decreasing the environmental impact and supporting sustainable farming practices.

A primary concern in the food industry is guaranteeing the safety and quality of food items because of the possible health dangers and financial ramifications of food-borne diseases and contamination [7]. Conventional techniques for identifying pollutants, such as chemical analysis and microbiological culture, can be laborious and may not yield the required sensitivity or fast findings [8]. Owing to their quick reaction times and ability to identify minuscule levels of pollutants such as toxins, allergies, pesticides, and microorganisms, nanobiosensors are effective instruments for monitoring food safety in real time [9]. This helps producers maintain strict quality control requirements while also improving consumer safety.

The combination of biosensor technology and nanotechnology has led to new opportunities for innovation in a number of industries, including agriculture and food. Biosensors are analytical devices that use a combination of biological recognition elements (such as enzymes, antibodies, or nucleic acids) and a transducer to detect specific biological or chemical substances [10]. Nanotechnology, on the other hand, involves manipulating materials at the nanoscale, giving these materials unique physical, chemical and biological properties, such as increased surface area, enhanced reactivity, and improved electrical conductivity [11].

The food and agricultural industries have experienced dramatic breakthroughs due to the rapid progress of nanotechnology [12]. The creation and use of nanobiosensors is one of the most exciting applications. Analytical tools known as nanobiosensors combine the special qualities of nanomaterials with the sensitivity and specificity of their ability to recognize biological components [13]. These tools are extremely useful in agricultural and food safety procedures because they provide a strong means of identifying and monitoring a broad variety of biological and chemical contaminants [14].

2 Fundamentals of Nanobiosensors

The three parts of a nanobiosensor are the biological probe, which includes affinity-based materials such as those found in interactions between antibodies and antigens, enzymes and substrates, nucleic acids, and cells; the transducer, which transforms biological signals into digital signals; and the data recording unit, which stores and transfers data [15, 16]. Biomimetics, aptamers, enzymes, antibodies, and DNA are examples of biological components [17]. Synthetic receptors play a role in naturally occurring biological components to imitate functions with a faster and more focused detection range [18]. Biological probes in conjunction with different nanoparticles, including magnetic, metallic, quantum dot, grapheme oxide, potentiometric, and optical signals such as calorimetric, surface plasmon resonance, metallic fluorescence, and optical fibers, are examples of transducers that use electrochemical signals [19, 20]. The development of nanosensors has made use of carbon nanotube-based nanomaterials, namely, single-walled carbon nanotubes (SWCNTs) and multiwalled carbon nanotubes (MWCNTs) [21, 22]. Carbon-based nanomaterials provide a superior surface for immobilizing biological components in biosensors [23]. It adds physical and mechanical qualities. However, the main drawback in this instance is the low solubility of nanomaterials in water. The creation of nanobiosensors, toxicity, and lack of selectivity in combination with the sustainability of nanostructures are additional difficulties [24, 25]. Silicon oxides, titanium, and fullerenes are among the other materials employed. For encapsulation, polymers, including polyacrylic acid, zeolites, and chitosan, can be utilized. Because of their safe qualities, certain metallic nanoparticles often made of silver and gold can be employed for nanobiosensors [26].

The most common materials utilized in food nanobiosensors to identify pollutants and pathogens in food and water are gold- and silver-based nanocompounds. One noble metal with intriguing nanoscale features is gold. One of the nanomaterials that has been the subject of active study over the past ten years is gold nanoparticles (AuNPs), also referred to as gold sols [27] (colloids in which solid particles are scattered in the liquid phase of countries). It consists of various sizes and is composed of gold nanomaterials, nanoclusters and nanorods. To create a lateral flow-based immunostrip test for multiple detection of *Escherichia coli* and *Shigella bodyii* in food samples, AuNPs were employed as optical markers [28, 29]. Secondary antibodies labeled with AuNPs were immobilized on the strip together with monoclonal antibodies specific to the aforementioned pathogens. The developed lateral flow strip has the benefits of being affordable, simple to use, sensitive, and useful in low-resource environments. Another study demonstrated the use of AuNP-based nanoELISA (enzyme-like antibody aptamer sandwich) for *Salmonella typhimurium* detection [30]. Salmonella-specific aptamers were rendered immobile on magnetic nanoparticles, which serve as a means of entrapment [31]. Using various biorecognition components, AuNPs work as catalysts in the enzyme-linked design of gold nanoparticles and quantum dot-based colorimetric and fluorescence sensors [32].

A unique material called the magnetoelastic material is made of magnets and gold nanoparticles. Easy separation and the avoidance of nonspecific binding of biological molecules, which lowers the signal-to-noise ratio, are two benefits that magnetic nanomaterials can offer. These include three groups, oxides, metals, and alloys, each with unique characteristics [33]. Among these developments is the microfluidic method, which incorporates biosensing, nanotechnology, and microsystems at microscale values [34]. These microfluidics techniques have been applied to detect pollutants, pathogens and food-borne allergies. Together with hydrodynamic and electrokinetic fluxes, manufacturing at a smaller scale still presents significant difficulty [35]. It is possible to circumvent these difficulties by incorporating DNA into microfluidic devices. To promote food safety, another tactic is to integrate the microfluidic technique with cell morphology. Nanobiosensors have been made using biological components in place of metals and other elements [36]. Bilayer lipid membranes are among these materials. The majority of lipid membrane-based nanobiosensors have been developed and used for the detection of microbes, poisons and metals from food [37, 38]. Nanobiosensors engage new tools to preserve agricultural practice sustainability, which can alleviate a few issues with traditional approaches, including the inability to handle a large number of samples at once, interference with different matrices and the length of the sample intake time for analysis [39, 40].

The unsustainability of agriculture and food was caused by the overuse of chemicals and pesticides. Increased yield is the outcome of the use of agricultural techniques based on nanotechnology methods, and nanobiosensors may be based on the larger-scale and widespread responses of a crop in an entire field. The mathematical models of sensors used in nanobionics are used to measure the fluxes of different phytohormones and signaling chemicals. Because nanobiosensors can sense, analyze, and identify changes in the environment, their usage has generally contributed to the establishment of smart agriculture and/or precise farming [41–43].

3 Application in the Agriculture Industry

3.1 *Analytical Tools in Agriculture*

Nanobiosensors are spherical sensors with diameters ranging from 20 to 200 nm made of sensor molecules encased by microemulsion polymerization in a chemically inert matrix [44]. Certain sensor molecules may be trapped, including those that sense fluorescence, pH, Ca2+ ions, optical transitions, or any combination of these [45]. In addition to being crucial for protein interactions, they control the intracellular imaging of ions and molecules and exhibit excellent stability and reversibility throughout photobinding [46]. Ions and molecules may be found via these sensors in real time. They have solid oxygen-sensing capabilities in human plasma that are less impacted by light dispersion and autofluorescence [47]. In complex biological settings, golden nanoshells are used in fast immunoassays to identify analytes without sample pretreatment. The near-infrared extinction spectra of the antibody/nanoshell conjugates are shown in the presence of analytics. Nanoshells can improve chemical sensitivity by a factor of ten billion [48, 49]. Compared with conventional biosensors, nanomaterial-based biosensors are becoming more intriguing. Owing to their exceptional sensitivity [50] and specificity for identification, they have enhanced biodefense capabilities, irrigation capabilities, food quality control, environmental and bioprocess management, and, most importantly, medical applications [51, 52]. To diagnose soil illness, it is crucial to quantify the variations in oxygen intake (relative activities) between bad microbes and good microbes in the soil environment [53]. The biosensing chip consists of two electrodes soaked in a soil sample suspension in a buffer solution, one loaded with positive microbes and the other loaded with negative microbes, and data on oxygen intake from two bacteria are computed [54]. The microorganisms that enhance the soil may be easily identified by comparing two sets of data. Additionally, we can predict when soil disease will be ready for soil breakdown [55]. The biosensor provides a semi-quantitative method for novel soil diagnostic procedures. Materials that contain and transfer nutrients inside nanoparticles, called nanofertilizers, are better solutions for enhancing the soil environment [44]. Nutrients can be distributed as minuscule particle matter or emulsions, wrapped in a thin coating, or integrated into nanomaterials such as nanotubes or nanoporous materials [56]. The use of nanofertilizers may reduce leaching, emission and long-term absorption of soil microbes, which could lead to monitoring soil quality for better farming and control pest movement around the field [57].

3.2 Precision Farming

With the very desirable objective of increasing the overall goal of increasing crop output while minimizing the use of fertilizers, pesticides, herbicides, etc., precision farming uses information technology to manage successful farming. The changes in the surroundings with sporadic duties need to be monitored [58]. Worldwide technologies such as remote sensing, geographic information systems, and systems make farming more economical and effective [59]. Measurements of the local environment are made via geographical methods involving the use of information systems to determine the locations of agricultural expansion distributions that might administer site-specific fertilizers and manage infections [60], both of which are essential for yield research, monitoring for pests, measuring parameters, including floods and drought, and predicting [61]. Every nano-based material has a distinct physicochemical characteristic that makes it quantifiable and appropriate for a variety of sensing techniques [62]. Furthermore, nano-based materials are analytical instruments or devices that are highly developed and enhanced biosensors that make it possible to identify biological components with the aid of signals produced by transducers [63]. With the use of these data, plant scientists can foresee disease signs and ascertain the nutritional and water needs of crops [64]. Composted plants, pesticides, and other very successful agronomic management techniques can all be made feasible by a well-designed system of nanobiosensors at the nanoscale [65]. The use of nanosensors and nanobiosensors benefits precision agriculture, including sense variations, intensity, polarity, time, and phase [66].

Moreover, near-infrared fluorescent H_2O_2 nanosensors are thought to be a revolutionary tool in precision farming, as they can detect early indicators of stress [67]. Additionally, red-color selenium nanoparticles produced by bioreduction may be employed as optical biosensors in tests for heavy metal toxicity [68]. With a larger range of applications in precision agriculture systems, the nanosensor-based electronic tongue eTongue and electronic nose eNose [69] are extremely precise, sensitive, redundant, selective, and dependable devices [70]. Their uses include the detection of a class of gases, taste, scents, and their fluctuating concentrations. Their functions are comparable to those of the human nose and tongue. Industries, including food and drink, agriculture, pharmaceuticals, and the engineering and processing of personal care products, have reached a key level that allows them to control both the amount and quality of material [71]. For example, eNose may be used to assess the scent of rice and thus stop its deprivation by acting in a timely manner [72]. The major force behind computerization in the field of precision agriculture is wireless sensor networks [73]. To guarantee productivity and sustainability, wireless nanosensors are deployed in fields to gather accurate data on crop, soil, and weather conditions [74].

Smart irrigation, smart fertilization, smart pest management, and greenhouse monitoring are a few of the more common applications of wireless sensor networks in precise farming [75]. To increase output, fertilizers must be applied in a very exact amount. Soil fertility is measured in real time via an automated wireless sensor-based fertilization system [76]. The agriculture business also uses electrochemical sensors for a variety of applications, identifying certain ions in the soil [77]; electrochemical sensors collect chemical data about the soil [78]. The pH and soil nutrient concentrations were used to collect these data. Electrochemical sensors are employed to detect MPK, pesticide residues, pathogenic *Staphylococcus aureus*, chloramphenicol, and nitrogen species, excluding spectroscopic detection [79, 80]. Many nanosensors have been developed and used to detect heavy field-effect transistors. Recent research has reported that optical nanosensors based on graphene perform better in detecting heavy metal ions. Electrochemical nanosensors can be used to monitor the concentration of heavy metal ions in the soil [81, 82]. Moreover, a DNA optical biosensor in combination with evanescent wave evaluation might be used for the identification of heavy metals via in situ detection to successfully measure heavy metal ions [83]. Beta-carotene from palm kernels might be utilized as a biological reporter in biosensors that use biopolymers to detect heavy metals such as aluminum [84]. Soil contains a multitude of nutrients in diverse forms, and each type emits a distinct signal. For example, high ammonia and nitrate levels in the soil indicate the presence of soil-feeding insects and microbial imbalances, respectively [85]. Numerous microorganisms are prevalent in soil and have a major influence on plant development and soil health [86]. The soil moisture content is another soil characteristic that is used to monitor and control crop production [87]. Recently, nanosensors have created rapid and accurate methods for measuring the moisture content of soil in the field via an integrated capacitive sensor [88]. Moreover, nitrite and urease concentrations in soil and water are detected via biosensors via microfluidic impedimetric and colorimetric assays [89–91].

3.3 *Environmental Monitoring*

In environmental monitoring, nanotechnology plays a crucial role in enhancing the sensitivity, accuracy and speed of detecting pollutants, as well as developing new methods for pollution control and environmental remediation. One of the key uses of nanotechnology in environmental monitoring is the creation of sophisticated nanosensors. These sensors are intended to detect and measure environmental toxins at extremely low concentrations, frequently down to parts per billion or even parts per trillion [92]. This degree of sensitivity is crucial for monitoring contaminants with hazardous effects even at low concentrations, such as heavy metals, insecticides, and persistent organic pollutants [93]. Carbon nanotubes, quantum dots, and metal oxide nanoparticles are commonly employed in the creation of these nanosensors because of their distinct electrical, optical and chemical capabilities [94]. Carbon nanotubes, e.g., may be functionalized with certain chemical groups to preferentially attach to certain contaminants, making them extremely effective at detecting harmful gases such as carbon monoxide and volatile organic compounds (VOCs) [95]. Another important use of nanomaterials in environmental monitoring is water quality evaluation. Nanotechnology enables the creation of portable, low-cost instruments for detecting toxins in water, including heavy metals, infections, and organic pollutants. Nanoscale materials, such as silver nanoparticles and titanium dioxide, are employed in water treatment procedures to decompose organic pollutants by photocatalysis or to disinfect water by killing bacteria and viruses [96]. Furthermore, nanoparticles may be included in membranes for water filtration, increasing the efficiency with which pollutants are removed compared with typical filtering procedures. This is especially crucial for supplying safe drinking water in locations with limited access to modern water treatment facilities [97].

Nanobiosensors, which combine nanomaterials and biological recognition components, are another effective technique for environmental monitoring. These biosensors use biological molecules such as enzymes, antibodies, and DNA to detect environmental contaminants with great selectivity [98]. The use of nanomaterials in these biosensors improves signal transduction, increases sensitivity, and allows real-time monitoring of environmental parameters [99]. For example, gold nanoparticles can be used to magnify signals in DNA biosensors, allowing for the identification of minute quantities of genetic material from dangerous microbes in water or soil samples [100]. Nanobiosensors are also being developed to detect endocrine-disrupting substances, which are causing increasing concern because of their potential to interact with hormone systems in wildlife and people [101]. In addition to being used for detection, nanobiosensors are being used to remediate environmental pollutants. Nanomaterials are being employed to create novel ways to clean contaminated soil and water. For example, iron nanoparticles are utilized to remediate groundwater contaminated with chlorinated hydrocarbons. These nanoparticles can react with contaminants, breaking them down into innocuous chemicals via a process known as reductive dechlorination [102]. Similarly, nanomaterials such as grapheme oxide are being investigated for their ability to absorb heavy metals from contaminated water, providing a potential option for minimizing the impacts of industrial pollution [103]. The use of nanotechnology in environmental monitoring also includes air quality evaluation. Nanoparticle-based sensors are being developed to accurately and continuously monitor air pollutants such as particulate matter, nitrogen oxides, and sulfur dioxide [104]. These sensors are not only more sensitive than traditional sensors but also smaller and more energy efficient, making them ideal for use in urban areas where air pollution is a major concern [105]. Additionally, nanoparticles are being studied for their ability to absorb and neutralize airborne contaminants, providing a proactive approach to air quality management [106]. While nanotechnology has intriguing uses in environmental monitoring, there are certain limitations and concerns about its utilization [107]. The possible environmental and health concerns associated with the discharge of engineered nanoparticles into the environment are not completely known [108]. As nanotechnology advances, it is critical to perform rigorous risk assessments and develop regulatory frameworks to guarantee that the benefits of new technologies are realized without jeopardizing the environment or public health.

3.4 Implications of Agrochemicals

One of the most common applications of nanobiosensors in agrochemicals is the precise detection and monitoring of pesticide residues. Pesticides, while necessary for crop protection from pests and illnesses, can have negative impacts on human health and the environment if not used properly [109]. Conventional pesticide residue detection methods frequently require complex and time-consuming laboratory tests. In contrast, nanobiosensors provide a quick, onsite solution. These sensors can be built to attach precisely to pesticide molecules via nanomaterials such as gold nanoparticles or carbon nanotubes, allowing real-time detection at extremely low concentrations [110]. This enables farmers to monitor pesticide levels in crops and soil, ensuring that they are within acceptable limits and lowering the danger of contamination [111]. Nanobiosensors are also being developed to target fertilizer applications. Fertilizers are necessary for agricultural growth, but excessive use can cause environmental problems such as soil deterioration, water pollution, and greenhouse gas emissions [112]. Nanosensors can monitor soil nutrient levels and provide real-time data to help with precision in fertilizer applications. For example, a biosensor may be used to monitor nitrogen levels in the soil, allowing farmers to apply fertilizer only when necessary and in the appropriate proportions [113]. This not only increases agricultural yields but also reduces the environmental impact of excessive fertilizer use. In addition to monitoring, nanobiosensors can be integrated into smart agrochemical delivery systems. These methods encapsulate agrochemicals with nanomaterials, which protects them from degradation and allows for regulated release [114]. Nanobiosensors can be programmed to release these compounds in response to certain environmental cues, such as the presence of pests or changes in soil nutrient levels [42]. This tailored delivery technique guarantees that agrochemicals are not only released when they are required but also decrease waste and increase the efficiency of their use. Another important application of nanobiosensors in agrochemicals is the detection and control of soil pathogens. Soil-borne illnesses can devastate agricultural output; therefore, early discovery is important for effective management [91].

Nanosensors can identify certain diseases in soil before they cause major damage. For example, a biosensor could be developed to detect the presence of fungal spores that cause root rot in crops [115]. Farmers who identify the pathogen early can use fungicides more effectively, minimizing disease spread and reducing the overall amount of pesticides required [116]. Nanobiosensors can help improve precision agriculture methods, which optimize the use of agrochemicals through data-driven decision making. Nanobiosensors, which provide real time, site-specific information on soil health, pest populations, and crop conditions, allow farmers to customize their pesticide use to the individual demands of their fields [117]. This not only increases crop output but also minimizes the environmental footprint of agriculture by limiting the use of chemicals [118]. Nanobiosensors are effective instruments for increasing the efficiency, precision, and sustainability of pesticide use in agriculture [119]. These sensors, which enable real-time monitoring and targeted administration of pesticides, fertilizers, and other agrochemicals, can help reduce environmental impacts, increase crop yields, and encourage the development of more sustainable agricultural practices [120].

4 Application in the Food Industry

4.1 Detection of Food Chemicals, Toxins, and Other Materials

Nanoparticles are ideal examples of biosensors because of their highly developed sensing capabilities and other advantageous qualities. Nanosensors or nanobiosensors are used to quantify the components of food that are readily available, identify pathogens in food and agricultural products, and notify distributors and customers about the safety of good foods [121]. The nanosensors perceive minute changes in the external environment, such as temperature fluctuations, humidity fluctuations, and microbial contamination in storage rooms [122]. The rapid reaction mechanisms further reveal these changes. Since biological and chemical pollutants cannot be readily detected as physical contaminants can, it is essential to detect their presence in food [123]. In addition to detecting sugars, alcohols, flavors, amino acids, and sweeteners, accurate detection of biological and chemical contamination is necessary for a variety of culinary applications [124].

The instruments used to detect chemical and chemical pollutants are known as chem-bio nanosensors. Chemical pollutants include pesticides, veterinary medications, and food additives; biological contaminants include toxicants and pathogens [125]. The main sources of biological pollutants include food-borne pathogenic bacteria; natural toxicants such as ochratoxins and aflatoxins; and other microorganisms that are created during the preparation, distribution, and preservation of food [126].

One type of food industry-targeted research is food preservation. A number of nanosensor-based food preservation systems have been implemented in the last few years [127]. These include the use of silicon-based microfluid systems and the inclusion of chemicals to improve the performance of various electrophoresis techniques in chip setup and NEMS assembled devices [128]. These device components, which range in size from the nanoscale to the millimeter scale and are employed in food preservation, have attracted much economic attention [129]. NEMS technology can be used to identify the presence of trans fat in food [130].

Even a tiny amount of heavy metal exposure can cause serious health issues in humans. One of the causes of soil pollution is the removal of water and waste products from industry [131, 132]. Notably, chromium contamination in certain regions leads to the build-up of chromium ions during local plant life [133]. When these ions reach above their threshold level, they have hazardous effects on both living things and people. They may also cause harm to the liver, skin, and nervous system. When biosensors are used, the detection sensitivity of nanomaterials can increase [134]. These include silica mesoporous/nanospheres, multiwall CNTs and other metal-based nanoparticles [135]. According to recent research, aminated magnetic iron oxide nanoparticles such as copper and zinc can operate as adsorbents to draw aqueous heavy metal ions out of a sample [136]. The magnesium oxide NPs produced via the solgel and calcination methods are capable of removing heavy metals from polluted water samples and reducing the impact of bacterial infection [137]. Reusable and recoverable synthetic iron oxides are stable when heavy metals are removed from a sample [138].

Steroids are lipid-containing natural hormones; more than thirty different steroids are generated in the adrenal cortex. The body's level of steroids is important since increased levels of this hormone can lead to endocrinological disorders that affect gonadal and adrenal function [139]. Adults have extremely low concentrations of steroid hormones, such as estradiol (110 nmol/L), whereas children have concentrations between 18 and 165 nmol/L. The concentration of progesterone is 3.2 nmol/L in women, whereas that of testosterone in males is 0.35 nmol/L [140]. Hormone detection is traditionally accomplished via immunoassays, such as radioimmunoassays. Nanosensors have used carbon nanotubes and silica nanowires doped with boron. Because the function of detection isomerization is well established [141], these compounds are utilized as receptors for steroids. The well-organized intermolecular binding of a charged ligand to mimic the total binding of the analyte to the protein residues is the fundamental idea behind the sensing process [142]. Large driving forces are created by hydrophobic effects to encourage appropriate protein binding to hydrophobic ligands [143]. The thermodynamics of protein–ligand binding may be effectively manipulated by adjusting the enthalpy [144].

4.2 Nanopackaging and Food Preservation

Nanobiosensors in nanopackage provide real-time monitoring of food freshness. Food deterioration is frequently caused by microbial growth, oxidation, or the production of toxic gases such as ammonia or sulfur compounds. Nanobiosensors incorporated in packaging materials can detect changes at the molecular level [145]. For example, a sensor may be built to detect the presence of particular gases that are emitted as food spoils, such as ethylene in fruits and vegetables. When sensors detect gas, they can display a visual or electronic alarm on the packaging, indicating that the product is no longer fresh [146]. This type of smart packaging allows consumers to make more educated judgments regarding the quality of the food they buy and consume, decreasing food waste and ensuring safety [147]. In addition to detecting spoilage, nanobiosensors can improve food safety by recognizing infections or pollutants [9]. Foodborne infections caused by bacteria, including *Salmonella, Escherichia coli* and *Listeria,* are major public concerns [148]. Nanosensors can be used to detect pathogen DNA or particular proteins in food goods [149]. For example, gold nanoparticles or carbon nanotubes can be utilized to increase the sensitivity of biosensors, allowing them to detect tiny levels of harmful microbes [150]. When incorporated in packaging, these sectors can continuously check food products for contamination and provide real-time data to consumers or retailers. This proactive approach to food safety can help prevent the spread of food-borne infections and reduce the need for costly recalls [151].

Nanobiosensors in nanopackaging help preserve food by controlling the release of preservatives and other active chemicals. In some circumstances, these sectors can be programmed to respond to specific environmental cues, such as temperature, humidity, or pH changes, which may hasten deterioration [152]. In active packaging systems, e.g., nanobiosensors can detect changes in food surroundings and release antibacterial or antioxidant compounds to avoid rotting [153]. This regulated release method guarantees that preservatives are utilized only when absolutely necessary, avoiding the need for synthetic additives while preserving the natural quality of the food [154]. Another novel application for nanobiosensors in nanopackaging is the detection of nutrient degradation. The contents of nutrients such as vitamins, antioxidants, and proteins in food can decline with age [155]. Nanosensors can be designed to track the amount of these nutrients and alert when they are below a given threshold. For example, a sensor could be used to monitor the degradation of vitamin C in fruit juices or the loss of omega 3 fatty acids in packaged fish [156]. They also involve environmentally friendly, sustainable packaging options, and help reduce the environmental effects of food packaging by optimizing preservative use and increasing food preservation efficiency [157]. Nanosensors are intended to detect specific markers that indicate the onset of spoilage, such as gas release (e.g., ethylene, ammonia) and pH changes [158]. Some nanobiosensors are fabricated to detect VOCs generated as food decays, such as breakdown of proteins in meat or fish, which produces ammonia or sulfur residues [159]. Metal oxides and quantum dots respond to specific spoiling indications. These sensors can be included in packaging materials to provide real time, onsite feedback on food [160]. Silver nanoparticles are recognized for their antibacterial characteristics and may be included in packaging to operate as controlled release agents when spoilage germs are discovered [161].

4.3 Application in the Fermentation Process and GMOs

Fermentation is a common technique in food production, particularly in the manufacture of dairy products, alcoholic drinks, soy products, and fermented vegetables. The fermentation process is extremely sensitive to microbial activity, temperature, pH and other environmental conditions, all of which must be carefully managed to ensure the quality and safety of the finished product [162]. Any variations may result in rotting, off-flavors, or the formation of toxic byproducts [163]. Traditionally, fermentation monitoring has depended on periodic sampling and laboratory testing, which take time and sometimes cause delays in discovering problems [164]. These sensors, which are made of nanomaterials such as carbon nanotubes, quantum dots, and metal nanoparticles, can detect small changes in the chemical environment with extraordinary accuracy [165]. Nanosensors can detect the concentrations of essential metabolites such as glucose, ethanol, and organic acids, all of which are important markers of fermentation progression [6]. By identifying these molecules at the nanoscale, companies can optimize fermentation conditions, ensuring consistent product quality and lowering the danger of spoiling [166]. In addition to monitoring metabolites, nanosensors can detect microbial activity during fermentation. Specific bacteria and yeasts must be present and active to ensure fermentation success [167]. Nanosensors equipped with biological recognition components, such as antibodies or enzymes, may identify individual bacteria by recognizing their distinct proteins or DNA. This enables manufacturers to ensure that the proper strains of bacteria are active and to identify contamination by undesired species. Real-time detection of microbial contamination enables prompt remedial measures, reducing spoilage and product waste [168, 169].

Another important use for nanosensors is the detection of genetically modified organisms (GMOs) in food items. As the use of GMOs in agriculture has increased, there is a need for precise, dependable technologies to identify and quantify GMOs in food [170]. Similarly, consumers and authorities are increasingly worried about the presence of GMOs in food products, especially in areas where GMO labeling is mandated by law [171]. Traditional methods of GMO identification, such as polymerase chain reaction (PCR) and enzyme-linked immunosorbent assay (ELISA), are very accurate but sometimes require specialized equipment and substantial laboratory work, making them time-consuming and expensive [172]. Nanosensors provide a faster and more efficient option for GMO detection. These sensors may be programmed to identify certain DNA sequences or proteins associated with genetically modified features. Nanosensors made from DNA or RNA probes can attach to complementary sequences present in GMO crops, such as those used to impart herbicide or insect resistance in genetically modified maize or soybeans [173]. Once the nanosensors attach to the target sequence, they produce a measurable signal, such as a change in fluorescence or electrical conductivity [174], allowing quick detection of the GMO content. In addition to identifying particular genetic markers, nanosensors may be incorporated into portable devices, allowing for onsite GMO testing in food processing plants or even in the field [175]. This real-time capacity greatly decreases the time and expense involved with standard GMO testing procedures. Nanosensor mobility and ease of use make them accessible to a diverse variety of users, from food manufacturers to regulatory bodies, guaranteeing effective GMO detection across the supply chain [176]. Nanosensors can be used to perform quantitative analyses of GMO content, assisting companies in meeting regulatory limitations on the proportion of GMO material permitted in non-GMO-labeled products. These sensors can provide extremely sensitive readings, detecting even tiny amounts of genetically modified DNA or proteins in food items, which is crucial for satisfying labeling requirements and addressing public concerns about GMO contamination [177].

5 Conclusion

Sensing platforms made of nanomaterials that exhibit remarkable physical, chemical, and biological capabilities have significantly improved upon traditional systems that use bulk materials. With the added benefit of having chemical and biological receptors, these nanosensors help to interact with target molecules in an effective and precise manner even in the presence of complex food matrices, enabling the quick detection of toxins in soil and food. The advantages of nanotechnology-based sensors over traditional methods are their speed in seconds, sensitivity in the picomolar range, selectivity, and ease of use for nontrained individuals. Food companies and environmental agencies can benefit from the use of nanosensing platforms to identify food pollutants, including infections and harmful substances. The use of these nanosensors to advance regular food and environmental monitoring is difficult. Food and environmental nanosensors will be developed with further research into hybrid nanomaterials with high responsiveness and room-temperature-stable bioreceptors.

References

1. Gomez-Zavaglia A, Mejuto JC, Simal-Gandara J (2020) Mitigation of emerging implications of climate change on food production systems. Food Res Int 134:109256. https://doi.org/10.1016/j.foodres.2020.109256
2. Mohsin M, Shad AA, Ali J, Sajjad-ur-Rahman E (2020) Antimicrobial resistance, food systems and climate change. In: Panwar H, Sharma C, Lichtfouse E (eds) Sustainable agriculture reviews 46. Cham, Springer International Publishing, vol 46, p 59–81. https://doi.org/10.1007/978-3-030-53024-2_3
3. Pawlak K, Kołodziejczak M (2020) The role of agriculture in ensuring food security in developing countries: considerations in the context of the problem of sustainable food production. Sustainability 12:5488. https://doi.org/10.3390/su12135488
4. Popescu GC, Popescu M, Khondker M, Clay DE, Pampana S, Umehara M (2022) Agricultural sciences and the environment: reviewing recent technologies and innovations to combat the challenges of climate change, environmental protection, and food security. Agron J 114:1895–1901. https://doi.org/10.1002/agj2.21164
5. Fan Y, Wang X, Funk T, Rashid I, Herman B, Bompoti N et al (2022) A critical review for real-time continuous soil monitoring: advantages, challenges, and perspectives. Environ Sci Technol 56:13546–13564. https://doi.org/10.1021/acs.est.2c03562
6. Kundu M, Krishnan P, Kotnala RK, Sumana G (2019) Recent developments in biosensors to combat agricultural challenges and their future prospects. Trends Food Sci Technol 88:157–178. https://doi.org/10.1016/j.tifs.2019.03.024
7. Fung F, Wang H-S, Menon S (2018) Food safety in the 21st century. Biomed J 41:88–95. https://doi.org/10.1016/j.bj.2018.03.003
8. Franco-Duarte R, Černáková L, Kadam S, S. Kaushik K, Salehi B, Bevilacqua A et al (2019) Advances in chemical and biological methods to identify microorganisms—from past to present. Microorganisms 7:130. https://doi.org/10.3390/microorganisms7050130
9. Raghu HV, Parkunan T, Kumar N (2020) Application of nanobiosensors for food safety monitoring. In: Dasgupta N, Ranjan S, Lichtfouse E (eds) Environmental nanotechnology Volume 4, vol 32, Cham, Springer International Publishing, pp 93–129. https://doi.org/10.1007/978-3-030-26668-4_3
10. Karunakaran C, Rajkumar R, Bhargava K (2015) Introduction to biosensors. Biosens Bioelectron 1–68. https://doi.org/10.1016/B978-0-12-803100-1.00001-3
11. Nagarajan R, Hatton TA (2008) Nanoparticles: synthesis, stabilization, passivation, and functionalization, vol 996. Washington, DC, American Chemical Society. https://doi.org/10.1021/bk-2008-0996
12. Sekhon B (2014) Nanotechnology in agri-food production: an overview. NSA, pp 31. https://doi.org/10.2147/NSA.S39406
13. Ramesh M, Janani R, Deepa C, Rajeshkumar L (2022) Nanotechnology-enabled biosensors: a review of fundamentals, design principles, materials, and applications. Biosensors 13:40. https://doi.org/10.3390/bios13010040
14. Fu X, Ying Y (2016) Food safety evaluation based on near infrared spectroscopy and imaging: a review. Crit Rev Food Sci Nutr 56:1913–1924. https://doi.org/10.1080/10408398.2013.807418
15. Nangare SN, Patil PO (2021) Affinity-based nanoarchitectured biotransducer for sensitivity enhancement of surface plasmon resonance sensors for in vitro diagnosis: a review. ACS Biomater Sci Eng 7:2–30. https://doi.org/10.1021/acsbiomaterials.0c01203
16. Prasad S (2014) Nanobiosensors: the future for diagnosis of disease? NDD 1. https://doi.org/10.2147/NDD.S39421
17. Huang Z, Qiu L, Zhang T, Tan W (2021) Integrating DNA nanotechnology with aptamers for biological and biomedical applications. Matter 4:461–489. https://doi.org/10.1016/j.matt.2020.11.002

18. Eersels K, Lieberzeit P, Wagner P (2016) A review on synthetic receptors for bioparticle detection created by surface-imprinting techniques—from principles to applications. ACS Sens 1:1171–1187. https://doi.org/10.1021/acssensors.6b00572
19. Munawar A, Ong Y, Schirhagl R, Tahir MA, Khan WS, Bajwa SZ (2019) Nanosensors for diagnosis with optical, electric and mechanical transducers. RSC Adv 9:6793–6803. https://doi.org/10.1039/C8RA10144B
20. Naresh Varnakavi, Lee N (2021) A review on biosensors and recent development of nanostructured materials-enabled biosensors. Sensors 21:1109. https://doi.org/10.3390/s21041109
21. Consales M, Cutolo A, Penza M, Aversa P, Giordano M, Cusano A (2008) Fiber optic chemical nanosensors based on engineered single-walled carbon nanotubes: a review. J Sens 2008:936074. https://doi.org/10.1155/2008/936074
22. Capek I (2009) Dispersions, novel nanomaterial sensors and nanoconjugates based on carbon nanotubes. Adv Coll Interface Sci 150:63–89. https://doi.org/10.1016/j.cis.2009.05.006
23. Eivazzadeh-Keihan R, Bahojb Noruzi E, Chidar E, Jafari M, Davoodi F, Kashtiaray A et al (2022) Applications of carbon-based conductive nanomaterials in biosensors. Chem Eng J 442:136183. https://doi.org/10.1016/j.cej.2022.136183
24. Yadav N, Garg VK, Chhillar AK, Rana JS (2021) Detection and remediation of pollutants to maintain ecosustainability employing nanotechnology: a review. Chemosphere 280:130792. https://doi.org/10.1016/j.chemosphere.2021.130792
25. Kumar S, Singh H, Feder-Kubis J, Nguyen DD (2023) Recent advances in nanobiosensors for sustainable healthcare applications: a systematic literature review. Environ Res 238:117177. https://doi.org/10.1016/j.envres.2023.117177
26. Kumar A, Maity D, Vyas G, Bhatt M, Bhatt S, Paul P (2021) Polyacrylic acid@zeolitic imidazolate framework-8 nanoparticles for detection and absorptive removal of cyanide from aqueous media with high efficiency. Colloids Surf A 617:126358. https://doi.org/10.1016/j.colsurfa.2021.126358
27. Khan MS, Vishakante GD, Siddaramaiah H (2013) Gold nanoparticles: a paradigm shift in biomedical applications. Adv Coll Interface Sci 199–200:44–58. https://doi.org/10.1016/j.cis.2013.06.003
28. Bahadır EB, Sezgintürk MK (2016) Lateral flow assays: principles, designs and labels. TrAC, Trends Anal Chem 82:286–306. https://doi.org/10.1016/j.trac.2016.06.006
29. Sohrabi H, Majidi MR, Fakhraei M, Jahanban-Esfahlan A, Hejazi M, Oroojalian F et al (2022) Lateral flow assays (LFA) for detection of pathogenic bacteria: a small point-of-care platform for diagnosis of human infectious diseases. Talanta 243:123330. https://doi.org/10.1016/j.talanta.2022.123330
30. Wu W, Li J, Pan D, Li J, Song S, Rong M et al (2014) Gold nanoparticle-based enzyme-linked antibody-aptamer sandwich assay for detection of *Salmonella* Typhimurium. ACS Appl Mater Interfaces 6:16974–16981. https://doi.org/10.1021/am5045828
31. Dehghani Z, Nguyen T, Golabi M, Hosseini M, Rezayan AH, Mohammadnejad J et al (2021) Magnetic beads modified with Pt/Pd nanoparticle and aptamer as a catalytic nanobioprobe in combination with loop mediated isothermal amplification for the on-site detection of *Salmonella Typhimurium* in food and fecal samples. Food Control 121:107664. https://doi.org/10.1016/j.foodcont.2020.107664
32. Li Q, Li H, Li K, Gu Y, Wang Y, Yang D et al (2023) Specific colorimetric detection of methylmercury based on peroxidase-like activity regulation of carbon dots/Au NPs nanozyme. J Hazard Mater 441:129919. https://doi.org/10.1016/j.jhazmat.2022.129919
33. Fernández-García M, Martínez-Arias A, Hanson JC, Rodriguez JA (2004) Nanostructured oxides in chemistry: characterization and properties. Chem Rev 104:4063–4104. https://doi.org/10.1021/cr030032f
34. Kim J, Junkin M, Kim D-H, Kwon S, Shin YS, Wong PK et al (2009) Applications, techniques, and microfluidic interfacing for nanoscale biosensing. Microfluid Nanofluid 7:149–167. https://doi.org/10.1007/s10404-009-0431-8

35. Gill RT, Harbottle MJ, Smith JWN, Thornton SF (2014) Electrokinetic-enhanced bioremediation of organic contaminants: a review of processes and environmental applications. Chemosphere 107:31–42. https://doi.org/10.1016/j.chemosphere.2014.03.019
36. Weng X, Zhang C, Jiang H (2021) Advances in microfluidic nanobiosensors for the detection of foodborne pathogens. LWT 151:112172. https://doi.org/10.1016/j.lwt.2021.112172
37. Nikoleli G-P, Nikolelis DP, Siontorou CG, Nikolelis M-T, Karapetis S (2018) The application of lipid membranes in biosensing. Membranes 8:108. https://doi.org/10.3390/membranes8040108
38. Harish V, Tewari D, Gaur M, Yadav AB, Swaroop S, Bechelany M et al (2022) Review on nanoparticles and nanostructured materials: bioimaging, biosensing, drug delivery, tissue engineering, antimicrobial, and agro-food applications. Nanomaterials 12:457. https://doi.org/10.3390/nano12030457
39. Andrade Korn MDG, Da Boa Morte ES, Batista Dos Santos DCM, Castro JT, Barbosa JTP, Teixeira AP, et al. Sample Preparation for the Determination of Metals in Food Samples Using Spectroanalytical Methods—A Review. Applied Spectroscopy Reviews 2008;43:67–92. https://doi.org/10.1080/05704920701723980.
40. Chiu ML, Lawi W, Snyder ST, Wong PK, Liao JC, Gau V (2010) Matrix effects—a challenge toward automation of molecular analysis. J Assoc Lab Autom 15:233–242. https://doi.org/10.1016/j.jala.2010.02.001
41. Butnariu M, Butu A (2019) Plant nanobionics: application of nanobiosensors in plant biology. In: Prasad R (ed) Plant Nanobionics. Cham, Springer International Publishing, pp 337–376. https://doi.org/10.1007/978-3-030-16379-2_12
42. Sharma P, Pandey V, Sharma MMM, Patra A, Singh B, Mehta S et al (2021) A review on biosensors and nanosensors application in agroecosystems. Nanoscale Res Lett 16:136. https://doi.org/10.1186/s11671-021-03593-0
43. Gangwar DS (2024) Recent trends in biosensor development for climate smart organic agriculture and their role in environmental sustainability. Environ Dev Sustain. https://doi.org/10.1007/s10668-024-05300-2
44. Ghaffar N, Farrukh MA, Naz S (2020) Applications of nanobiosensors in agriculture. In: Javad S (ed) Nanoagronomy. Cham: Springer International Publishing, pp 179–196. https://doi.org/10.1007/978-3-030-41275-3_10
45. Carter KP, Young AM, Palmer AE (2014) Fluorescent sensors for measuring metal ions in living systems. Chem Rev 114:4564–4601. https://doi.org/10.1021/cr400546e
46. Zeng L, Gupta P, Chen Y, Wang E, Ji L, Chao H et al (2017) The development of anticancer ruthenium(II) complexes: from single molecule compounds to nanomaterials. Chem Soc Rev 46:5771–5804. https://doi.org/10.1039/C7CS00195A
47. Hang Y, Boryczka J, Wu N (2022) Visible-light and near-infrared fluorescence and surface-enhanced Raman scattering point-of-care sensing and bioimaging: a review. Chem Soc Rev 51:329–375. https://doi.org/10.1039/C9CS00621D
48. Jain KK (2005) Nanotechnology in clinical laboratory diagnostics. Clin Chim Acta 358:37–54. https://doi.org/10.1016/j.cccn.2005.03.014
49. Sharma D, Teli G, Gupta K, Bansal G, Gupta GD, Chawla PA (2022) Nanobiosensors from agriculture to nextgen diagnostic tools. CNM 7:110–138. https://doi.org/10.2174/2405461507666220131104843
50. Petrucci R, Bortolami M, Di Matteo P, Curulli A (2022) Gold nanomaterials-based electrochemical sensors and biosensors for phenolic antioxidants detection: recent advances. Nanomaterials 12:959. https://doi.org/10.3390/nano12060959
51. Rai V, Acharya S, Dey N (2012) Implications of nanobiosensors in agriculture. JBNB 03:315–324. https://doi.org/10.4236/jbnb.2012.322039
52. Jan N, Qazi HA, Ramzan S, John R (2018) Developing stress-tolerant plants through in vitro tissue culture: family brassicaceae. In: Gosal SS, Wani SH (eds) Biotechnologies of crop improvement, vol. Cham, Springer International Publishing, pp 327–72. https://doi.org/10.1007/978-3-319-78283-6_10

53. Fierer N, Wood SA, Bueno De Mesquita CP (2021) How microbes can, and cannot, be used to assess soil health. Soil Biol Biochemistr 153:108111. https://doi.org/10.1016/j.soilbio.2020.108111
54. Jang A, Zou Z, Lee KK, Ahn CH, Bishop PL (2011) State-of-the-art lab chip sensors for environmental water monitoring. Meas Sci Technol 22:032001. https://doi.org/10.1088/0957-0233/22/3/032001
55. Van Bruggen AHC, Semenov AM (2000) In search of biological indicators for soil health and disease suppression. Appl Soil Ecol 15:13–24. https://doi.org/10.1016/S0929-1393(00)00068-8
56. Mousa RMA, El-Hady DA (2018) Nanofood technology and nutrition. In: Dasgupta N, Ranjan S, Lichtfouse E (eds) Environmental nanotechnology, vol 14. Cham, Springer. International Publishing, pp 59–74. https://doi.org/10.1007/978-3-319-76090-2_2
57. Subramanian KS, Thirunavukkarasu M (2017) Nanofertilizers and nutrient transformations in soil. In: Ghorbanpour M, Manika K, Varma A (eds) Nanoscience and plant–soil systems, vol 48. Cham, Springer International Publishing, pp 305–319. https://doi.org/10.1007/978-3-319-46835-8_11.
58. Mainwaring A, Culler D, Polastre J, Szewczyk R, Anderson J (2002) Wireless sensor networks for habitat monitoring. In: Proceedings of the 1st ACM international workshop on Wireless sensor networks and applications, Atlanta Georgia USA: ACM, pp 88–97. https://doi.org/10.1145/570738.570751
59. Nath SS, Bolte JP, Ross LG, Aguilar-Manjarrez J (2000) Applications of geographical information systems (GIS) for spatial decision support in aquaculture. Aquacult Eng 23:233–278. https://doi.org/10.1016/S0144-8609(00)00051-0
60. Dyck VA, Hendrichs J, Robinson AS (2005) Sterile insect technique: principles and practice in area-wide integrated pest management. Dordrecht, Springer Netherlands. https://doi.org/10.1007/1-4020-4051-2
61. Olatinwo R, Hoogenboom G (2014) Weather-based pest forecasting for efficient crop protection. Integrated Pest Management, Elsevier, pp 59–78. https://doi.org/10.1016/B978-0-12-398529-3.00005-1
62. Bashir O, Bhat SA, Basharat A, Qamar M, Qamar SA, Bilal M et al (2022) Nanoengineered materials for sensing food pollutants: technological advancements and safety issues. Chemosphere 292:133320. https://doi.org/10.1016/j.chemosphere.2021.133320
63. George Kerry R, Ukhurebor KE, Kumari S, Maurya GK, Patra S, Panigrahi B et al (2021) A comprehensive review on the applications of nanobiosensor-based approaches for noncommunicable and communicable disease detection. Biomater Sci 9:3576–3602. https://doi.org/10.1039/D0BM02164D
64. Mahlein A-K, Oerke E-C, Steiner U, Dehne H-W (2012) Recent advances in sensing plant diseases for precision crop protection. Eur J Plant Pathol 133:197–209. https://doi.org/10.1007/s10658-011-9878-z
65. Yadav A, Yadav K (2018) Nanoparticle-based plant disease management: tools for sustainable agriculture. In: Abd-Elsalam KA, Prasad R (eds) Nanobiotechnology applications in plant protection. Cha Springer International Publishing, pp 29–61. https://doi.org/10.1007/978-3-319-91161-8_2
66. Srivastava AK, Dev A, Karmakar S. Nanosensors for Food and Agriculture. In: Ranjan S, Dasgupta N, Lichtfouse E (eds) Nanoscience in food and agriculture, vol 26. Cham, Springer International Publishing, pp 41–79. https://doi.org/10.1007/978-3-319-58496-6_3
67. Ansari MHD, Lavhale S, Kalunke RM, Srivastava PL, Pandit V, Gade S et al (2020) Recent advances in plant nanobionics and nanobiosensors for toxicology applications. CNANO 16:27–41. https://doi.org/10.2174/1573413715666190409101305
68. Rievaj M, Culková E, Šandorová D, Durdiak J, Bellová R, Tomčík P (2023) A review of analytical techniques for the determination and separation of silver ions and its nanoparticles. Nanomaterials 13:1262. https://doi.org/10.3390/nano13071262
69. Balasurya S, Kokilavani S, Janani B, Sudheer Khan S (2023) Recent advances in bioelectronic noses based on olfactory receptors. In: Nanotechnology-Based E-noses, Elsevier, pp 125–141. https://doi.org/10.1016/B978-0-323-91157-3.00002-7

70. Kumar P, Chugh P, Ali SS, Chawla W, Sushmita S, Kumar R et al (2024) Trends of nanobiosensors in modern agriculture systems. Appl Biochem Biotechnol. https://doi.org/10.1007/s12010-024-05039-6
71. Tan J, Xu J (2020) Applications of electronic nose (e-nose) and electronic tongue (e-tongue) in food quality-related properties determination: a review. Artific Intell Agricult 4:104–115. https://doi.org/10.1016/j.aiia.2020.06.003
72. Feng Y, Cao H, Song H, Huang K, Zhang Y, Zhang Y et al (2024) The formation mechanism, analysis strategies and regulation measures of cereal aroma: a review. Trends Food Sci Technol 147:104452. https://doi.org/10.1016/j.tifs.2024.104452
73. Abd El-kader SM, Mohammad El-Basioni BM (2013) Precision farming solution in Egypt using the wireless sensor network technology. Egyptian Inform J 14:221–233. https://doi.org/10.1016/j.eij.2013.06.004
74. Kumar SA, Ilango P (2018) The impact of wireless sensor network in the field of precision agriculture: a review. Wireless Pers Commun 98:685–698. https://doi.org/10.1007/s11277-017-4890-z
75. Soussi A, Zero E, Sacile R, Trinchero D, Fossa M (2024) Smart sensors and smart data for precision agriculture: a review. Sensors 24:2647. https://doi.org/10.3390/s24082647
76. GL, CR, PG (2020) An automated low cost IoT based Fertilizer Intimation System for smart agriculture. Sustain Comput Inform Syst 28:100300. https://doi.org/10.1016/j.suscom.2019.01.002
77. Nadporozhskaya M, Kovsh N, Paolesse R, Lvova L (2022) Recent advances in chemical sensors for soil analysis: a review. Chemosensors 10:35. https://doi.org/10.3390/chemosensors10010035
78. Chugh B, Thakur S, Singh AK, Joany RM, Rajendran S, Nguyen TA (2022) Electrochemical sensors for agricultural application. Nanosens Smart Agricult 147–164. https://doi.org/10.1016/B978-0-12-824554-5.00018-5
79. Manisha H, Priya Swetha PD, Shim Y-B, Prasad KS (2019) Revisiting fluorescent carbon nanodots for environmental, biomedical applications and puzzle about fluorophore impurities. Nano-Structures & Nano-Objects 20:100391. https://doi.org/10.1016/j.nanoso.2019.100391
80. Omara AE-D, Elsakhawy T, Alshaal T, El-Ramady H, Kovács Z, Fári M (2019) Nanoparticles: a novel approach for sustainable agro-productivity. EBSS 3:30–40. https://doi.org/10.21608/jenvbs.2019.7478.1050
81. Numan A, Gill AAS, Rafique S, Guduri M, Zhan Y, Maddiboyina B et al (2021) Rationally engineered nanosensors: a novel strategy for the detection of heavy metal ions in the environment. J Hazard Mater 409:124493. https://doi.org/10.1016/j.jhazmat.2020.124493
82. Rasheed T, Shafi S, Sher F (2022) Smart nanoarchitectures as potential sensing tools for detecting heavy metal ions in aqueous matrices. Trends Environ Anal Chemistr 36:e00179. https://doi.org/10.1016/j.teac.2022.e00179
83. Long F, Zhu A, Shi H, Wang H, Liu J (2013) Rapid on-site/in situ detection of heavy metal ions in environmental water using a structure-switching DNA optical biosensor. Sci Rep 3:2308. https://doi.org/10.1038/srep02308
84. Knaus J, Palzenberger M (2018) PARMA. A full text search based method for matching nonpatent literature citations with scientific reference databases. a pilot study. 1188419, 36536, 779370, 1639, 1199. https://doi.org/10.17617/2.2540157
85. Brune A, Ohkuma M (2010) Role of the termite gut microbiota in symbiotic digestion. In: Bignell DE, Roisin Y, Lo N (eds) Biology of termites: a modern synthesis, Dordrecht, Springer Netherlands, pp 439–475. https://doi.org/10.1007/978-90-481-3977-4_16
86. Tarkka M, Schrey S, Hampp R (2008) Plant Associated Soil Microorganisms. In: Nautiyal CS, Dion P, editors. Molecular Mechanisms of Plant and Microbe Coexistence, vol. 15, Berlin, Heidelberg: Springer Berlin Heidelberg; 2008, p. 3–51. https://doi.org/10.1007/978-3-540-75575-3_1.
87. Ochsner TE, Cosh MH, Cuenca RH, Dorigo WA, Draper CS, Hagimoto Y et al (2013) State of the Art in Large-Scale Soil Moisture Monitoring. Soil Sci Soc Am J 77:1888–1919. https://doi.org/10.2136/sssaj2013.03.0093

88. Umasankareswari T, Dharshana V, Saravanadevi K, Dorothy R, Sasilatha T, Nguyen TA et al (2022) Soil moisture nanosensors. Nanosens Smart Agricult 185–216. https://doi.org/10.1016/B978-0-12-824554-5.00019-7
89. Mahmud MAP, Ejeian F, Azadi S, Myers M, Pejcic B, Abbassi R et al (2020) Recent progress in sensing nitrate, nitrite, phosphate, and ammonium in aquatic environment. Chemosphere 259:127492. https://doi.org/10.1016/j.chemosphere.2020.127492
90. Alhalaili B, Popescu IN, Rusanescu CO, Vidu R (2022) Microfluidic devices and microfluidics-integrated electrochemical and optical (Bio)sensors for pollution analysis: a review. Sustainability 14:12844. https://doi.org/10.3390/su141912844
91. Salouti M, Khadivi Derakhshan F (2020) Biosensors and nanobiosensors in environmental applications. In: Ghorbanpour M, Bhargava P, Varma A, Choudhary DK (eds) Biogenic nanoparticles and their use in agro-ecosystems, Singapore, Springer Singapore, pp 515–591. https://doi.org/10.1007/978-981-15-2985-6_26.
92. Tyagi D, Wang H, Huang W, Hu L, Tang Y, Guo Z et al (2020) Recent advances in two-dimensional-material-based sensing technology toward health and environmental monitoring applications. Nanoscale 12:3535–3559. https://doi.org/10.1039/C9NR10178K
93. El-Shahawi MS, Hamza A, Bashammakh AS, Al-Saggaf WT (2010) An overview on the accumulation, distribution, transformations, toxicity and analytical methods for the monitoring of persistent organic pollutants. Talanta 80:1587–1597. https://doi.org/10.1016/j.talanta.2009.09.055
94. Hassan D, Bakhsh H, Khurram AM, Bhutto SA, Jalbani NS, Ghumro T et al (2022) Fluorescent nanotechnology: an evolution in optical sensors. CAC 18:176–185. https://doi.org/10.2174/1573411017666201215121420
95. Gulati S, Lingam BHN, Kumar S, Goyal K, Arora A, Varma RS (2022) Improving the air quality with functionalized carbon nanotubes: sensing and remediation applications in the real world. Chemosphere 299:134468. https://doi.org/10.1016/j.chemosphere.2022.134468
96. Bhardwaj AK, Sundaram S, Yadav KK, Srivastav AL (2021) An overview of silver nanoparticles as promising materials for water disinfection. Environ Technol Innov 23:101721. https://doi.org/10.1016/j.eti.2021.101721
97. Silva JA (2023) Water supply and wastewater treatment and reuse in future cities: a systematic literature review. Water 15:3064. https://doi.org/10.3390/w15173064
98. Wang X, Lu X, Chen J (2014) Development of biosensor technologies for analysis of environmental contaminants. Trends Environ Anal Chemistr 2:25–32. https://doi.org/10.1016/j.teac.2014.04.001
99. Justino C, Duarte A, Rocha-Santos T (2017) Recent progress in biosensors for environmental monitoring: a review. Sensors 17:2918. https://doi.org/10.3390/s17122918
100. Shanker R, Singh G, Jyoti A, Dwivedi PD, Singh SP (2020) Nanotechnology and detection of microbial pathogens. Anim Biotechnol 593–611. https://doi.org/10.1016/B978-0-12-811710-1.00026-4
101. Encarnação T, Pais AA, Campos MG, Burrows HD (2019) Endocrine disrupting chemicals: Impact on human health, wildlife and the environment. Sci Prog 102:3–42. https://doi.org/10.1177/0036850419826802
102. Marcon L, Oliveras J, Puntes VF (2021) In situ nanoremediation of soils and groundwaters from the nanoparticle's standpoint: a review. Sci Total Environ 791:148324. https://doi.org/10.1016/j.scitotenv.2021.148324
103. Dayana Priyadharshini S, Manikandan S, Kiruthiga R, Rednam U, Babu PS, Subbaiya R et al (2022) Graphene oxide-based nanomaterials for the treatment of pollutants in the aquatic environment: Recent trends and perspectives—a review. Environ Pollut 306:119377. https://doi.org/10.1016/j.envpol.2022.119377
104. Ramaiyan KP, Mukundan (Mukund) R (2019) Electrochemical sensors for air quality monitoring. Electrochem Soc Interface 28:59–63. https://doi.org/10.1149/2.F08193IF
105. Kumar P, Morawska L, Martani C, Biskos G, Neophytou M, Di Sabatino S et al (2015) The rise of low-cost sensing for managing air pollution in cities. Environ Int 75:199–205. https://doi.org/10.1016/j.envint.2014.11.019

106. Yunus IS, Harwin, Kurniawan A, Adityawarman D, Indarto A (2012) Nanotechnologies in water and air pollution treatment. Environ Technol Rev 1:136–148. https://doi.org/10.1080/21622515.2012.733966
107. Corsi I, Winther-Nielsen M, Sethi R, Punta C, Della Torre C, Libralato G et al (2018) Ecofriendly nanotechnologies and nanomaterials for environmental applications: key issue and consensus recommendations for sustainable and ecosafe nanoremediation. Ecotoxicol Environ Saf 154:237–244. https://doi.org/10.1016/j.ecoenv.2018.02.037
108. Hristozov D, Malsch I (2009) Hazards and risks of engineered nanoparticles for the environment and human health. Sustainability 1:1161–1194. https://doi.org/10.3390/su1041161
109. Özkara A, Akyil D, Konuk M (2016) Pesticides, environmental pollution, and health. In: Larramendy M, Soloneski S (eds) Environmental health risk—hazardous factors to living species. https://doi.org/10.5772/63094
110. Mirres ACDM, Silva BEPDMD, Tessaro L, Galvan D, Andrade JCD, Aquino A et al (2022) Recent advances in nanomaterial-based biosensors for pesticide detection in foods. Biosensors 12:572. https://doi.org/10.3390/bios12080572
111. Damalas CA, Eleftherohorinos IG (2011) Pesticide exposure, safety issues, and risk assessment indicators. IJERPH 8:1402–1419. https://doi.org/10.3390/ijerph8051402
112. Xuejun L, Fusuo Z (2011) Nitrogen fertilizer induced greenhouse gas emissions in China. Curr Opin Environ Sustain 3:407–413. https://doi.org/10.1016/j.cosust.2011.08.006
113. Guerrero A, De Neve S, Mouazen AM (2021) Current sensor technologies for in situ and on-line measurement of soil nitrogen for variable rate fertilization: a review. Adv Agron 168:1–38. https://doi.org/10.1016/bs.agron.2021.02.001
114. Singh H, Sharma A, Bhardwaj SK, Arya SK, Bhardwaj N, Khatri M (2021) Recent advances in the applications of nanoagrochemicals for sustainable agricultural development. Environ Sci: Processes Impacts 23:213–239. https://doi.org/10.1039/D0EM00404A
115. Ray M, Ray A, Dash S, Mishra A, Achary KG, Nayak S et al (2017) Fungal disease detection in plants: Traditional assays, novel diagnostic techniques and biosensors. Biosens Bioelectron 87:708–723. https://doi.org/10.1016/j.bios.2016.09.032
116. Corkley I, Fraaije B, Hawkins N (2022) Fungicide resistance management: Maximizing the effective life of plant protection products. Plant Pathol 71:150–169. https://doi.org/10.1111/ppa.13467
117. Srinivasan S, Raajasubramanian D, Ashokkumar N, Vinothkumar V, Paramaguru N, Selvaraj P et al (2024) Nanobiosensors based on on-site detection approaches for rapid pesticide sensing in the agricultural arena: a systematic review of the current status and perspectives. Biotech Bioeng 121:2585–2603. https://doi.org/10.1002/bit.28764
118. Khan S, Hanjra MA (2009) Footprints of water and energy inputs in food production—global perspectives. Food Policy 34:130–140. https://doi.org/10.1016/j.foodpol.2008.09.001
119. Thakur M, Wang B, Verma ML (2022) Development and applications of nanobiosensors for sustainable agricultural and food industries: recent developments, challenges and perspectives. Environ Technol Innov 26:102371. https://doi.org/10.1016/j.eti.2022.102371
120. Mazuryk J, Klepacka K, Kutner W, Sharma PS (2023) Glyphosate separating and sensing for precision agriculture and environmental protection in the era of smart materials. Environ Sci Technol 57:9898–9924. https://doi.org/10.1021/acs.est.3c01269
121. Maishu SP, Atemenkeh NC (2022) Applications of nanosensors as pathogen detectors in packaged food. In: Annu, Bhattacharya T, Ahmed S (eds) Nanotechnology in intelligent food packaging. 1st ed., Wiley, pp 291–312. https://doi.org/10.1002/9781119819011.ch12
122. Tshabalala ZP, Mokoena TP, Motaung DE (2022) Current commercial nanosensors and devices/products used in agriculture. In: Nanosensors for smart agriculture. Elsevier, p. 165–181. https://doi.org/10.1016/B978-0-12-824554-5.00034-3
123. McGrath TF, Elliott CT, Fodey TL (2012) Biosensors for the analysis of microbiological and chemical contaminants in food. Anal Bioanal Chem 403:75–92. https://doi.org/10.1007/s00216-011-5685-9

124. Castro-Muñoz R, Correa-Delgado M, Córdova-Almeida R, Lara-Nava D, Chávez-Muñoz M, Velásquez-Chávez VF et al (2022) Natural sweeteners: sources, extraction and current uses in foods and food industries. Food Chem 370:130991. https://doi.org/10.1016/j.foodchem.2021.130991
125. Thompson LA, Darwish WS (2019) Environmental chemical contaminants in food: review of a global problem. J Toxicol 2019:1–14. https://doi.org/10.1155/2019/2345283
126. Garvey M (2019) Food pollution: a comprehensive review of chemical and biological sources of food contamination and impact on human health. Nutrire 44:1. https://doi.org/10.1186/s41110-019-0096-3
127. Patel G, Pillai V, Bhatt P, Mohammad S. Application of nanosensors in the food industry. In: Nanosensors for smart cities, Elsevier, pp 355–68. https://doi.org/10.1016/B978-0-12-819870-4.00020-7
128. Lei KF (2012) Microfluidic systems for diagnostic applications: a review. SLAS Technol 17:330–347. https://doi.org/10.1177/2211068212454853
129. Ravichandran R (2010) Nanotechnology applications in food and food processing: innovative green approaches, opportunities and uncertainties for global market. Int J Green Nanotechnol Phys Chem 1:P72-96. https://doi.org/10.1080/19430871003684440
130. Dahlan NA, Thiha A, Ibrahim F, Milić L, Muniandy S, Jamaluddin NF et al (2022) Role of nanomaterials in the fabrication of bioNEMS/MEMS for biomedical applications and toward pioneering food waste utilization. Nanomaterials 12:4025. https://doi.org/10.3390/nano12224025
131. Mishra S, Bharagava RN, More N, Yadav A, Zainith S, Mani S et al (2019) Heavy metal contamination: an alarming threat to environment and human health. In: Sobti RC, Arora NK, Kothari R (eds) Environmental biotechnology: for sustainable future, Singapore: Springer Singapore, pp 103–125. https://doi.org/10.1007/978-981-10-7284-0_5
132. Chopra AK, Pathak C, Prasad G (2009) Scenario of heavy metal contamination in agricultural soil and its management. JANS 1:99–108. https://doi.org/10.31018/jans.v1i1.46
133. Zulfiqar U, Haider FU, Ahmad M, Hussain S, Maqsood MF, Ishfaq M et al (2023) Chromium toxicity, speciation, and remediation strategies in soil-plant interface: a critical review. Front Plant Sci 13:1081624. https://doi.org/10.3389/fpls.2022.1081624
134. Pirzada M, Altintas Z (2019) Nanomaterials for healthcare biosensing applications. Sensors 19:5311. https://doi.org/10.3390/s19235311
135. Wang Y, Du X, Liu Z, Shi S, Lv H (2019) Dendritic fibrous nanoparticles (DFNPs): rising stars of mesoporous materials. J Mater Chem A 7:5111–5152. https://doi.org/10.1039/C8TA09815H
136. Hemmati M, Rajabi M, Asghari A (2018) Magnetic nanoparticle based solid-phase extraction of heavy metal ions: a review on recent advances. Microchim Acta 185:160. https://doi.org/10.1007/s00604-018-2670-4
137. Jagadeeshan S, Parsanathan R (2019) Nanometal oxides for antibacterial activity. In: Naushad Mu, Rajendran S, Gracia F (eds) Advanced nanostructured materials for environmental remediation, vol 25. Cham: Springer International Publishing, pp 59–90. https://doi.org/10.1007/978-3-030-04477-0_3
138. Abdullah NH, Shameli K, Abdullah EC, Abdullah LC (2019) Solid matrices for fabrication of magnetic iron oxide nanocomposites: synthesis, properties, and application for the adsorption of heavy metal ions and dyes. Compos B Eng 162:538–568. https://doi.org/10.1016/j.compositesb.2018.12.075
139. Handa RJ, Weiser MJ (2014) Gonadal steroid hormones and the hypothalamo–pituitary–adrenal axis. Front Neuroendocrinol 35:197–220. https://doi.org/10.1016/j.yfrne.2013.11.001
140. Elmlinger MW, Kühnel W, Ranke MB (2002) Reference Ranges for Serum Concentrations of Lutropin (LH), Follitropin (FSH), Estradiol (E2), Prolactin, Progesterone, Sex Hormone-Binding Globulin (SHBG), Dehydroepiandrosterone Sulfate (DHEAS), Cortisol and Ferritin in Neonates, Children and Young Adults. Clinic Chemistr Lab Med 40. https://doi.org/10.1515/cclm.2002.202

141. De Cesare F, Macagnano A (2015) Electrospinning-based nanobiosensors. In: Macagnano A, Zampetti E, Kny E (eds) Electrospinning for high performance sensors. Cham, Springer International Publishing, pp 225–279. https://doi.org/10.1007/978-3-319-14406-1_10
142. Paleček E, Tkáč J, Bartošík M, Bertók T, Ostatná V, Paleček J (2015) Electrochemistry of nonconjugated proteins and glycoproteins. toward sensors for biomedicine and glycomics. Chem Rev 115:2045–2108. https://doi.org/10.1021/cr500279h
143. Du X, Li Y, Xia Y-L, Ai S-M, Liang J, Sang P et al (2016) Insights into Protein-Ligand interactions: mechanisms, models, and methods. IJMS 17:144. https://doi.org/10.3390/ijms17020144
144. Perozzo R, Folkers G, Scapozza L (2004) Thermodynamics of protein-ligand interactions: history, presence, and future aspects. J Recept Sign Transduct 24:1–52. https://doi.org/10.1081/RRS-120037896
145. Sharma A, Ranjit R, Pratibha, Kumar N, Kumar M, Giri BS (2023) Nanoparticles based nanosensors: principles and their applications in active packaging for food quality and safety detection. Biochem Eng J 193:108861. https://doi.org/10.1016/j.bej.2023.108861
146. Shaalan NM, Ahmed F, Saber O, Kumar S (2022) Gases in food production and monitoring: recent advances in target chemiresistive gas sensors. Chemosensors 10:338. https://doi.org/10.3390/chemosensors10080338
147. Verghese K, Lewis H, Lockrey S, Williams H (2015) Packaging's role in minimizing food loss and waste across the supply chain: packaging's role in minimizing food waste across the supply chain. Packag Technol Sci 28:603–620. https://doi.org/10.1002/pts.2127
148. Abebe E, Gugsa G, Ahmed M (2020) Review on major food-borne zoonotic bacterial pathogens. J Trop Med 2020:1–19. https://doi.org/10.1155/2020/4674235
149. Kumar H, Kuča K, Bhatia SK, Saini K, Kaushal A, Verma R et al (2020) Applications of nanotechnology in sensor-based detection of foodborne pathogens. Sensors 20:1966. https://doi.org/10.3390/s20071966
150. Sutarlie L, Ow SY, Su X (2017) Nanomaterials-based biosensors for detection of microorganisms and microbial toxins. Biotechnol J 12:biot.201500459. https://doi.org/10.1002/biot.201500459
151. Ahmed I, Lin H, Zou L, Li Z, Brody AL, Qazi IM et al (2018) An overview of smart packaging technologies for monitoring safety and quality of meat and meat products. Packag Technol Sci 31:449–471. https://doi.org/10.1002/pts.2380
152. Isapour G, Lattuada M (2018) Bioinspired stimuli-responsive color-changing systems. Adv Mater 30:1707069. https://doi.org/10.1002/adma.201707069
153. Sundramoorthy AK, Vignesh Kumar TH, Gunasekaran S (2018) Graphene-based nanosensors and smart food packaging systems for food safety and quality monitoring. Graphene Bioelectron 267–306. https://doi.org/10.1016/B978-0-12-813349-1.00012-3
154. Rout S, Tambe S, Deshmukh RK, Mali S, Cruz J, Srivastav PP et al (2022) Recent trends in the application of essential oils: the next generation of food preservation and food packaging. Trends Food Sci Technol 129:421–439. https://doi.org/10.1016/j.tifs.2022.10.012
155. Obrenovich E, Li Y, Parvathaneni K, Yendluri B, Palacios H, Leszek J et al (2011) Antioxidants in health, disease and aging. CNSNDDT 10:192–207. https://doi.org/10.2174/187152711794480375
156. Yenipazar H, Şahin-Yeşilçubuk N (2023) Effect of packaging and encapsulation on the oxidative and sensory stability of omega-3 supplements. Food Sci Nutr 11:1426–1440. https://doi.org/10.1002/fsn3.3182
157. Han J, Ruiz-Garcia L, Qian J, Yang X (2018) Food packaging: a comprehensive review and future trends. Comp Rev Food Sci Food Safe 17:860–877. https://doi.org/10.1111/1541-4337.12343
158. Das J, Mishra HN (2023) A comprehensive review of the spoilage of shrimp and advances in various indicators/sensors for shrimp spoilage monitoring. Food Res Int 173:113270. https://doi.org/10.1016/j.foodres.2023.113270
159. Joshi N, Pransu G, Adam C-J (2023) Critical review and recent advances of 2D materials-Based gas sensors for food spoilage detection. Crit Rev Food Sci Nutr 63:10536–10559. https://doi.org/10.1080/10408398.2022.2078950

160. Khan S, Monteiro JK, Prasad A, Filipe CDM, Li Y, Didar TF (2024) Material breakthroughs in smart food monitoring: intelligent packaging and on-site testing technologies for spoilage and contamination detection. Adv Mater 36:2300875. https://doi.org/10.1002/adma.202300875
161. Gorup LF, Neto FNS, Kubo AM, Souza JAS, Fernandes RA, Fernandes GL et al (2017) Nanostructured functional materials: silver nanoparticles in polymer for the generation of antimicrobial characteristics. In: Longo E, La Porta FDA (eds) Recent advances in complex functional materials. Cham, Springer International Publishing, pp 271–292. https://doi.org/10.1007/978-3-319-53898-3_11
162. Motarjemi Y (2002) Impact of small scale fermentation technology on food safety in developing countries. Int J Food Microbiol 75:213–229. https://doi.org/10.1016/S0168-1605(01)00709-7
163. Guadayol M, Cortina M, Guadayol JM, Caixach J (2016) Determination of dimethyl selenide and dimethyl sulphide compounds causing off-flavors in bottled mineral waters. Water Res 92:149–155. https://doi.org/10.1016/j.watres.2016.01.016
164. Cinar A, Parulekar SJ, Undey C, Birol G (2003) Batch fermentation. 0 ed. CRC Press. https://doi.org/10.1201/9780203911358
165. Aragay G, Pino F, Merkoçi A (2012) Nanomaterials for sensing and destroying pesticides. Chem Rev 112:5317–5338. https://doi.org/10.1021/cr300020c
166. Hashim N, Abdullah S, Yusoh K (2022) Graphene nanomaterials in the food industries: quality control in promising food safety to consumers. Graphene 2D Mater 7:1–29. https://doi.org/10.1007/s41127-021-00045-5
167. Tien T, Saccomano SC, Martin PA, Armstrong MS, Prud'homme RK, Cash KJ (2022) Sensors in a flash! oxygen nanosensors for microbial metabolic monitoring synthesized by flash nanoprecipitation. ACS Sens 7:2606–2614. https://doi.org/10.1021/acssensors.2c00859
168. Møretrø T, Langsrud S (2017) Residential bacteria on surfaces in the food industry and their implications for food safety and quality. Comp Rev Food Sci Food Safe 16:1022–1041. https://doi.org/10.1111/1541-4337.12283
169. Alegbeleye O, Odeyemi OA, Strateva M, Stratev D (2022) Microbial spoilage of vegetables, fruits and cereals. Appl Food Res 2:100122. https://doi.org/10.1016/j.afres.2022.100122
170. Anklam E, Gadani F, Heinze P, Pijnenburg H, Van Den Eede G (2002) Analytical methods for detection and determination of genetically modified organisms in agricultural crops and plant-derived food products. Eur Food Res Technol 214:3–26. https://doi.org/10.1007/s002170100415
171. Vecchione M, Feldman C, Wunderlich S (2015) Consumer knowledge and attitudes about genetically modified food products and labeling policy. Int J Food Sci Nutr 66:329–335. https://doi.org/10.3109/09637486.2014.986072
172. Salisu IB, Shahid AA, Yaqoob A, Ali Q, Bajwa KS, Rao AQ et al (2017) Molecular approaches for high throughput detection and quantification of genetically modified crops: a review. Front Plant Sci 8:1670. https://doi.org/10.3389/fpls.2017.01670
173. Kumar V, Arora K (2020) Trends in nanoinspired biosensors for plants. Mater Sci Energy Technol 3:255–273. https://doi.org/10.1016/j.mset.2019.10.004
174. Perdomo SA, Marmolejo-Tejada JM, Jaramillo-Botero A (2021) Review-bio-nanosensors: fundamentals and recent applications. J Electrochem Soc 168:107506. https://doi.org/10.1149/1945-7111/ac2972
175. Kumar V, Guleria P (2020) Application of DNA-nanosensor for environmental monitoring: recent advances and perspectives. Curr Pollution Rep. https://doi.org/10.1007/s40726-020-00165-1
176. Maksimović M, Omanović-Mikličanin E, Badnjević A (2019) Nanofood and internet of nano things: for the next generation of agriculture and food sciences. Cham: Springer International Publishing. https://doi.org/10.1007/978-3-030-15054-9
177. Mayer M, Baeumner AJ (2019) A megatrend challenging analytical chemistry: biosensor and chemosensor concepts ready for the internet of things. Chem Rev 119:7996–8027. https://doi.org/10.1021/acs.chemrev.8b00719

Emerging Trends in Nanobiosensors Development for Quality Control

Abhishant Nigam, Suhani Sharma, Anushka Puri, Sarang Gajare, and Shadab Ahmed

Abstract The global food industry today is valued at more than nine trillion USD and is growing annually at a rate of 6.47%. Food adulteration is a common practice, prevalent in many high-value, high-consumption items such as milk, honey, spices, popular beverages, etc., which necessitates the importance of rapid detection of food adulterants. Various nanoparticles such as metal oxide, carbon nanotubes (CNTs), graphene, carbon nanofibers (CNFs), magnetic nanoparticles, and electrospun nanofibers are widely used to design nanobiosensors that can detect adulteration and spoilage in specific food items. The intrinsic characteristic of nanomaterials allows it to interact with target analytes more efficiently facilitating the development of sensors capable of detecting minute quantities of adulterants with remarkable sensitivity. Nanobiosensors in the food industry can be employed for detection of various contaminants (Pb, Cd, As, Hg, NaNO3, Etc.) toxins, adulterants (detergents, starch, preservatives), dyes (artificial food colors), and pesticides (organophosphates, carbamates, pyrethroids, organochlorines, triazines, etc.) which are commonly found in food items. Compared to conventional sensors, nanobiosensors offer significantly faster, more precise, and reliable detection of these contaminants, enabling real-time monitoring and rapid response to food safety threats. Additionally, the improved electrical and optical characteristics of nanomaterials hold great promise to revolutionize food quality control monitoring mechanisms.

Keywords Nanobiosensors · Nanomaterials · Nanoparticles · Quality control · Heavy metals · Carbon nanotubes

1 Introduction

Biosensor technology is fundamentally composed of three integral components that work in concert to detect specific biological or chemical substances. The first of these components is the biological detection system, commonly referred to as the

A. Nigam · S. Sharma · A. Puri · S. Gajare · S. Ahmed (✉)
Department of Biotechnology, Savitribai Phule Pune University, Pune, India
e-mail: shadab@iitg.ernet.in

D. Sillu et al. (eds.), *Nanobiosensors for the Food Industry*, Smart Nanomaterials Technology, https://doi.org/10.1007/978-981-95-0136-6_4

bioreceptor. The bioreceptor plays a crucial role in the biosensor's functionality and can take various forms, including enzymes or antibodies, which are specifically designed to interact with target analytes. This interaction occurs through a unique binding process which can be noncovalent in nature, where the bioreceptor recognizes and binds to the analyte, triggering a response that is critical for further signal processing. Next in-line component is the transducer, which serves as the intermediary that transforms a biochemically generated signal from the interaction between the analyte and the bioreceptor into a measurable signal. This transformation is vital, as it converts events at the molecular level into a format that can be quantified and analyzed. The transducer can operate through various mechanisms, including optical, electrochemical, or piezoelectric methods, depending on the specific application and desired output. The final component of a biosensor is the signal processing system. This system is responsible for converting the raw output from the transducer into a workable form, which is often presented as a digital readout or graphical representation. Signal processing involves amplifying the signal, filtering out noise, and interpreting the data to provide meaningful results that can be easily understood by the end user (Fig. 1).

The advancements in biosensor technology have led to the integration of bioreceptors with nanoparticles, such as gold nanoparticles. This integration can be achieved through various techniques, enhancing the sensitivity and specificity of the biosensor. Nanobiosensors, which employ these nanoscale materials, benefit from increased surface area and improved electronic properties, allowing for more accurate detection of analytes at lower concentrations. This development not only enhances the performance of traditional biosensors but also expands the potential applications within medical diagnostics, environmental monitoring, and food safety.

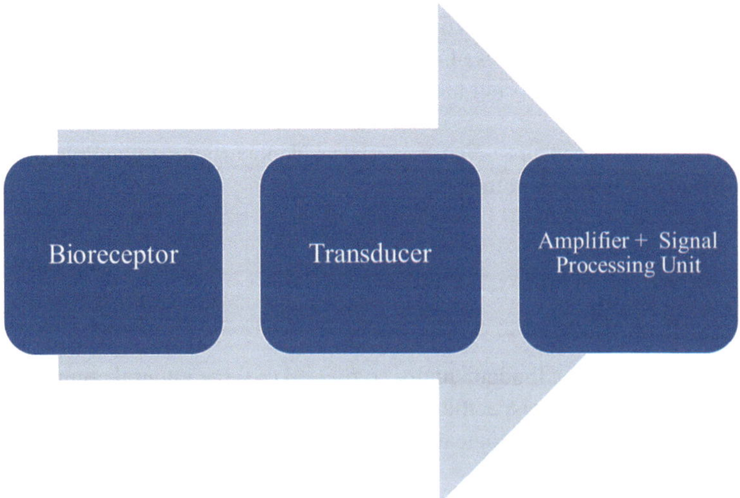

Fig. 1 Basic design of a biosensor

2 Need of Biosensors for Quality Control

Biosensors today have a crucial role in quality control, particularly in the food and pharmaceutical industries, diagnostics, and ecological studies. They are employed to provide a rapid and reliable detection method that can detect contaminants, adulterants, pathogens, or any chemical imbalances in the provided samples [1]. Biosensors offer high sensitivity, which can be helpful for the detection of even very low concentrations of the target analytes. Along with this aspect, they can be highly specific ensuring accurate results [2, 3]. Nanobiosensors are a revolutionary step toward the existing biosensor industry as they provide higher sensitivity and precision by incorporating their large surface area for the ease of smooth detection. Nanomaterials allow for extensive miniaturization that can enable them to be portable making them an ideal choice for on-site testing [4]. They have improved specificity and can be optimized for consuming less power compared to the traditional biosensors [5]. Being miniaturized, they can often react more quickly and can even offer real-time monitoring.

3 Characteristics of Nanobiosensor

Nanobiosensors represent a significant advancement in the realm of sensor technology, merging the principles of nanotechnology with biosensing capabilities to detect biological molecules and changes within environmental conditions at remarkably small scales. These sophisticated devices exhibit several key characteristics that enhance their functionality in various applications. One of the most notable features of nanobiosensors is their high sensitivity like sensitivity for *E.coli* and *S. aureus* 0.048 and 0.046 (μA/(CFU/mL))/cm^2, respectively [6]. The nanoscale dimensions of these sensors enable them to detect extremely low concentrations of analytes, which is crucial in many diagnostic and monitoring scenarios [7]. The large surface area-to-volume ratio inherent in nanoscale materials greatly amplifies their ability to interact with target molecules, thus improving sensitivity and detection limits [8].

Another important characteristic is the high specificity of the biosensor. Nanobiosensors can be meticulously engineered to identify distinct biological molecules or pathogens through the incorporation of biological recognition elements [9]. These elements can include antibodies, enzymes, or nucleic acids specifically designed to bind to their target substances [10, 11]. This targeted interaction allows for precise detection of specific biological entities, which is essential for applications in medical diagnostics and pathogen detection.

The miniaturization of these devices is also a significant advantage. The compact size of nanobiosensors facilitates their integration into portable or wearable technology, thereby making them highly suitable for point-of-care diagnostics and real-time monitoring applications. This portability ensures that users can conduct tests in various environments, improving accessibility to critical health information. In addition to miniaturization, nanobiosensors exhibit rapid response times. This is attributed to their nanoscale size and high surface area, which promote swifter interactions between the sensor and the target molecules. As a result, these sensors can provide quicker detection times, which is especially valuable in emergency medical situations where timely results are crucial.

Moreover, nanobiosensors are designed to function with low sample volumes. They can often operate effectively with only minute amounts of a sample, which is particularly beneficial when dealing with situations where sample availability is restricted, such as in pediatric patients or rare biological specimens. Versatility is another key characteristic of nanobiosensors. These devices can be tailored to detect an extensive range of biological and chemical targets, which may include proteins, DNA, small molecules, and various pathogens. This breadth of application enhances their utility across different fields, from clinical diagnostics to environmental assessments. The high resolution of nanobiosensors enables them to provide detailed information about molecular interactions. This capability is critical for researchers seeking to understand complex biological processes, as it allows for the observation and analysis of interactions at exquisite detail.

Furthermore, nanobiosensors can be integrated with a variety of detection technologies, such as optical, electrochemical, or piezoelectric methods. This integration not only enhances their functionality but also broadens the potential applications of these sensors in diverse fields, including healthcare, environmental science, and food safety. Real-time monitoring is an additional strength of many nanobiosensors. They can deliver continuous data streams, facilitating the dynamic observation of biological systems. This real-time capability is invaluable in situations where immediate feedback on biological processes is essential for decision-making [12].

Finally, the cost-effectiveness of nanobiosensors has improved significantly due to advancements in fabrication techniques, such as nanolithography [13]. These improvements have played a crucial role in reducing manufacturing costs, making nanobiosensors more accessible for a variety of applications across different sectors (Fig. 2).

In conclusion, the distinctive characteristics of nanobiosensors—high sensitivity, specificity, miniaturization, rapid response times, low sample volume requirements, versatility, high resolution, integration with other technologies, real-time monitoring capabilities, and cost-effectiveness—underscore their immense value in critical fields such as medical diagnostics, environmental monitoring, and food safety. Their continued development and application promise to enhance diagnostic accuracy and efficiency, ultimately contributing to improved health outcomes and environmental stewardship. Some other characteristics of a good nanobiosensor can depend on reproducibility of results, range of function, lifetime of materials, detection limit.

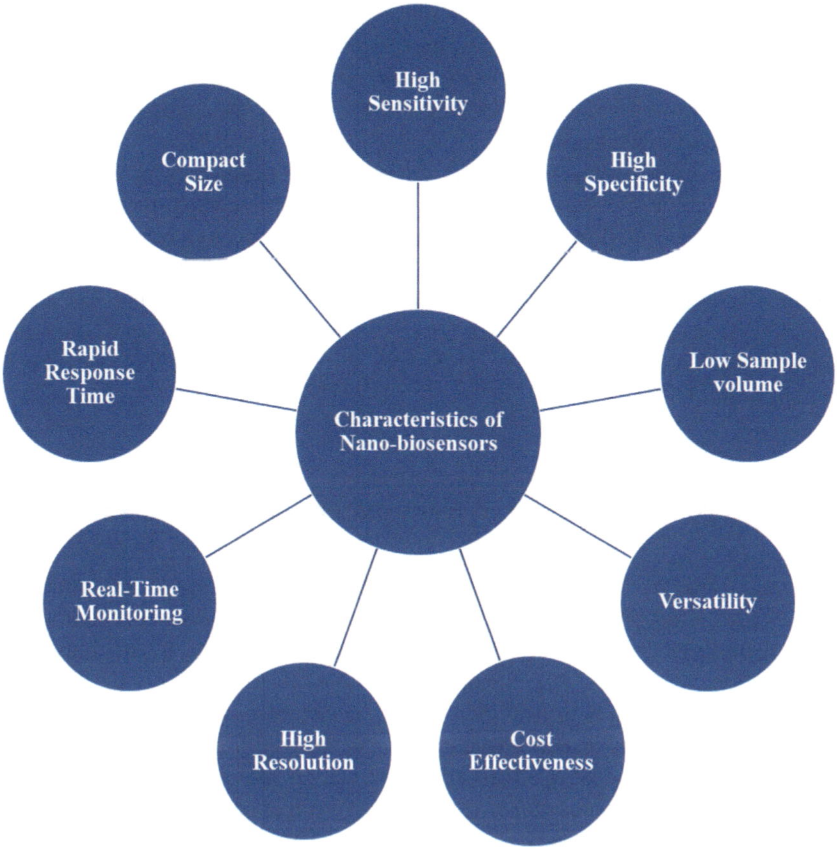

Fig. 2 Characteristics of nanobiosensors

4 Generations of Biosensors

Biosensors, popular analytical devices, are categorized into three main generations. This classification is based on the range of integration of separate components, such as the different methods of linking the bioreceptor molecules or their biorecognition to the core transducer element. These are categorized as:

First-Generation Biosensors: The first-generation biosensors consist of biosensors where the bioreceptor's molecule is physically entrapped in the area of the base sensor followed by a discriminating membrane like dialysis membrane. In the next generations, immobilization can be achieved by properly customizing the transducer interface through covalent bonding or otherwise by its inclusion into a polymer matrix on the surface of transduction [3].

Second-Generation Biosensors: The second generation of biosensors differs from other generations as the electronics, biomolecule, and electrode components are placed separately and away from each other properly and discreetly. Enzymes are immobilized on the electrode surface where their interaction with specific substrates generates a detectable signal. This immobilization is often achieved through physical entrapment or covalent bonding or by the inclusion of substrates into a polymer matrix on the surface of transduction [14].

Third-Generation Biosensors: This generation of biosensors is distinguished where the bioreceptor molecule becomes a crucial component of the base sensing element. This concept proved to have wider applications even in biosensors despite being originally designed for enzyme electrode systems. These use molecularly imprinted polymers as recognition elements. MIPs are synthetic polymers engineered to have specific binding sites for target analytes, providing high selectivity and stability. The following table enumerates some more differences between the three generations (Table 1).

5 Use of Nanoparticles and Nanomaterials in Biosensors

Nanoscale particles (NPs) are commonly utilized in biosensors to bridge the gap between the bioreceptor and the converter. The integration of NPs with electrochemical methods has led to the development of biosensors with enhanced sensitivity, superior detection capabilities (19). Nanomaterials are a broad category of materials and have revolutionized the biosensor industry due to their ultra-small dimensions. They are opted for their rapid detection quality and can be customized for a wider range of applications. Development of advanced nanostructures into various biosensing platforms is a promising approach for wide applications like opioid detection, resulting in swift and precise identification.

Various popular nanomaterials which are used in biosensors can be categorized as:

5.1 Nanoparticle-Based Biosensor

Gold nanoparticles (AuNPs) are widely used for incorporation into biosensors as they enhance visibility and provide optical advantages. Moreover, they enhance signals that can be easily detected using fluorometric and colorimetric assays. Similarly, silver nanoparticles (AgNPs) are also used majorly due to their plasmonic properties increased sensitivity toward environmental changes, flexible optical properties, enhancement of local electromagnetic field, and improved efficiency for light harvesting devices [18, 19]. Apart from these, platinum-based nanoparticle biosensors are also used.

Table 1 Generations of biosensors

Distinguishing factor	First generation	Second generation	Third generation	References
Relying factor	It relies on enzymes like oxygenases and dehydrogenases	It relies on enzymes as biorecognition elements. Through this, the biosensors use electrochemical methods to detect and measure the biological interactions	These biosensors rely on utilizing direct electron transfer between the bioreceptor (which could be an enzyme) and the electrode reducing the need for mediators and enhancing the efficiency of the electron transfer process	[15]
Technology used	It uses electrochemical which is mainly enzyme-based technology	It uses electrochemical technology by improving the enzyme and mediator use	Optical, piezoelectric and other advanced transduction methods are used in the third-generation biosensors	[14, 16, 17]
Transducer	Simple electrodes like the Clark-type electrodes are used in the first generation	In this generation, the electrodes are improved by using mediators	This generation sees the use of advanced transducers including surface plasmon resonance (SPR) and quartz crystal microbalance (QCM)	[14, 16, 17]
Applications	Major applications are basic clinical diagnostics	Some of the applications include the use of more complex biochemical assays and environmental monitoring	The applications for third-generation biosensors include complex and multianalyte assays, real-time and point-of-care diagnostics	[14, 16, 17]

5.2 *Metal-Oxide Nanoparticle-Based Biosensors*

Some examples like zinc-oxide biosensors and titanium dioxide nanoparticles are used due to their enhanced electrochemical properties allowing for swift detection. They are also employed due to their photocatalytic properties which can be advantageous for optical biosensors [20].

5.3 Carbon-Based Nanomaterial Biosensors

These can be further categorized into carbon nanotubes, graphene, and graphene oxide biosensors. These are employed due to their large surface area and high electrical conductivity. These properties make them fit for various electrochemical biosensors [18]. They exhibit a well-ordered arrangement of carbon atoms and are connected through sp^2 bonds, which imparts great strength and rigidity to these materials. These nanomaterials are among the most extensively studied for use in biosensors across various diagnostic applications in medical and research fields and also prove to be effective scaffolds for the immobilization of biomolecules on their surfaces [21].

5.4 Quantum Dot Biosensors

Quantum dots (QDs) are tiny semiconductor crystals ranging from 2.0 to 10.0 nm in size [21]. They are used for their unique optical properties, such as emitting light at different colors based on their size, which makes them great for fluorescence-based detection of organic compounds and macromolecules [20]. The categorization is as 5.0–6.0 nm dots are orange/red, while 2.0–3.0 nm dots are blue/green [21].

5.5 Nanowire Biosensors

Nanowires (NWs)-based biosensors are tiny, wire-like structures made from materials including semiconducting metal oxides, metal nanotubes, and carbon [21]. One such widely used nanowire-based biosensor is a silicon nanowire biosensor. These biosensors are beneficial due to their small size and exceptional thermal, mechanical, chemical, optical, and various electronic properties that are difficult to find in bulk materials. They have heightened sensitivity which can be exploited advantageously in the detection of proteins and DNA material. A stark application is the study pursued by Ivanov et al., 2021 focusing on the use of silicon nanowire field-effect transistors (Si-NW FETs) to study cancer-related genetic markers.

6 Mechanisms and Types of Nanobiosensors

6.1 Electrochemical Nanobiosensors

Electrochemical biosensor platforms offer numerous advantages, such as affordability, speed, specificity, and portability. They are also useful for enhancing the sensitivity and various other features of cell phones and other analytical methods [22]. Electrochemical biosensors are analytical tools that transform biological interactions—such as those between an enzyme and a substrate or an antibody and an antigen—into electrical signals like current, voltage, or impedance. A major part of these sensors is the electrode, which provides stable platform for electron flow and biomolecule immobilization. The use of various nanoparticles with extensive surface areas enhances the loading capacity and mass transfer of reactants, leading to superior analytical sensitivity. The selection of the working electrode material is crucial for sensor and biosensor performance. Although traditional materials like metals and carbon-based electrodes have been commonly used, nanomaterials offer distinct advantages [23] (Table 2).

Table 2 Common working electrode materials and their characteristics in electrochemical biosensor [22]

Working electrode material	Advantages	Disadvantages	Applications
Metallic electrodes	High conductivity, biocompatibility, good mechanical properties	Can be expensive, susceptible to corrosion, may require surface modification	Voltammetry, amperometry, cyclic voltammetry
Carbon-based electrodes	Low cost, wide potential window, chemically inert	May have lower conductivity than metallic electrodes, can be prone to fouling	Voltammetry, amperometry, cyclic voltammetry
Mercury electrodes	High sensitivity, low background current	Toxic, limited potential window, requires specialized equipment	Polarography, stripping voltammetry
Nanomaterials	High surface area, enhanced electron transfer, improved sensitivity, high mechanical strength, high electrical conductivity, catalytic activity Biocompatibility	Can be challenging to synthesize and characterize	Electrochemical biosensors, nano-sensors, energy storage devices

6.2 Optical Nanobiosensors

Optical nanobiosensors generate biosignals by emitting light signals when target analytes bind to their respective receptors. These sophisticated analytical devices are engineered for the selective identification of specific biomolecules, utilizing an immobilized bioreceptor and a physicochemical transducer to quantify analyte concentrations. Recent advancements in optical nanobiosensor technologies have introduced various signal readout methods, including fluorescence, chemiluminescence, colorimetric, surface-enhanced Raman spectroscopy (SERS), and surface plasmon resonance (SPR) [24].

Nanomaterials exhibit unique and tuneable physical, chemical, optical, electrical, and magnetic properties, such as a high surface area, increased bandgap, enhanced binding energy, biocompatibility, and remarkable stability [25]. Among these attributes, their optical properties stand out as the most significant and advantageous [26]. The optical characteristics of nanomaterials allow for the straightforward modification and adjustment of their shape, surface functionalities, composition, and size. This capability leads to the development of optically active nanomaterials, which hold promise for innovative nanosensor systems aimed at detecting chemicals and biomolecules [27] (Table 3).

Table 3 Comparison of colorimetric, fluorimetric, and SERS-based nanosensors [28]

Feature (s)	Colorimetric sensors	Fluorimetric sensors	SERS-based sensors
Detection principle	Color change	Fluorescence emission	Raman scattering enhancement
Signal readout	UV–Vis spectrophotometry, digital image analysis	Fluorescence microscopy	Raman spectroscopy
Advantages	Simple, direct visual detection, low cost	High sensitivity, low background noise, multiplexed detection	Highly sensitive, specific, label-free detection
Disadvantages	Less sensitive for low concentrations, susceptible to interference	Requires specialized equipment, may suffer from photobleaching	Requires complex sample preparation, may be sensitive to matrix effects
Applications	Detection of small molecules, ions, pathogens	Detection of pathogens, biomolecules, contaminants	Detection of pathogens, biomolecules, contaminants

6.3 Piezoelectric Nanobiosensors

Piezoelectric biosensors offer a promising platform for detecting various analytes due to their ability to directly measure affinity interactions without the need for additional reagents. However, they require careful consideration of factors such as sensitivity and fragility. Different types of piezoelectric nanobiosensors are explained by [29].

Piezoelectric immunosensors are analytical devices that utilize antibodies as biorecognition elements to detect specific macromolecular compounds or microorganisms. The specificity of the immunosensor depends largely on the specificity of the antibody used.

The principle of piezoelectric immunosensors:

- Immunological reaction: The antibody or antigen on the immunosensor surface binds to its specific target molecule.
- Mass change: This interaction results in a change in mass on the piezoelectric material.
- Frequency shift: The alteration in mass affects the resonant frequency of the piezoelectric material.
- Detection: Frequency shift is measured and correlated to the concentration of the target molecule.

Molecularly imprinted polymers (MIPs) are artificial materials designed to selectively recognize and bind target analytes, similar to the way antibodies interact with antigens in biosensing applications. They are synthesized by polymerizing a mixture of monomers around a template molecule, which is subsequently extracted, resulting in the formation of specific binding sites that match the shape, size, and functional groups of the template.

The principle of molecularly imprinted polymer (MIP)-based piezoelectric biosensors.

- MIP synthesis: The MIP is synthesized in the presence of a template molecule, creating a cavity with a complementary shape and binding sites for the target analyte.
- Template removal: The template molecule is removed from the MIP, leaving behind a porous structure with specific binding sites.
- Analyte binding: When the target analyte interacts with the MIP, it attaches to the specific binding sites located within the cavities.
- Piezoelectric signal: The binding of the analyte to the MIP causes a change in mass on the piezoelectric material. This change in mass alters the resonant frequency of the material.
- Detection: The variation in frequency is recorded and related to the concentration of the target analyte.

Magnetic nanoparticles (MNPs), mainly superparamagnetic nanoparticles, are highly esteemed among smart materials due to their distinct and tunable characteristics. When exposed to an external magnetic field, MNPs' great magnetic reactivity enables them to be quickly retrieved from intricate sample matrices [30]. Because most (MNPs) can be recovered and reused several times to avoid contamination, they are also generally considered ecologically favorable. Utilizing these efficient magnetic nanoaids, several food safety analysis techniques have been developed, including magnetic solid-phase extraction (MSPE) [31, 32].

Microextraction methods based on MNPs have become increasingly popular for the analysis of various contaminants in food systems; by modifying the chemical entities or molecular fragments covalently bound or adsorbed on the surface of magnetic nanoparticles, researchers can achieve careful and efficient entrapment of trace-level contaminants. The magnetic properties of MNPs allow for rapid separation, eliminating the need for traditional filtration and centrifugation steps. This leads to faster preconcentration procedures and improved detection efficiency. Advantages of MNP-based microextraction methods include reduced solvent consumption which is MNPs can be used with minimal or no organic solvents, making them more environmentally friendly. Simplified procedures help in the magnetic separation process streamline the analysis process. Enhanced sensitivity in which MNPs can effectively capture and concentrate trace-level contaminants, and last one is versatility where MNPs can be modified with various functional groups to target different analytes in various sample matrices [33].

7 Applications

Nanobiosensors have expanded tremendously in the food industry over the last few years. This expansion is due to nanomaterials' unique properties, which offer higher sensitivity and selectivity in detecting contaminants. We highlight some specific applications of nanobiosensors for quality control (Table 4):

- Detection of heavy metals: These metals can accumulate in the body and lead to severe health risks. Nanobiosensors have proven effective in detecting toxic heavy metals, including lead (Pb), cadmium (Cd), arsenic (As), and mercury (Hg), in various food products. Gold and magnetic nanoparticles are commonly employed for identifying these contaminants in foods such as rice, vegetables, and seafood [34].

Table 4 Applications of different nanobiosensors

Application	Nanomaterials used	Contaminants detected	References
Heavy metal detection	Gold nanoparticles	Pb, Cd, As, Hg	[34]
Adulteration in milk	CNTs, graphene	Starch, detergents, sugars	[35]
Pesticide detection	CNTs, graphene	Organophosphates, carbamates, pyrethroids	[36]
Microbial contamination	Magnetic nanoparticles	*E. coli, Salmonella*	[37]

- Adulterant detection in milk and honey: Milk and honey are frequently adulterated with additives such as starch, detergents, and artificial sugars. Carbon nanotube (CNT)-based sensors can rapidly detect these contaminants at low concentrations. The use of graphene oxide and electrospun nanofibers has further improved the sensitivity and detection speed [35].
- Pesticide detection: Pesticides, such as organophosphates, carbamates, and organochlorines, are common contaminants in food crops. Nanobiosensors, particularly those utilizing CNTs and graphene, provide rapid and reliable detection of these pesticides in fruits, vegetables, and cereals [36].
- Microbial contamination monitoring: The detection of bacterial pathogens (e.g., *E. coli, Salmonella*) in food has also been made efficient through nanobiosensors. Magnetic nanoparticles combined with DNA aptamers are capable of identifying microbial contaminants in food matrices like meat and dairy [37].

8 Future Prospects

With ongoing advancements in material science and sensor design, nanobiosensor technology is expected to further revolutionize food quality control. Several emerging trends offer promising prospects:

- Integration with AI and IoT: The integration of nanobiosensors with artificial intelligence (AI) and Internet of Things (IoT) technologies will facilitate real-time monitoring of food products across the entire supply chain. This can lead to predictive quality control, minimizing waste, and ensuring food safety at all stages [38].
- Hybrid nanomaterials providing enhanced sensitivity: The combination of different nanomaterials like metal oxides and carbon nanofibers (CNFs) is expected to create hybrid sensors with enhanced sensitivity and selectivity. These advanced sensors can detect lower levels of contaminants more accurately, providing a competitive edge over conventional methods [39].
- Point-of-Care (POC) Nanobiosensors: The development of POC devices based on nanobiosensors will allow rapid on-site testing of food samples for contamination or spoilage, which is particularly useful in remote areas with limited access to sophisticated lab infrastructure [40].
- Biodegradable Nanobiosensors: With the growing demand for sustainable technologies, biodegradable and eco-friendly nanobiosensors made from cellulose nanofibers and other green materials are gaining attention. These sensors will reduce environmental impact while providing accurate food quality assessments [9].

9 Conclusion

Nanobiosensors promise transformative potential for quality control, enabling the detection of a multitude of contaminants with remarkable precision and accuracy. Their ability to detect minute levels of adulterants, toxins, and microbial contaminants ensures better food safety and consumer protection. The integration of nanobiosensors with IoT and AI technologies will drive future developments, ensuring real-time monitoring and predictive control of food safety across the global supply chain. As technology evolves, there is immense potential for improving food quality control mechanisms, resulting in safer and healthier food for consumers worldwide.

References

1. Li X et al (2011) Biosynthesis of nanoparticles by microorganisms and their applications. J Nanomater 2011:1–16. https://doi.org/10.1155/2011/270974
2. Liu H, Wang J (2007) Electrochemical sensors and biosensors based on nanomaterials and nanostructures. J Nanosci Nanotechnol 7(12):4271–4280
3. Wang J (2008) Electrochemical biosensors for biomedical applications. Anal Chem 80(8):2262–2271
4. Sood A, Kumar V (2016) Nanomaterial-based portable biosensors for point-of-care applications. Sens Actuators, B Chem 231:1184–1197
5. Choi J, Park H (2018) Low-power nanoscale biosensors for wearable and portable applications. IEEE Trans Nanotechnol 17(2):276–285
6. Gangwar R et al (2022) Label-free miniaturized electrochemical nanobiosensor triaging platform for swift identification of the bacterial type. Anal Chim Acta 1233:340482. https://doi.org/10.1016/j.aca.2022.340482
7. Kumari M et al (2023) Microfluidics-Based nanobiosensors for healthcare monitoring. Mol Biotechnol 66(3):378–401. https://doi.org/10.1007/s12033-023-00760-9
8. Hanif A et al (2019) Aptamer based nanobiosensors: promising healthcare devices. Saudi Pharmaceutic J 27(3):312–319. https://doi.org/10.1016/j.jsps.2018.11.013
9. Singh N et al (2023) Nanobiosensors design using 2D materials: implementation in infectious and fatal disease diagnosis. Biosensors 13(2):166. https://doi.org/10.3390/bios13020166
10. Amarante AM et al (2014) Modeling the coverage of an AFM tip by enzymes and its application in nanobiosensors. J Mol Graph Model 53:100–104. https://doi.org/10.1016/j.jmgm.2014.07.009
11. Negahdary M, Angnes L (2022) Electrochemical nanobiosensors equipped with peptides: a review. Microchimica Acta 189(3). https://doi.org/10.1007/s00604-022-05184-x
12. Ye C, Wang M, Min J et al (2024) A wearable aptamer nanobiosensor for non-invasive female hormone monitoring. Nat Nanotechnol 19(3):330–337. https://doi.org/10.1038/s41565-023-01513-0
13. Tebbe M et al (2017) Homopolymer nanolithography. Small (Weinheim an der Bergstrasse, Germany) 13(37). https://doi.org/10.1002/smll.201702043
14. Heller A, Feldman B (1987) Design and performance of glucose sensors based on glucose oxidase and redox mediators. Anal Chem 59(6):1067–1074
15. Parthasarathy P, Vivekanandan S (2018) A comprehensive review on thin film-based nanobiosensor for uric acid determination: arthritis diagnosis. World Rev Sci Technol Sustain Develop. https://doi.org/10.1504/WRSTSD.2018.092824
16. Clark LC, Lyons C (1962) Electrode systems for continuous monitoring in cardiovascular surgery. Ann N Y Acad Sci 102(1):29–45

17. Li Y, Zhang S, Sun L (2020) Recent advances in optical biosensors for point-of-care applications. Biosens Bioelectron 148:111806
18. Abdel-Karim R (2024) Nanotechnology-enabled biosensors: a review of fundamentals, materials, applications, challenges, and future scope. Biomed Mater Dev 2:759–777. https://doi.org/10.1007/s44174-023-00147-z
19. Lee YM, Kim SE, Park JE (2023) Strong coupling in plasmonic metal nanoparticles. Nano Convergence 10:34. https://doi.org/10.1186/s40580-023-00383-5
20. Gençdağ Şensoy K, Muti M (2021) The novel nanomaterials based biosensors and their applications. IntechOpen. https://doi.org/10.5772/intechopen.94930
21. Malik S, Singh J, Goyat R, Saharan Y, Chaudhry V, Umar A, Ibrahim AA, Akbar S, Ameen S, Baskoutas S (2023) Nanomaterials-based biosensor and their applications: a review. Heliyon 9(9):e19929, ISSN 2405–8440
22. Jaffrezic-Renault N, Ariño C, Hassan RY (2022) Advances in electrochemical nano-biosensors for biomedical and environmental applications: from current work to future perspectives. Sensors 22:7539. https://doi.org/10.3390/S22197539
23. Cho IH, Kim DH, Park S (2020) Electrochemical biosensors: perspective on functional nanomaterials for on-site analysis. Biomater Res 24(1). https://doi.org/10.1186/s40824-019-0181-y
24. Vairaperumal T, Huang CC, Liu PY (2023) Optical nanobiosensor-based point-of-care testing for cardiovascular disease biomarkers. ACS Appl Bio Mater 6(7):2591–2613
25. Asha AB, Narain R (2020) Nanomaterials properties. In: Elsevier eBooks, pp 343–359. https://doi.org/10.1016/b978-0-12-816806-6.00015-7
26. Zhang JZ (2009) Optical properties and spectroscopy of nanomaterials. Optic Proper Spectroscop Nanomater 1–383. https://doi.org/10.1142/7093/SUPPL_FILE/7093_CHAP03
27. Sahoo SK (2022) Sensing and biosensing with optically active nanomaterials: a note. Sens Biosens Optical Active Nanomater 1–7. https://doi.org/10.1016/B978-0-323-90244-1.00006-9
28. Li X, Cui H, Zeng Z (2018) A simple colorimetric and fluorescent sensor to detect organophosphate pesticides based on adenosine triphosphate-modified gold nanoparticles. Sensors 18(12):4302. https://doi.org/10.3390/S18124302
29. Pohanka M (2018) Overview of piezoelectric biosensors, immunosensors and DNA sensors and their applications. Materials 11(3):448. https://doi.org/10.3390/MA11030448
30. Gao Q, Luo D, Bai M, Chen ZW, Feng YQ (2011) Rapid determination of estrogens in milk samples based on magnetite nanoparticles/polypyrrole magnetic solid-phase extraction coupled with liquid chromatography-tandem mass spectrometry. J Agricult Food Chemistr 59(16):8543–8549. https://doi.org/10.1021/JF201372R/ASSET/IMAGES/LARGE/JF-2011-01372R_0006
31. Hubetska TS, Kobylinska NG, Menendez JRG (2020) Application of hydrophobic magnetic nanoparticles as cleanup adsorbents for pesticide residue analysis in fruit, vegetable, and various soil samples. J Agric Food Chem 68(47):13550–13561. https://doi.org/10.1021/ACS.JAFC.0C00601/ASSET/IMAGES/LARGE/JF0C00601_0007.JPEG
32. Zhou P, Zheng R, Zhang W, Liu W, Li Y, Wang H, Wang X (2019) Development of an effervescent tablet microextraction method using $NiFe_2O_4$-based magnetic nanoparticles for preconcentration/extraction of heavy metals prior to ICP-MS analysis of seafood. J Anal At Spectrom 34(3):598–606. https://doi.org/10.1039/C8JA00331A
33. Yu X, Zhong T, Zhang Y, Zhao X, Xiao Y, Wang L, Liu X, Zhang X (2022) Design, preparation, and application of magnetic nanoparticles for food safety analysis: a review of recent advances. J Agric Food Chem 70(1):46–62. https://doi.org/10.1021/ACS.JAFC.1C03675/ASSET/IMAGES/LARGE/JF1C03675_0003.JPEG
34. Smith J et al (2023) Detection of heavy metals in food using nanoparticles. J Food Chemistr
35. Jones A et al (2022) Adulteration detection in dairy products using CNT-based sensors. Sens Actuat
36. Liu T et al (2021) Nanotechnology in pesticide detection for food safety. Food Anal Meth
37. Kumar R et al (2020) Magnetic nanoparticle-based biosensors for pathogen detection. Biosens Bioelectron

38. Wang Y et al (2024) IoT-enabled nanobiosensors for real-time food safety monitoring. Adv Sens
39. Patel S et al (2021) Hybrid nanomaterials in food quality monitoring. Nanotechnol Rev
40. Zhang F et al (2022) Point-of-care nanobiosensors for on-site food testing. Front Nanotechnol

Nanobiosensors for Food Pathogen Detection: Trends and Limitations

Abhinay Thakur, Valentine C. Anadebe, Abdelkader Zarrouk, Omar Dagdag, Ashish Kumar, and Virat Khanna

Abstract Nanobiosensors have pushed the pace for the rapid detection of foodborne pathogens by providing essential public health strategies oriented to food safety. The incorporation of nanomaterials including nanoparticles, nanotubes, and nanowires has greatly improved sensitivity and specificity of detection in pathogenic bacteria and viruses at very low concentrations. This chapter discusses nanobiosensors and their working, categories with special reference to their use in the food industry. However, the chapter also looks at the drawbacks and difficulties encountered by nanobiosensors with regard to their usage in real-life situations. Issues such as matrix effects, stability of the nanomaterials, and the need to develop stable calibration methods pose major barriers to the realization of the targeted performance. Furthermore, the latest findings about multifunctional nanobiosensors are presented

A. Thakur (✉)
Division of Research and Development, Lovely Professional University, Phagwara, Punjab 144411, India
e-mail: thakurabhinay96@gmail.com

V. C. Anadebe
Department of Chemical Engineering, Alex Ekwueme Federal University Ndufu Alike, P. M. B. 1010, Abakaliki, Ebonyi State, Nigeria

A. Zarrouk
Laboratory of Materials, Nanotechnology and Environment, Faculty of Sciences, Mohammed V University in Rabat, Av. Ibn Battouta, P.O. Box. 1014, Agdal-Rabat, Morocco

Research Centre, Manchester Salt and Catalysis, Unit C, 88-90 Chorlton Rd, Manchester M15 4AN, UK

O. Dagdag
Department of Mechanical Engineering, Gachon University, Seongnam 13120, Republic of Korea

A. Kumar
Department of Science, Technology and Technical Education, Nalanda College of Engineering, Bihar Engineering University, Government of Bihar, Bihar Sharif, Bihar 803108, India

V. Khanna
University Centre for Research and Development, Chandigarh University, Mohali, Punjab 140413, India

D. Sillu et al. (eds.), *Nanobiosensors for the Food Industry*, Smart Nanomaterials Technology, https://doi.org/10.1007/978-981-95-0136-6_5

with examples for sensing more than a pathogen at a time and the use of smart technologies for easy operation. Possible future developments are indicated, including the incorporation of nanobiosensors with portable systems for real-time analysis of food safety which would be a drastic change from current practices.

Keywords Nanobiosensors · Detection · Pathogens · Health · Toxicity

1 Introduction

Pathogenic microorganisms in foods have remained a major menace to human health and are estimated to cause several million infections globally annually [1–3]. Foodborne pathogens are microorganisms such as bacteria, viruses, and parasites that can get into food at any point in its chain from production through processing, distribution, and preparation. The increase in foodborne diseases due to globalization, increased imports, and changes in human diet explains why efficient detectors for foods are important. New pathogens make a new appearance and old strains, once controlled, are also on the scene, the situation aggravates the problem more due to the fact that it requires constant probing and monitoring [4–7]. For instance, microorganisms including *Salmonella*, *Listeria*, and *Escherichia coli* are known to cause major diseases and sometimes death. This has made conventional methods such as identifying microorganisms by their ability to grow in different environments and resist food preservation methods ineffective. The safety of foods is of uttermost importance for public health, the economy, and the security of food supply chains. Food that has been contaminated produces serious health effects, effects that cause hospitalization and in some instances, death. The WHO states that food, water and other related diarrheal diseases claim an estimated 2 million lives per year with the affected population group being children below five years of age. In addition to health effects, foodborne illnesses impact the economy in relation to costs of treatment, revenue loss due to loss in productivity and brand value diminution due to loss in consumers' confidence in food products [8, 9]. It causes losses affecting producers, processors, and retailers to the level they experience devastating losses from food safety incidents that span through supply chain links. Recently, consumers have become more concerned with increased food safety standards due to shocking incidences of foodborne diseases associated with tainted foods. This has boosted capital investment in new technologies and practices with a view to enhancing detection and prevention approaches. Conventional procedures in food safety diagnosis entail elaborate laboratory tests, which take considerable time before the contamination sources can be identified, and the outbreak contained. Consequently, there is a growing importance of methodologies that are efficient, selective, and accurate enough to be used in real-time environments. Recent advances in nanotechnology have attracted attention as the solution to the challenges faced regarding the detection of foodborne pathogens. In particular, the application of nanoscale devices including

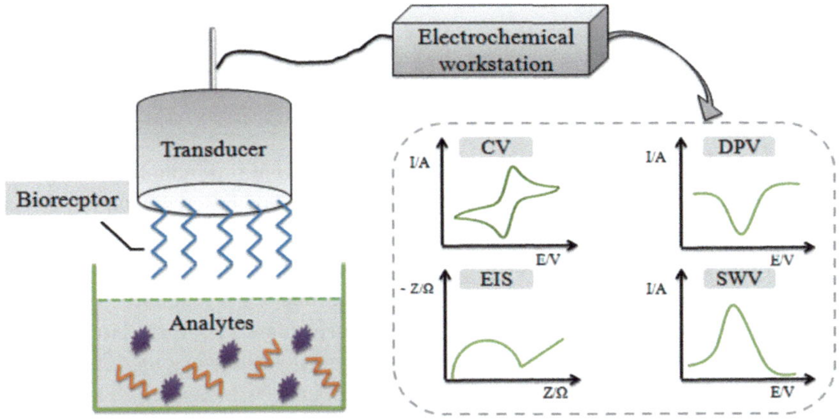

Fig. 1 Detecting mechanism of the biosensor for the detection of the pathogens. Adapted from Ref. [10] under CCBY 4.0

nanoparticles, nanotubes, and nanowires results in higher sensitivity and /or specificity of nanobiosensors. These devices can sample pathogens right down to trace level and so can be used in assessing food safety in different circumstances. At the same time, they can be used for monitoring food safety problems in real time which will allow to prevent large-scale outbreaks of foodborne illness. Figure 1 shows the use of biosensors for the rapid detection of food pathogen.

The application of nanobiosensors in food safety is a radical shift in the way that pathogen detection is considered. Thus, by utilizing the characteristics of nanomaterials, SO researchers and other specialists can create new approaches and tools that, aside from being effective, can easily evolve to fit the requirements of the constantly developing sphere. The implementation of these technologies can be envisaged to bring about change in protecting food from the risks posed by pathogenic bacteria [11, 12]. As the knowledge regarding nanobiosensors increases, it is imperative to discuss their potential as well as the existing problems of their use. Given the character and intricacy of foodborne pathogens, intervention measures require a systems approach with rapid detection technologies being vital components. This chapter seeks to explore the existing trends in nanobiosensor systems and analyze their ability to identify foodborne pathogens, together with the challenges that researchers and practitioners come across. Through presenting an evaluation of the contemporary state of development of nanobiosensors, this chapter aims to advance the discussion on increasing the efficiency of making foods safer and thus benefiting people's health.

2 Nanobiosensors

2.1 Definition and Functionality

Nanobiosensors are very innovative tools of analysis that link nanotechnology and biotechnology to measure biological analytes. They are based on nanomaterials which due to their size—between one and one hundred nanometers—have specific physical and chemical properties. Such properties include a higher surface area to volume ratio, better electrical and optical properties, and better reactivity which when incorporated into nanobiosensors improve their performance compared to traditional biosensors [13, 14]. The use of nanobiosensors rests on the capacity of the system to convert a biological response to a signal that can be measured. This process typically involves two key components: a biorecognition element and a transducer are provided. Biorecognition element is essential to selectively recognize the target analyte and it can be any pathogen toxin or biomolecule such as protein, nucleic acid, or antibody. In more detail, the biorecognition element can be obtained from enzymes, antibodies, DNA, RNA, and even whole cells [15, 16]. These biorecognition elements make it possible for nanobiosensors to render high selectivity and specificity for the sensed analyte, which makes it possible to sense the target securely in wealthy biological solutions. In specific, when the target analyte interacts with the biorecognition element, a biochemical change takes place resulting in an alteration of the transducer. The transducer is also bounded by the task of converting the biochemical signal originating from the interaction into a measure. There is variability in the transduction techniques that can be applied in nanobiosensors; electrochemical, optical, and piezoelectric transducers differ in their modes of operation.

2.2 Types of Nanobiosensors

The variety of nanobiosensors gives a clear distinction of its types as each one is built to detect on the premise of the principles shown above. This classification is largely based on the transduction mechanisms that are used by such sensors. There are four main classes of nanobiosensors namely: electrochemical nanobiosensors; optical nanobiosensors, piezoelectric nanobiosensors, and finally surface plasmon resonance nanobiosensors. Both have their peculiarities of application, strengths, and weaknesses and can be used in very many occasions for the job of detection across various disciplines.

Electrochemical nanobiosensors are one of the oldest and most commonly used types of nanobiosensors in the present context. There are two types, namely the current monitors where changes in current, voltage, or impedance measurements signify biochemical interactions at the surface of the electrode [17, 18]. The basic concept of electrochemical detection is based on the electrochemical reactions that happen when an analyte comes into contact with a biorecognition material placed on an electrode. Electrochemical nanobiosensors are known to have high sensitivity, and this is because of the nanomaterials incorporated in the fabrication of bioelectronics. It was reported that the gold nanoparticles, carbon nanotubes, and graphene have sought material to cause the electrochemical response of the sensors to improve due to their increased surface area [19]. In this case, when the target analyte binds to the biorecognition element, it can cause either the redox reaction or a change in electron transfer rate, which in turn brings about a measurable electrical signal. This makes the electrochemical nanobiosensors capable of rapid detection of low-abundance analytes and is especially beneficial in areas such as food safety and clinical diagnostics.

Some of the methods used in electrochemical nanobiosensors are the following methods Amperometry; Potentiometry; and Impedance Spectroscopy. Amperometric sensors determine the current responses of redox reactions on the electrode surface, therefore directly relating the concentration analyte to the current response [20–22]. Due to the capability of this technique to monitor the concentration of target biomolecules in real time, it is commonly applied for pathogen, toxins, and other biomolecule detection. Solid-state potentiometric sensors track the change in voltage upon the change in the concentration of an ionic species in the solution and come in handy in tracking the density of ionic species in nutrient-rich solutions. In contrast, impedance spectroscopy monitors the change in electrical impedance upon analyte binding and yields information concerning binding kinetics and all the processes taking place in the electrode layer. Electrochemical nanobiosensors have therefore been widely used in food safety, especially for the detection of foodborne pathogens. For example, sensors for detecting *Salmonella*, *E. coli*, or *Listeria* monocytogenes have been produced so that food screenings for potential foodborne diseases can be done quickly [23, 24]. Moreover, electrochemical detection is very flexible to enable multiplex systems allowing for the detection of two or more pathogens or analytes in a solitary test. Such multiplexing capability is pivotal in food indicating that it is highly desirable to analyze a range of compounds in the shortest time possible for the sake of public health. In spite of these qualities, electrochemical nanobiosensors have been known to have certain drawbacks also. Peculiarities in the stabilization of the sensors' response depend on the variability of the environmental conditions, fouling of the electrodes, and the degradation of biorecognition elements. Furthermore, the matrices from samples for instance food products or bio fluids present challenges toward acquiring precise and reproducible measurements. These challenges are being met by current researchers through the invention of new nanomaterials, better designs of sensors, and enhanced signal processing methods.

Optical nanobiosensors are one of the largest groups of sensor systems that use optical techniques for the detection of biological analytes. These sensors are especially prized for their high selectivity, and the opportunity to monitor processes in real time as well as the possibility of multi-analyte detection. Optical nanobiosensors act through changes in the optical characteristics of the target analyte after interacting with light. An important feature of optical nanobiosensors is their ability to quantify target analytes without the use of reagents for sample preparation or conjugation. This makes detection easy and minimizes the time needed for analysis than other laboratory diagnostic methods. Optical sensors work based on fluorescent, surface plasmon resonance (SPR), changes in absorbencies and reflected light [25]. All of these principles are useful in their way for specific applications in food safety as well as other disciplines. Optical nanobiosensors based on florescence are produced from fluorophores that are linked to the biorecognition molecule. Upon binding of the target analyte to the biorecognition element, the characteristics of the fluorescence signal change, and this change can be measured. This method is an effective indicator to identify the low concentrations of pathogens, and in food safety tests where early detection is important. It also makes fluorescent labeling increase the specificity of the detection because the binding event is distinguished from the nonspecific signal. The second important optical method, which is used in nanobiosensing, is surface plasmon resonance (SPR). SPR quantifies variations in the refractive index at the sensor surface due to the binding of the analytes to the biorecognition element. This technique makes it possible to monitor binding events on antioxidants and other molecules in real time without labeling, thus improving the ease and efficiency of detection. In general, it is possible to obtain high sensitivity and specificity with SPR-based sensor systems to detect various biological targets associated with foodborne pathogens [26, 27].

Optical nanobiosensors can also be based on absorbance methods in which variation in the absorbance of the light at selective wavelength is monitored once the target analyte has bonded to the receptor. This approach is simple and can be employed in a number of methods such as the discovery of contaminants in foods. Reflectance-based sensors also work with this concept only, using changes in the amount of light reflected from the sensor surface on the binding to the analyte. Optical nanobiosensors depend on nanomaterials, which enable them to improve their performance. The integration of gold and silver nanoparticles can provide enhancement of optical signal through localized surface plasmon resonance effects, thus improving sensitivity. Apart from improving the signal detected, these nanomaterials also allow for a large area to be coated with biorecognition elements leading to better sensor performance [28, 29]. Optical nanobiosensors have many uses in the safety of food the identification of pathogens, food toxins, chemical contaminants and much more. For instance, rice Cyclers that are used to identify specific food pathogens such as *E. coli* and *Salmonella* can give results within a short span that will enable food producers and the authorities to contain incidences of the diseases. Besides, one can introduce portable optical sensors to monitor food quality and safety indicators directly at the place of their application. However, it needs to be noted that even with all these advantages in mind, the application of optical nanobiosensors meets certain difficulties. The optical detection techniques are generally selective; however, their sensitivities depend on the conditions normally present in the environment, background noises, and matrix effects that are normally present in complex samples. Researchers are keen to overcome such limits by refining new sensor characteristics, optimizing the biorecognition layer and inventing new signal processing methods.

Piezoelectric nanobiosensors belong to a unique class of sensors that utilize the piezoelectric phenomenon for the detection of biomolecular reactions [30, 31]. Most of these sensors incorporate elements like quartz crystals or piezoelectric polymers to determine variations in mass or mechanical characteristics when target analytes interact with the biorecognition element. Piezoelectricity is defined simply as a phenomenon where an electrical charge arises from applied mechanical stress on piezoelectric materials or where the material changes shape when an electric field has been applied. The mode of piezoelectric nanobiosensors works according to the law that the analyte binding to the biorecognition element causes a change in the mass/ mechanical parameter of the sensor which in turn affects the frequency or amplitude of the acoustic wave. Some of the piezoelectric nanobiosensors are able to detect these changes at the picogram level, due to the highly sensitive nature of this method. Therefore, the capacity to obtain such high sensitivity is especially beneficial in instances when the concentrations of the analyzed analyte are low, for example in the analysis of food and water samples. Quartz crystal microbalance (QCM) is one of the most common existing types of piezoelectric sensors. The QCMs comprise a thin layer of quartz crystal that vibrates at a particular frequency when subjected to an oscillating voltage. On binding of the analyte to the immobilized biorecognition element on the crystal surface, there is a change in mass of the system, which, via the Sauerbrey equation, can be measured in terms of frequency change. This correlation between the mass change and the shift in the frequency indicates that the binding events occurring in real-time make the QCM more appropriate for detecting foodborne pathogens and other bio-molecules. The other category of piezoelectric nanobiosensors is the surface acoustic wave (SAW) device [32, 33]. SAW devices also work on a similar concept as QCMs, employing surface acoustic waves that traverse at the surface of the piezoelectric layer. When an analyte is attracted to the biorecognition element, the properties of the acoustic wave change causing a shift in its frequency or amplitude. SAW devices are characterized by a high sensitivity and short response time, which makes them attractive for real-time applications.

Piezoelectric nanobiosensors have several advantages over other sorts of sensors. On this list, one of the major advantages consists in their ability to function in the matrices of the multi-component variety without the necessity to undertake a great deal of preparation of samples. This characteristic makes piezoelectric sensors particularly useful in the applications of food safety since the samples are often heterogeneous and normally contain different interferences. Consequently, the fast response times and the real-time monitoring also allow prompt decision-making in many cases which are crucial for maintaining the quality of food. However, there are still issues with piezoelectric nanobiosensors as well. The general and consistent performance of these sensors depends on external factors, parameters of the piezoelectric material, and the biorecognition elements tethered onto the QCM. Further, signals arising due to matrix effects and nonspecific binding can pose a significant challenge by giving false positive results or incorrect measurements and, therefore, require optimization of the design of sensors and conditions under which they will be operated.

3 Advancements in Nanobiosensor Technology

The present development of nanobiosensor technology has seen a great boost in the recent past because of the deposition of new nanomaterials, highly sensitized response, specificity, and real-time analysis of biological analytes. These enhancements have enhanced the performance of nanobiosensors and expanded the flow area to embrace several industries such as the medical, environmental, and food industries [34, 35]. Constant development and improvement of nanobiosensor technology are indispensable to achieving modern socio-functional objectives related to health and safety, the main advantages of which are high sensitivity and effectiveness of pathogen and toxic substance detection. In this field, another remarkable development in nanobiosensor technology is the incorporation of different forms of nanomaterials into nanoscale biosensors. These nanomaterials are anything that has a size close to or less than 100 nm, and the enhanced properties of such materials are beneficial to biosensors. When it comes to nanomaterials of nanoscale biosensors, nanoparticles, nanotubes, and nanowires are most notable and prominent because of their unique properties and uses [36]. Nanoparticles have become the most widely used nanomaterial in nanobiosensor development because of their large surface area/ volume ratio, size, and surface chemistry. These properties allow them to interface with biomolecules in a fashion that improves the sensitivity and selectivity of the sensor. For instance, gold nanoparticles are quite important in optical and electrochemical biosensors owing to the optimized characteristics of surface plasmon resonance and biocompatibility. When conjugated with antibodies or other biorecognition elements, their capacity to enhance the signal has been particularly useful in the identification of low-concentration pathogens. The application also has benefits in that the antimicrobial action of silver nanoparticles can reduce nonspecific binding and contribute to the increased dependability of the sensor.

Another class of nanomaterials, carbon nanotubes (CNTs), has also given paramount support to nanobiosensor science [37, 38]. CNTs are known for their excellent electrical conductivity, high mechanical properties, and large surface area, which qualifies them for improving electrochemical sensors. Incorporation of CNTs into nanobiosensors can enhance the electron transfer activities and hence increase signal responses. Also, such structure provides the opportunity to immobilize the biorecognition elements here giving more interactions with the target analyte. Over the years, scientists have proven that CNT-based nanobiosensors can closely and selectively capture biomarkers and pathogens; therefore, CNT-based nanobiosensors are good candidates for disease diagnosis and food safety monitoring. Carbon nanotubes and another type of nanomaterial named nanowires also have enormous potential to raise the efficiency of nanobiosensors. They are single-scaled structures with high width-to-thickness ratios and can be made from metals, semiconductors, and oxides. Nanowires can be applied to numerous sensory structures, but field effect transistors (FET) and conductance sensors found to have the ability to incorporate nanowires. Because of the large surface area and chemical reactivity of nanotubes, they can enhance the sensitivity and selectivity of nanobiosensors through chemical interaction with biological molecules. For instance, silicon nanowire-based biosensors have shown the possibility of identifying low concentrations of cancer biomarkers in different body fluids, hence the possibility of medical diagnosis [39–41]. The incorporation of these nanomaterials into the synthesis of nanobiosensor systems has advanced treatments concerning sensitivity and specificity. Nanostructured materials have been employed in the reduction of the detection limits thus it became possible to identify target analytes at nanoscale concentrations. The modifications occurring on the surface due to reinforcements from nanomaterials also enhance selectivity which minimizes the inclusion of other unnecessary binding forces which are nonspecific. These are even more significant in food safety applications because low levels of pathogens or initial contamination agents are normally lethal. Determination of such trace levels with selective characteristics improves the efficiency and effectiveness of the sensor hence improving the public's confidence in food safety analysis. In an experiment, Tessaro et al. [42] developed a surface plasmon resonance (SPR) nanobiosensor based on gold nanoparticles (AuNPs) for the rapid detection of SARS-CoV-2 on food surfaces. Given the heightened concern over potential virus transmission through food, particularly during the COVID-19 pandemic, this research addresses the urgent need for fast, accurate, and cost-effective diagnostic tools. Traditional detection methods such as quantitative reverse transcription polymerase chain reaction (RT-qPCR) are highly sensitive but can be time-consuming, require specialized equipment, and may not be feasible for large-scale or on-site testing. In contrast, biosensors offer the potential to overcome these limitations by providing rapid, point-of-care detection with simplified protocols. The gold nanoparticles used in this SPR nanobiosensor are functionalized with specific oligonucleotides designed to bind to the viral RNA of SARS-CoV-2. The choice of AuNPs is pivotal due to their unique optical properties, which allow for sensitive detection based on surface plasmon resonance. When light interacts with the nanoparticles, it causes oscillations in the conduction electrons, leading to a characteristic absorption peak that can be measured through ultraviolet–visible (UV–Vis) spectroscopy. In this case, the presence or absence of the virus is indicated by shifts in this absorption peak. Specifically, when viral RNA binds to the functionalized AuNPs, the peak shifts from its baseline position at 525 nm (indicative of the functionalized AuNPs alone) to a region near 300 nm, signaling the presence of the virus (Fig. 2). This change in the absorption spectrum serves as a clear indicator of viral detection.

One of the key innovations of this study is the use of loop-mediated isothermal amplification (LAMP) for the amplification of viral genetic material. Unlike PCR, LAMP does not require complex thermal cycling and can amplify nucleic acids at a constant temperature, making the process faster and more suitable for field applications. This method allows the detection of the virus from raw samples with minimal preparation, streamlining the overall testing procedure and making it more accessible for use in various settings, including food safety testing. The nanobiosensor is designed to detect a specific region of the SARS-CoV-2 N gene, which is essential for the virus's replication. The sensor's dynamic range is broad, spanning from 0.1 to 50×10^3 ng/mL, with a detection limit as low as 1 ng/mL (equivalent to

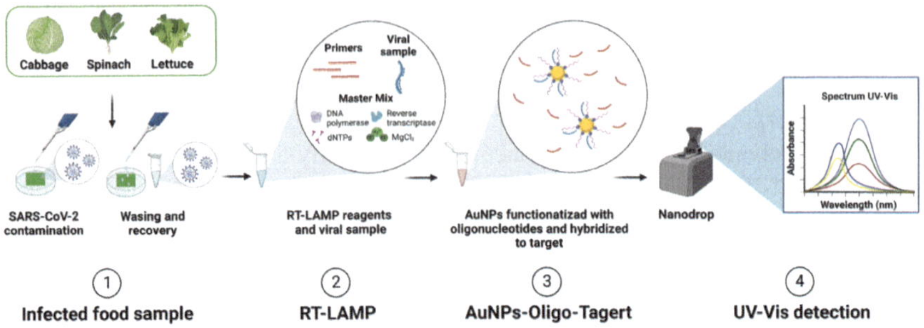

Fig. 2 Schematic of key stages in the development and application of an AuNP-based nanobiosensor functionalized with oligonucleotides for detecting SARS-CoV-2 on food surfaces. Adapted from Ref. [42] under CCBY 4.0

2.7×10^3 copies per μL) (Fig. 3a, b). This level of sensitivity makes the sensor highly effective for identifying even low concentrations of viral particles, which is crucial for ensuring the safety of food products that may come into contact with the virus. Testing the nanobiosensor on food surfaces, specifically on vegetables, further highlights its practical applications. The sensor demonstrated 100% accuracy when compared to the gold-standard RT-qPCR technique, reinforcing its reliability. This high level of precision suggests that the nanobiosensor could be deployed in various food safety scenarios, providing a valuable tool for monitoring and mitigating the risk of viral contamination in the food supply chain.

Lang et al. [43] introduced an advanced WaveFlex biosensor, a groundbreaking tool designed for the rapid and sensitive detection of Staphylococcus aureus (S. aureus), a bacterium frequently associated with foodborne illnesses. This innovative

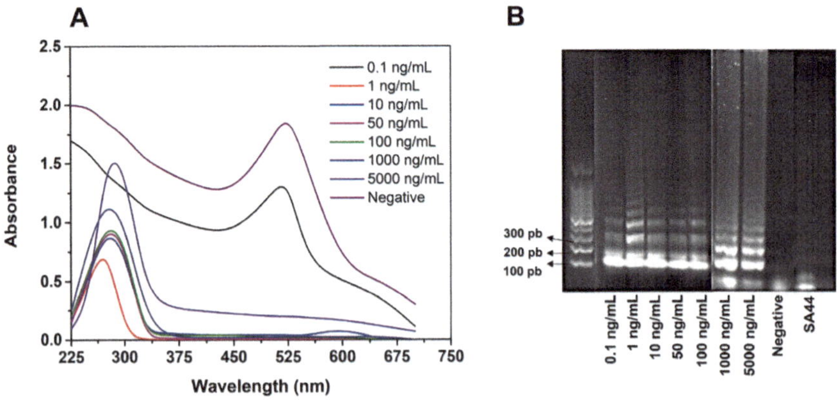

Fig. 3 **a** UV–Vis spectra for SARS-CoV-2 RNA concentrations (0.1 to 50×10^3 ng/mL), negative control, and betacoronavirus SA44. **b** 1% agarose gel electrophoresis confirming RNA amplification and fragment identification. Adapted from Ref. [42] under CCBY 4.0

biosensor is based on optical fiber technology, specifically utilizing a flexible tapered single-mode fiber structure. This design not only enhances its sensitivity but also facilitates the integration of various nanomaterials that are crucial for detecting low concentrations of pathogens. At the core of the sensor's functionality is the incorporation of gold nanoparticles (AuNPs), which play a pivotal role in inducing localized surface plasmon resonance (LSPR). This phenomenon significantly amplifies the optical signals, thereby enabling the detection of S. aureus with high precision. In addition to AuNPs, the biosensor employs chitosan-coated iron (III) oxide nanoparticles (Fe_3O_4–CS NPs) and tungsten disulfide quantum dots (WS2-QDs), further enhancing its sensitivity and specificity. The chitosan coating not only improves biocompatibility but also assists in the immobilization of the iron oxide nanoparticles, contributing to the overall effectiveness of the sensor. Meanwhile, the WS2-QDs provide unique optical properties that complement the plasmonic effects of the AuNPs. The integration of these advanced materials ensures that the biosensor can detect very low levels of S. aureus, which is critical given that foodborne pathogens often exist in low concentrations. To achieve targeted detection, the optical fiber surface of the biosensor is functionalized with specific antibodies that selectively bind to S. aureus. This functionalization is essential for ensuring that the sensor can distinguish the pathogen from other microbial entities present in food samples. The results from testing demonstrate that the biosensor reaches a maximum sensitivity of 2.74 nm per logarithmic increase in S. aureus concentration, translating to a remarkably low detection limit of just 6.67 CFU/mL. Such sensitivity is crucial for food safety applications, where even minute levels of contamination can pose significant health risks. The potential applications of this WaveFlex biosensor extend beyond food safety; it holds promise for disease diagnostics and other medical applications. Its rapid response time and high sensitivity position make it as a valuable tool for real-time monitoring of bacterial contamination, thereby aiding in the prevention of foodborne outbreaks and enhancing public health safety. The successful integration of LSPR technology with advanced nanomaterials, like AuNPs, Fe_3O_4–CS NPs, and WS2-QDs, not only highlights the biosensor's versatility but also underscores its significance in advancing detection technologies across various fields. Overall, this research presents a significant step forward in biosensor development, paving the way for innovative solutions in food safety and health monitoring. Similarly, Song et al. [44] introduced an innovative biosensing platform that harnesses the capabilities of CRISPR/Cas12a, specifically targeting Staphylococcus aureus, a common and significant foodborne pathogen. While conventional CRISPR-based biosensors are lauded for their high specificity, they often suffer from limited sensitivity, necessitating the incorporation of nucleic acid amplification techniques to bolster their detection capabilities. However, this requirement for amplification typically confines the functionality of the biosensors to binary outcomes, which can significantly hinder their practical applications in real-world scenarios where nuanced detection is critical. To overcome these limitations, the researchers developed a novel CRISPR/Cas12a-powered Evanescent Wave Fluorescence Nanobiosensing Platform (CREAT) that facilitates ultrasensitive quantitative detection of pathogens without the need for nucleic acid amplification. The CREAT platform stands out by integrating multiple signal enhancement strategies, which collectively improve detection sensitivity and accuracy. A cornerstone of this innovation is the use of a nanophotonic structure that leverages evanescent wave fluorescence, a technique that significantly amplifies fluorescence signals beyond what traditional methods can achieve. Additionally, the platform employs various other enhancement mechanisms, including magnesium ion (Mg^{2+}) or DNA-mediated fluorescence amplification and a novel air-displacement fluorescence enhancement strategy. The latter, in particular, represents a simple yet effective method to significantly increase the fluorescence signal, thereby enhancing the overall sensitivity of the biosensor. When evaluated with real-world samples containing S. aureus, the CREAT platform exhibited remarkable performance, achieving detection limits of 592 CFU/mL within 45 min and an even lower limit of 13.2 CFU/mL in 90 min. These detection limits are comparable to those achieved through real-time quantitative polymerase chain reaction (RT-qPCR), a widely recognized gold standard for pathogen detection. The rapid and accurate identification of low concentrations of S. aureus is crucial in various fields, including food safety, clinical diagnostics, and environmental monitoring, where timely intervention is essential for public health. The advancement presented by the CREAT platform signifies a pivotal shift in biosensing technology, merging the precision of CRISPR with innovative signal enhancement techniques. This combination not only amplifies sensitivity but also streamlines the detection process, rendering it robust and adaptable for real-world applications. Furthermore, the implications of this research extend beyond S. aureus, suggesting that similar methodologies could be effectively applied to the detection of a wide range of pathogens. This adaptability enhances the platform's potential impact on public health initiatives, providing a reliable tool for rapid pathogen detection in diverse environments. Overall, the study underscores the transformative power of CRISPR/Cas12a technology in developing advanced biosensors capable of addressing the pressing demands for sensitive and rapid pathogen detection in today's global landscape.

Manzano et al. [45] developed an innovative biosensor that employs localized surface plasmon resonance (LSPR) technology, which offers a highly sensitive method for the detection of this specific yeast in wine samples. LSPR is a phenomenon that occurs when light interacts with noble-metal nanoparticles, such as gold, that are smaller than the wavelength of the incident light. This interaction induces the oscillation of free electrons on the surface of the nanoparticles, leading to significant enhancements in optical properties. The basis of the developed biosensor lies in a gold nanostructured surface designed for the immobilization of a thiol-modified DNA probe at the 5' end. This specific DNA probe is tailored to target Brettanomyces bruxellensis, ensuring that the biosensor can accurately identify the presence of this spoilage organism. In the construction of the biosensor, a thin layer of gold was deposited to a precise thickness of 4 nm, optimizing its plasmonic characteristics for improved sensitivity. To evaluate the performance of the LSPR biosensor, DNA samples from Brettanomyces bruxellensis and a negative control, Saccharomyces cerevisiae, were tested at various concentrations. The methodology involved observing the changes in the optical properties of the gold nanoparticles as a result of DNA probe binding, which manifested as shifts in the LSPR extinction maximum (Δλmax). These shifts provide critical information about the binding events and enable the detection of the target organism. The results of this research demonstrated that the LSPR biosensor could detect concentrations as low as 0.1 ng/μL of target DNA, underscoring its high sensitivity. This capability highlights the potential of this biosensing system to identify microbial pathogens present in trace amounts within food and beverage samples. Furthermore, the specificity of the DNA probe ensures that the biosensor can differentiate between Brettanomyces bruxellensis and nontarget organisms, such as Saccharomyces cerevisiae, effectively minimizing false positives and improving the reliability of detection. Similarly, Banerjee et al. [46] developed a novel tunable magneto-plasmonic nanosensor (MPnS) by integrating surface plasmon resonance (SPR) capabilities with spin–spin magnetic relaxation (T2 MR) technology. This sensor was engineered by encapsulating gold nanoparticles (GNPs) within a polymer coating on iron oxide nanoparticles (IONPs), creating a multifunctional platform capable of rapid and sensitive pathogen detection. The synthesis process involved coating IONPs with polyacrylic acid (PAA) using a solvent precipitation method, followed by the encapsulation of GNPs using gold chloride solution in a single step. Magnetic separation techniques were employed to purify the MPnS, and various characterization methods, including transmission electron microscopy (TEM) and energy-dispersive X-ray spectroscopy (EDS), were used to confirm the successful incorporation of GNPs within the IONPs. The MPnS demonstrates enhanced magnetic relaxation (T2 MR) properties alongside amplified optical capabilities compared to conventional GNPs, making it an ideal tool for detecting *E. coli* O157. By combining SPR and T2 MR techniques with colorimetric readout, this nanosensor provides a rapid and ultrasensitive detection platform. Experimental tests, conducted both in simple buffer solutions and in complex media such as milk, demonstrated that the MPnS-based assay could detect as few as 10 CFUs of *E. coli* O157 within minutes, with no cross-reactivity observed with other pathogens. This high sensitivity, coupled with the ability to work in complex media, highlights the robustness and versatility of the MPnS platform. Figure 4 presents the evaluation of the analytical sensitivity of the magneto-plasmonic nanosensor in detecting *E. coli* O157 within complex media, specifically spiked milk samples. In panels A and B, UV–Vis measurements illustrate the changes in surface plasmon resonance (SPR) signals, indicating the sensor's capability to detect the pathogen even in a complex medium like milk. The distinct shifts in the absorption peaks reflect the interaction between the sensor and the bacterial target, demonstrating that the sensor can reliably distinguish the presence of *E. coli* O157 from the background matrix of the milk. Panels C and D depict the results of T2 magnetic resonance (MR) experiments, where the sensor's magnetic properties were measured to assess its response to *E. coli* O157 in the same milk samples. The T2 MR data further corroborates the sensitivity of the nanosensor, showing clear differences in relaxation times that correlate with the concentration of the pathogen. Both the UV–Vis and T2 MR experiments yielded statistically significant results ($p < 0.0001$), confirming the sensor's ability to detect even low levels of *E. coli* O157 in complex media. The marked changes in both SPR and T2 MR data points underscore the robustness and accuracy of the nanosensor, making it a highly effective tool for pathogen detection in real-world food matrices like milk.

Moving to another facet of the nanobiosensor advancement, researchers were also able to develop methods for real-time monitoring of samples based on the presented nanobiosensor technology. Here, real-time monitoring is defined as the ability of sensors to indicate immediately whether or not the target analytes of interest are present so that timely action and judgments can be made in a wide range of applications. The improvement of nanomaterial sensitization, especially of the responsive type, has led to the establishment of real-time sensing systems. For example, electrochemical nanobiosensors have been shown to give a response immediately after Sample introduction. The real-time response is due to the swift electron transfer processes that the nanomaterials bring about resulting in immediate signal formation. This characteristic is especially valuable in medical diagnostics where the identification of the pathogen or biomarkers frequently has considerable implications for treatment outcomes. In food safety, monitoring the level of contamination makes

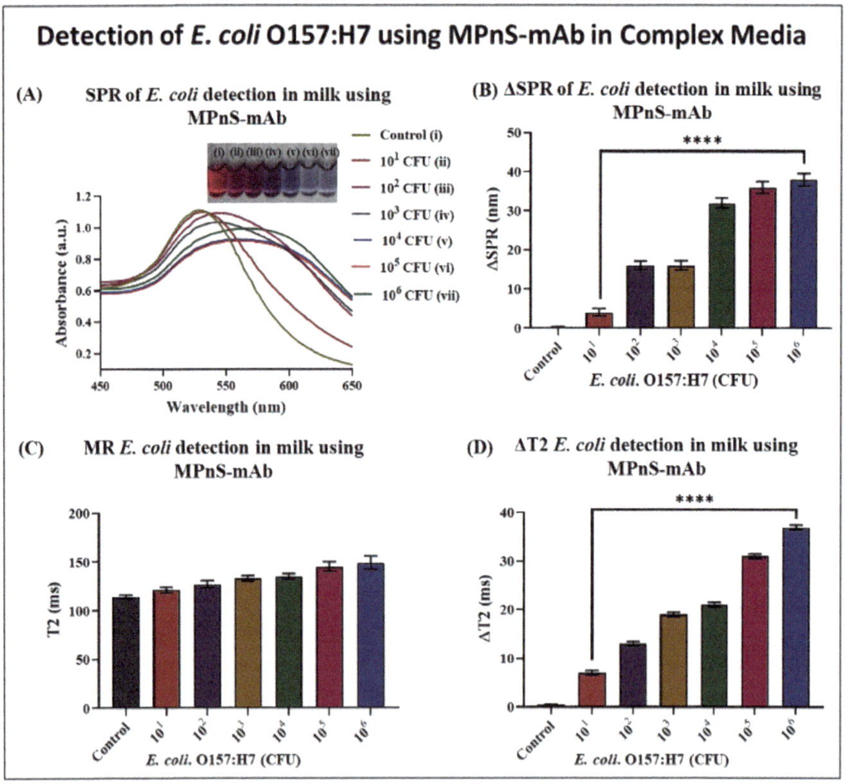

Fig. 4 Analytical sensitivity evaluation in complex media: **a**, **b** UV–Vis measurements, and **c**, **d** T2 MR experiments for detecting *E. coli* O157 in spiked milk. Significant differences ($p < 0.0001$) were observed in ΔSPR and ΔT2 data points. Adapted from Ref. [46] under CCBY 4.0

swift decisions in any dangers hence minimizing illness caused by food contamination. Other nanobiosensors, that employ fluorescence or surface plasmon resonance, are also capable of real-time detection. Such a configuration allows the integration of these sensors in a manner where changes in the optical signal can be quantified as target analytes interact with or bind to the biorecognition elements. The possibility of visualizing these interactions in real-time goes beyond merely complementing the assessment of the sensor's performance: it also helps in the detection of problems with sample preservation or the sensor itself [47, 48]. Besides, improvements in wireless communication technology and the miniaturization of nanobiosensors enable the fabrication of portable sensing gadgets with real-time capabilities. Such devices can be effectively used in food industries, healthcare facilities, and field research to obtain instant indications of the existence of bacteria or other contaminants. Portable and mobile nanobiosensors prove to be advantageous in this regard since they offer more functions and are easily usable by planning stakeholders in order to achieve more effective decision-making on safety and quality. For instance, in an experiment, an endocrinologist handled patients with diabetes and performed hormonal treatments on them. The developments made in the field of nanobiosensor technology based on the integration of nanomaterials, increased sensor sensitivity and selective coefficients, and real-time monitoring have revolutionized analytical detection. These innovations provide new approaches to solving some of the most important questions in public health, food safety, and environmental diagnostics [49, 50]. Therefore, as research in this field continues to advance, the design, materials, and functioning of nanobiosensors will likely be still more refined and efficient means of detection are expected to be developed. Ultimately, the integration of advanced nanobiosensor technology into routine monitoring practices holds the promise of safeguarding public health and ensuring the safety and quality of food products.

4 Challenges and Limitations

4.1 Matrix Effects on Detection

Matrix interferences involve the ability of a sample matrix to interfere with the ability to detect a target analyte in a specific sample. With reference to nanobiosensors, matrix effects can cause a positive or negative result to be encountered, thus challenging the validity and stability of the sensor. These effects are wanted most where there are many interfering substances, such as in food products, clinical samples, and environmental samples [51, 52]. That is why matrix effects in the analysis of real samples for food safety applications may originate from the interaction of fats, proteins, and sugars and other macromolecules with the biorecognition elements of the nanobiosensor. For example, when identifying foodborne pathogens, the background of the food sample under examination can have compounds that interact with the biorecognition element or interferants that can occupy the active sites of the target analyte. Such interference may result in changes in sensitivity and response characteristics thus providing low probabilities of identifying pathogens of low concentration. Consequently, the true risk assessment and measures put in place for food safety can fail or be inefficient and unsafe for the general population. Most researchers that use mass spectrometry have been trying to avoid matrix effects through various sample pretreatment procedures, the use of matrix-matching reagents, and internal standards. Some of the pretreatment techniques you may likely encounter may be dilution, filtration, or extraction in order to work on interferences prior to analysis. Matrix-matching methods involve having standards in the matrix as the sample being analyzed this helps to neutralize the effects of sample constituents. Furthermore, internal standards may offer correction of the signal damage caused by matrix effects, as well as offering a solitary point of reference that can improve the precision of the measurements [53–56]. However, eradicating matrix effects may still pose a significant problem and thus there is still extensive work that needs to be done in the development of nanobiosensors that could adequately mitigate complexity in interfering sample matrices. Recent studies have shown that designing new sensors is a key challenge: for instance, the application of biorecognition elements with higher selectivity, or, a combination of different types of detection systems. In the long run, the stability of the nanobiosensors to detachment from the matrix will be crucial in their successful application in the field.

4.2 Stability of Nanomaterials

The stability of materials employed in the fabrication of the nanobiosensors is equally another major concern since it is likely to compromise the efficiency and reliability of the nanobiosensors after some time. Despite the excellent properties of nanomaterials like nanoparticles, nanotubes, and nanowires in improving the sensitivity and selectivity of sensors, they usually exhibit varying stability under operating conditions depending on parameters such as the environmental conditions, presence of impurities, and characteristics of the nanomaterials used. The first important apprehensive connected with the stability of the nanomaterial is their ability to come together, or agglomerate, in solution. This aggregation can cause the surface area and functionality to be reduced and hence lead to poor sensor performance. For example, gold nanoparticles which have broad applications in optical biosensors can segregate into larger aggregates in specific conditions which alter their optical characterizing parameters and thus diminish their effectiveness in interacting with target species of interest [57, 58]. Likewise, the structure of the individual carbon nanotubes can be changed or the functional groups added to the carbon atom network that affect the electrical conductivity and electrochemical reactivity of the carbon nanotubes and thus of the electrochemical nanobiosensors. In addition, the stability of nanomaterials depends on the surrounding environment conditions including temperature, pH, ionic strength, and others [59, 60]. For instance, temperature fluctuations can hinder the conformation of biorecognition elements interacting with the sensor surface, which consequently reduces the binding capacity of the elements and changes the performance of the sensors. Likewise, the changes in pH or ionic force may affect the stability of electrostatic forces that are essential for the stability and functionality of nanomaterials.

In an effort to counter the stability problems of nanomaterials, scholars take a number of measures to enhance the reliability of such structures. This concerns the incorporation of stabilizing agents or coatings that can hinder aggregation and preserve the nanomaterials from unfavorable environmental factors. Furthermore, the synthesis of novel nanocomposites wherein the properties of different forms of nanomaterials are incorporated and can improve the general properties of the hybrids. For instance, the dispersion of metal nanoparticles in polymer matrices can protect the nanoparticles from degradations while at the same time maintaining themselves as sensitive sensors. Nonetheless, these developments have left the problem of stability of nanomaterials as one of the most critical issues facing nanobiosensors [61, 62]. More studies are required to determine the factors that define the stability of nanomaterials and to find new approaches to provide stable performance of nanomaterials that can be applied in practice. The issues with nanomaterials' stability must be met in order to further progress the use of nanobiosensors in practice.

4.3 Calibration Methods and Consistency

When dealing with the development of nanobiosensors, there is a crucial aspect which is calibration method as they are the ones that help to relate the sensor signals to target analyte concentration. Nonetheless, this is not always the case when it comes to samples obtained from real-world applications as many hurdles make it difficult to maintain uniformity in the way the calibrations are done. Since calibrations are affected and/or affect the quantification of analytes, such variability must always be avoided in order to avoid losing the reliability of the sensor [63–65]. One of the fundamental limitations posed by calibration relies on the availability of appropriate reference materials that are homologous, either wholly or partly, to the samples under consideration. Often, the commercially obtainable calibration standards do not simulate the complicated matrices that are encountered in real-world cases which leads to a difference between the calibration done and what is measured [66, 67]. Serious problems arise in the form of failure to maintain the correct concentration and presence of target pathogens in the appropriate food matrices. Even when one considers the matrix in which the pathogen exists, tempering enervating data compromises pathogen cultivability, or the concentration of pathogens in the matrix is compromised. Apart from that, even the calibration can be influenced by others such as temperature sensing, humidity sensing, and sample preparation. Changing these conditions is highly likely to affect how the sensors respond and thus affect the calibration results. For example, it is possible that temperature changes may occur at certain points during the calibration process and thus altering the kinetics of the biorecognition elements binding to the analytes, and therefore affecting the accuracy of the calibration curve [68].

In order to enhance the preciseness of the calibrations and maintain the reproducibility of results, numerous other techniques are being studied. One such method is the employment of matrix-matched standards that are composed of the matrix of the samples under analysis. This type of preparation should help reduce errors and improve the accuracy of the measurements. Along with that, applying proper procedures for quality authentication of the results, including repeated calibration for specified time intervals or use of internal controls, could largely eradicate the relative difference in the performance of the sensor system throughout use [69–71]. Also, new techniques in calibration, for instance, portable gadgets for calibration at the place of need, would be more useful. These techniques enable real-time calibration of nanobiosensors, reducing the reliance on laboratory-based calibration procedures and enhancing the practicality of the sensors for field applications.

5 Future Directions

5.1 *Integration with Mobile Platforms*

Incorporating nanobiosensors into mobile devices brings a revolution in the accessibility and simplicity of the operation of biosensing systems. It is very rare today to find people who do not carry mobile devices like tablets. These users have instant access to a lot of information and tools. If mobile or portable devices are used with nanobiosensors, the nanobiosensors can be utilized as portable appliances that can process and analyze information instantly with connectivity solutions available [72–74]. Along with numerous diagnostic advantages, integrating nanobiosensors with mobile platforms provides a unique capability of off-site monitoring and data gathering. For instance, residents may be able to utilize the mobile application for a nanobiosensor to test for pathogenic organisms in food samples, any hazardous materials within their environment and clinical samples for the above-mentioned biomarkers. The application may be able to capture the sensor's data intended for the app, accomplish the sensor's analysis, or provide the app's analysis results. This integration not only improves the interaction of users with the work system but also makes it feasible to obtain important data immediately, which is especially helpful in cases that require urgent action, such as food safety checks, or clinical diagnostics. Likewise, mobile devices can also act as media for storing or sharing information in the network, thereby enabling stakeholders to share and work together effectively. For example, the information obtained from several users can be bundled together and used to search for patterns, track epidemic occurrences, or check whether a certain food item is safe within the span of a few seconds [75, 76]. Such powers can help to improve public health efforts significantly and also assist regulatory bodies and food safety agencies with rapid developments [77–82]. Apart from this, the analysis and collection of data in real time, the incorporation of nanobiosensors into mobile devices goes a step further to facilitate the implementation of machine learning and artificial intelligence (AI) approaches. These technologies can complement the functionality of the nanobiosensors by using the advanced approach of historical data seeking and learning how to use the data in predicting results. Training machine learning models to be able to recognize the signatures of specific pathogens enables rapid identification and threat assessment. When considering the possibilities that emerge from combining the power of nanobiosensors with mobile platforms and AI, these diagnostic tools may be transformed into even more intelligent and responsive systems for better health and safety decisions [83–85]. In general, the combination of nanobiosensors within mobile platforms is a great opportunity to improve some properties of biosensing technologies. Development of the nanobiosensor–nanotechnology approach in mobile biosensing platforms may prove a valuable technology for monitoring and detecting diseases with emerging new trends in mobile tech.

5.2 Potential for On-Site Testing

Another important development in the realm of biosensors technology is the possibility of testing on-site via nanobiosensors. Background on-site testing involves the execution of analyses directly in the environment where the originating sample was collected, so avoiding time-consuming transportation of samples in order to reach a laboratory for analysis. For applications including food safety monitoring, clinical diagnostics, and environmental assessments (where quick decisions are critical), this ability is especially important. Three-Day On-Site Test Another thing that sets us apart from other bio-testing companies if the ability to have on-site testing and an instant result. For example, the immediacy of on-site testing allows food producers and regulators to respond quickly to potential risks associated with the presence of pathogens or contaminants in food [86, 87]. A quicker response will make it less likely that foodborne illness cases may expand to larger outbreaks, increasing consumer confidence in the safety of our food supply. Also, in healthcare scenarios, nanobiosensors may also be used for on-field testing to avert the conditions where precious hours are lost cricketing over easily curable diseases [88, 89]. These types of portable nanobiosensors could enable healthcare workers to perform fast screenings for infectious diseases so that they can immediately manage and start treating different conditions. This is particularly important in the context of remote settings or sites with limited access to laboratory facilities [7]. Not only is on-site testing convenient and efficient but this type of environmental monitoring can also be included within the same test group. With the ability to follow-up on contamination of air, water, or soil that occurs in real-time using nanobiosensors, regulatory departments and environmental associations can promptly respond to emerging dangers. This proactive approach to environmental monitoring can lead to more effective management of public health risks and contribute to sustainable environmental practices.

Furthermore, the development of portable, user-friendly nanobiosensors designed for on-site testing can empower nonexperts to conduct analyses without the need for specialized training. By simplifying the testing process and providing intuitive interfaces, these devices can be utilized by a broader range of users, including farmers, food industry professionals, and community health workers [90, 91]. This democratization of testing can enhance public engagement in health and safety initiatives and promote a culture of proactive monitoring. Despite the numerous advantages of on-site testing, several challenges must be addressed to fully realize its potential [92, 93]. These challenges include ensuring the reliability and accuracy of test results in varying environmental conditions and addressing concerns related to the stability of nanomaterials used in portable devices [94–96]. Ongoing research and innovation in sensor design, materials science, and data analysis will be essential for overcoming these obstacles and maximizing the effectiveness of on-site testing.

6 Conclusion

In summary, this chapter has explored the emerging field of nanobiosensors, highlighting their significance in detecting foodborne pathogens and ensuring food safety. The integration of nanomaterials such as nanoparticles, nanotubes, and nanowires has propelled the development of these sensors, enhancing their sensitivity, specificity, and overall performance. By utilizing advanced detection mechanisms, nanobiosensors are positioned to address critical public health concerns related to food safety and environmental monitoring. The advancements in nanobiosensor technology have opened up new avenues for rapid and accurate detection of pathogens, providing stakeholders with timely information to make informed decisions. The chapter discussed various types of nanobiosensors, including electrochemical, optical, and piezoelectric sensors, each offering unique advantages and functionalities. The ability to achieve real-time monitoring through these devices represents a significant step forward in ensuring food safety, as it allows for immediate identification of potential hazards and enables prompt intervention. However, despite the promising advancements, the chapter also highlighted the challenges and limitations associated with nanobiosensors. Matrix effects, stability of nanomaterials, and the need for robust calibration methods pose significant hurdles to the consistent performance of these devices. Addressing these challenges is essential for maximizing the effectiveness of nanobiosensors in real-world applications. Continuous research efforts are needed to improve the resilience of nanobiosensors to matrix interference, enhance the stability of nanomaterials, and develop standardized calibration protocols that ensure reliability across diverse sample types. Looking ahead, the integration of nanobiosensors with mobile platforms holds immense potential for revolutionizing the field of biosensing. This integration enables real-time data analysis, remote monitoring, and collaboration among stakeholders, ultimately enhancing the accessibility and usability of biosensing technology. Furthermore, the potential for on-site testing can empower users to conduct analyses directly in the field, facilitating rapid decision-making in food safety, healthcare, and environmental monitoring. This democratization of testing promotes proactive engagement in health and safety initiatives, fostering a culture of vigilance, and responsibility.

References

1. Maldonado J, González-Guerrero AB, Domínguez C, Lechuga LM (2016) Label-free bimodal waveguide immunosensor for rapid diagnosis of bacterial infections in cirrhotic patients, Biosens. Bioelectron 85:310–316. https://doi.org/10.1016/j.bios.2016.04.095
2. Awasthi G, Singh T, Tiwari, Y, Awasthi A, Tripathi RD, Shrivastava S, Vajpayee P, Kumar A, Awasthi KK (2020) A review on nanotechnological interventions for plant growth and production. Mater Today Proc 31:685–693. https://doi.org/10.1016/j.matpr.2020.07.255
3. Ali Q, Zheng H, Rao MJ, Ali M, Hussain A, Saleem MH, Nehela Y, Sohail MA, Ahmed AM, Kubar KA, Ali S, Usman K, Manghwar H, Zhou L (2022) Advances, limitations, and prospects of biosensing technology for detecting phytopathogenic bacteria. Chemosphere 296:133773. https://doi.org/10.1016/j.chemosphere.2022.133773
4. Shivashakarappa K, Reddy V, Tupakula VK, Farnian A, Vuppula A, Gunnaiah R (2022) Nanotechnology for the detection of plant pathogens. Plant Nano Biol 2:100018. https://doi.org/10.1016/j.plana.2022.100018
5. Jin L, Li T, Wu B, Yang T, Zou D, Liang X, Hu L, Huang G, Zhang J (2020) Rapid detection of *Salmonella* in milk by nuclear magnetic resonance based on membrane filtration superparamagnetic nanobiosensor. Food Control 110:107011. https://doi.org/10.1016/j.foodcont.2019.107011
6. Ranslow A, Crompton D, Mehta K, Butler P, Adair J (2015) Empowering community health workers with inkjet-printed diagnostic test strips. Procedia Eng 107:205–214. https://doi.org/10.1016/j.proeng.2015.06.075
7. Otari S, Bapat VA, Lakkakula J, Kadam US, Suprasanna P (2024) Advancements in bionanotechnological applications for climate-smart agriculture and food production. Biocatal Agric Biotechnol 57:103117. https://doi.org/10.1016/j.bcab.2024.103117
8. Ko S, Gunasekaran S, Yu J (2010) Self-indicating nanobiosensor for detection of 2,4-dinitrophenol. Food Control 21:155–161. https://doi.org/10.1016/j.foodcont.2009.05.006
9. Joung CK, Kim HN, Im HC, Kim HY, Oh MH, Kim YR (2012) Ultra-sensitive detection of pathogenic microorganism using surface-engineered impedimetric immunosensor. Sensors Actuat B Chem 161:824–831. https://doi.org/10.1016/j.snb.2011.11.041
10. Wu Q, Zhang Y, Yang Q, Yuan N, Zhang W (2019) Review of electrochemical DNA biosensors for detecting food borne pathogens. Sensors (Switzerland) 19. https://doi.org/10.3390/s19224916
11. Kaushal S, Pinnaka AK, Soni S, Singhal NK (2021) Antibody assisted graphene oxide coated gold nanoparticles for rapid bacterial detection and near infrared light enhanced antibacterial activity. Sens Actuat B Chem 329:129141. https://doi.org/10.1016/j.snb.2020.129141
12. Pandit C, Roy A, Ghotekar S, Khusro A, Islam MN, Bin Emran T, Lam SE, Khandaker MU, Bradley DA (2022) Biological agents for synthesis of nanoparticles and their applications. J King Saud Univ Sci 34:101869. https://doi.org/10.1016/j.jksus.2022.101869
13. Preethi M, Roy L, Lahkar S, Borse V (2022) Outlook of various diagnostics and nanodiagnostic techniques for COVID-19, Biosens. Bioelectron X 12:100276. https://doi.org/10.1016/j.biosx.2022.100276
14. Mahapatra S, Chandra P (2020) Clinically practiced and commercially viable nanobio engineered analytical methods for COVID-19 diagnosis. Biosens Bioelectron 165:112361. https://doi.org/10.1016/j.bios.2020.112361

15. Franco AJDM, Merca FE, Rodriguez MS, Balidion JF, Migo VP, Amalin DM, Alocilja EC, Fernando LM (2019) DNA-based electrochemical nanobiosensor for the detection of Phytophthora palmivora (Butler) Butler, causing black pod rot in cacao (Theobroma cacao L.) pods. Physiol Mol Plant Pathol 107:14–20. https://doi.org/10.1016/j.pmpp.2019.04.004
16. Nisha Raj S, Anooj ES, Rajendran K, Vallinayagam S (2021) A comprehensive review on regulatory invention of nano pesticides in agricultural nano formulation and food system. J Mol Struct 1239:130517. https://doi.org/10.1016/j.molstruc.2021.130517
17. Biswas P, Mukherjee A, Goyal P, Bhattacharya P, Dutta G, Chakraborty S (2024) A rapid diagnostic technology for isolating rare blood group patients under medical emergency using a three-fold paper-polymer microfluidic kit. Sens Actuat B Chem 409:135650. https://doi.org/10.1016/j.snb.2024.135650
18. Narmani A, Kamali M, Amini B, Kooshki H, Amini A, Hasani L (2018) Highly sensitive and accurate detection of Vibrio cholera O1 OmpW gene by fluorescence DNA biosensor based on gold and magnetic nanoparticles. Process Biochem 65:46–54. https://doi.org/10.1016/j.procbio.2017.10.009
19. Suaifan GARY, Shehadeh MB, Darwish RM, Alterify M, Abu Jbara W, Abu Jbara F, Alaridah N, Zourob M (2024) Magnetic beads-based nanozyme for portable colorimetric biosensing of Helicobacter pylori. Biosens Bioelectron 20:100517. https://doi.org/10.1016/j.biosx.2024.100517
20. Shen Y, Jia F, He Y, Fu Y, Fang W, Wang J, Li Y (2022) A CRISPR-Cas12a-powered magnetic relaxation switching biosensor for the sensitive detection of *Salmonella*. Biosens Bioelectron 213:114437. https://doi.org/10.1016/j.bios.2022.114437
21. Bashir O, Bhat SA, Basharat A, Qamar M, Qamar SA, Bilal M, Iqbal HMN (2022) Nano-engineered materials for sensing food pollutants: technological advancements and safety issues. Chemosphere 292:133320. https://doi.org/10.1016/j.chemosphere.2021.133320
22. Yi Z, Zhang Y, Gao X, Li S, Li K, Xiong C, Huang G, Zhang J (2024) Sensitive electrochemical immunosensor for rapid detection of *Salmonella* in milk using polydopamine/CoFe-MOFs@Nafion modified gold electrode. Int J Food Microbiol 425:110870. https://doi.org/10.1016/j.ijfoodmicro.2024.110870
23. Kumar S, Singh H, Feder-Kubis J, Nguyen DD (2023) Recent advances in nanobiosensors for sustainable healthcare applications: a systematic literature review. Environ Res 238:117177. https://doi.org/10.1016/j.envres.2023.117177
24. Al-Awwal N, Masjedi M, El-Dweik M, Anderson SH, Ansari J (2022) Nanoparticle immuno-fluorescent probes as a method for detection of viable *E. coli* O157:H7. J Microbiol Methods 193:106403. https://doi.org/10.1016/j.mimet.2021.106403
25. Warghane A, Saini R, Shri M, Andankar I, Ghosh DK, Chopade BA (2024) Application of nanoparticles for management of plant viral pathogen: current status and future prospects. Virology 592:109998. https://doi.org/10.1016/j.virol.2024.109998
26. Dam P, Paret ML, Mondal R, Mandal AK (2023) Advancement of noble metallic nanoparticles in agriculture: a promising future. Pedosphere 33:116–128. https://doi.org/10.1016/j.pedsph.2022.06.026
27. Elattar KM, Al-Otibi FO, El-Hersh MS, Attia AA, Eldadamony NM, Elsayed A, Menaa F, Saber WIA (2024) Multifaceted chemical and bioactive features of Ag@TiO2 and Ag@SeO2 core/shell nanoparticles biosynthesized using *Beta vulgaris L.* extract. Heliyon 10:e28359. https://doi.org/10.1016/j.heliyon.2024.e28359
28. Suaifan GARY, Alhogail S, Zourob M (2017) Paper-based magnetic nanoparticle-peptide probe for rapid and quantitative colorimetric detection of Escherichia coli O157:H7. Biosens Bioelectron 92 (2017) 702–708. https://doi.org/10.1016/j.bios.2016.10.023
29. Kamat S, Kumari M, Jayabaskaran C (2021) Nano-engineered tools in the diagnosis, therapeutics, prevention, and mitigation of SARS-CoV-2. J Control Release 338:813–836. https://doi.org/10.1016/j.jconrel.2021.08.046
30. Song D, Xu W, Han X, Wang H, Zhuo Y, Liu J, Zhu A, Long F (2023) CRISPR/Cas12a-powered evanescent wave fluorescence nanobiosensing platform for nucleic acid amplification-free detection of Staphylococcus aureus with multiple signal enhancements. Biosens Bioelectron 225:115109. https://doi.org/10.1016/j.bios.2023.115109

31. Lee J, Kim IS (2011) Microbial community in seawater reverse osmosis and rapid diagnosis of membrane biofouling, Desalination 273:118–126. https://doi.org/10.1016/j.desal.2010.12.005
32. Ghasemi S, Bari MR, Pirsa S, Amiri S (2020) Use of bacterial cellulose film modified by polypyrrole/TiO_2-Ag nanocomposite for detecting and measuring the growth of pathogenic bacteria. Carbohydr Polym 232:115801. https://doi.org/10.1016/j.carbpol.2019.115801
33. Alarcon-Angeles G, Álvarez-Romero GA, Merkoçi A (2016) Chapter 7-emerging nanomaterials for analytical detection. In: Scognamiglio V, Rea G, Arduini F, Palleschi G (eds) Bio sensing sustainable food—new opportunity technical challenges, Elsevier, pp 195–246. https://doi.org/10.1016/bs.coac.2016.03.022
34. Chand Mali S, Raj S, Trivedi R (2020) Nanotechnology a novel approach to enhance crop productivity. Biochem Biophys Rep 24:100821. https://doi.org/10.1016/j.bbrep.2020.100821
35. Baetsen-Young AM, Vasher M, Matta LL, Colgan P, Alocilja EC, Day B (2018) Direct colorimetric detection of unamplified pathogen DNA by dextrin-capped gold nanoparticles. Biosens Bioelectron 101:29–36. https://doi.org/10.1016/j.bios.2017.10.011
36. Singh KP, Dhek NS, Nehra A, Ahlawat S, Puri A (2017) Applying graphene oxide nanofilm over a polycarbonate nanoporous membrane to monitor E. coli by infrared spectroscopy, Spectrochim. Acta Part A Mol Biomol Spectrosc 170:14–18. https://doi.org/10.1016/j.saa.2016.06.053
37. Kaushal S, Priyadarshi N, Pinnaka AK, Soni A, Deep NK, Singhal K (2019) Glycoconjugates coated gold nanorods based novel biosensor for optical detection and photothermal ablation of food borne bacteria. Sens Actuat B Chem 289:207–215. https://doi.org/10.1016/j.snb.2019.03.096
38. Zheng XS, Jahn IJ, Weber K, Cialla-May D, Popp J (2018) Label-free SERS in biological and biomedical applications: recent progress, current challenges and opportunities. Spectrochim Acta Part A Mol Biomol Spectrosc 197:56–77. https://doi.org/10.1016/j.saa.2018.01.063
39. Xu J, Guo J, Maina SW, Yang Y, Hu Y, Li X, Qiu J, Xin Z (2018) An aptasensor for staphylococcus aureus based on nicking enzyme amplification reaction and rolling circle amplification. Anal Biochem 549:136–142. https://doi.org/10.1016/j.ab.2018.03.013
40. Guliy O, Dykman L (2024) Nanozyme-based sensors for cancer diagnosis. Biosens Bioelectron X 19:100512. https://doi.org/10.1016/j.biosx.2024.100512
41. Shokri E, Hosseini M, Sadeghan AA, Bahmani A, Nasiri A, Hosseinkhani S (2020) Virus-directed synthesis of emitting copper nanoclusters as an approach to simple tracer preparation for the detection of Citrus Tristeza Virus through the fluorescence anisotropy immunoassay. Sens Actuat B Chem 321:128634. https://doi.org/10.1016/j.snb.2020.128634
42. Tessaro L, Aquino A, Panzenhagen P, Ochioni AC, Mutz YS, Raymundo-Pereira PA, Vieira NKR, Belem CA (2022) Conte-junior, development and application of an SPR nanobiosensor based on AuNPs for the detection of SARS-CoV-2 on food surfaces. Biosensors 12. https://doi.org/10.3390/bios12121101
43. Lang X, Singh R, Zeng Q, Li J, Tosi D, Nedoma J, Marques C, Zhang B, Kumar S (2024) Chitosan-coated iron(III) oxide nanoparticles and tungsten disulfide quantum dots-immobilized Fiber-based WaveFlex Biosensor for Staphylococcus Aureus bacterial detection in real food samples. Sens Actuat Rep 8:100239. https://doi.org/10.1016/j.snr.2024.100239
44. Song D, Xu W, Han X, Wang H, Zhuo Y, Liu J, Zhu A, Long F (2023) CRISPR/Cas12a-powered evanescent wave fluorescence nanobiosensing platform for nucleic acid amplification-free detection of Staphylococcus aureus with multiple signal enhancements. Biosens Bioelectron 225:115109. https://doi.org/10.1016/j.bios.2023.115109
45. Manzano M, Vizzini P, Jia K, Adam PM, Ionescu RE (2016) Development of localized surface plasmon resonance biosensors for the detection of Brettanomyces bruxellensis in wine. Sens Actuat B Chem 223:295–300. https://doi.org/10.1016/j.snb.2015.09.099
46. Banerjee T, Panchal N, Sutton C, Elliott R, Patel T, Kajal K, Arogunyo E, Koti N, Santra S (2023) Tunable magneto-plasmonic nanosensor for sensitive detection of foodborne pathogens. Biosensors 13. https://doi.org/10.3390/bios13010109
47. Ramnani P, Saucedo NM, Mulchandani A (2016) Carbon nanomaterial-based electrochemical biosensors for label-free sensing of environmental pollutants. Chemosphere 143:85–98. https://doi.org/10.1016/j.chemosphere.2015.04.063

48. Ghasemi R, Mirahmadi-zare SZ, Allafchian A, Behmanesh M (2022) Fast fluorescent screening assay and dual electrochemical sensing of bacterial infection agent (Streptococcus agalactiae) based on a fluorescent-immune nanofibers. Sensors Actuat B Chem 352:130968. https://doi.org/10.1016/j.snb.2021.130968
49. Zeng Y, Zhu Z, Du D, Lin Y (2016) Nanomaterial-based electrochemical biosensors for food safety. J Electroanal Chem 781:147–154. https://doi.org/10.1016/j.jelechem.2016.10.030
50. Zhao X, Wang N, Chen H, Bai L, Xu H, Wang W, Yang H, Wei D, Yang L, Cheng, Preparation of a novel sandwich-type electrochemical immunosensor for AFP detection based on an ATRP and click chemistry technique11Electronic supplementary information (ESI) available. Polym Chem 11:900–908. https://doi.org/10.1039/c9py01279f
51. Mohankumar P, Ajayan J, Mohanraj T, Yasodharan R (2021) Recent developments in biosensors for healthcare and biomedical applications: a review. Measurement 167:108293. https://doi.org/10.1016/j.measurement.2020.108293
52. Sadik OA, Mwilu AK, Aluoch A (2010) Smart electrochemical biosensors: From advanced materials to ultrasensitive devices. Electrochim Acta 55:4287–4295. https://doi.org/10.1016/j.electacta.2009.03.008
53. Nodoushan SM, Nasirizadeh N, Sedighian H, Kachuei R, Azimzadeh-Taft M, Fooladi AAI (2022) Detection of Staphylococcal Enterotoxin A (SEA) using a sensitive nanomaterial-based electrochemical aptasensor. Diam Relat Mater 127:109042. https://doi.org/10.1016/j.diamond.2022.109042
54. Liu J, Lu L, Xu S, Wang W (2015) One-pot synthesis of gold nanoclusters with bright red fluorescence and good biorecognition Abilities for visualization fluorescence enhancement detection of E. coli. Talanta 134:54–59. https://doi.org/10.1016/j.talanta.2014.10.058
55. Lang X, Singh Q, Zeng Q, Li J, Tosi D, Nedoma J, Marques C, Zhang B, Kumar S (2024) Chitosan-coated iron(III) oxide nanoparticles and tungsten disulfide quantum dots-immobilized fiber-based waveflex biosensor for staphylococcus aureus bacterial detection in real food samples. Sens Actuat Rep 8:100239. https://doi.org/10.1016/j.snr.2024.100239
56. Ding S, Chen X, Yu B, Liu Z (2024) Electrochemical biosensors for clinical detection of bacterial pathogens: advances, applications, and challenges. Chem Commun 60:9513–9525. https://doi.org/10.1039/d4cc02272f
57. Krishnan S, Chinnasamy T, Veerappan S, Senthilkumar K, Kannaiyan D (2014) Dual labeled Ag@SiO2 core–shell nanoparticle based optical immunosensor for sensitive detection of E. coli. Mater Sci Eng C 45:337–342. https://doi.org/10.1016/j.msec.2014.09.028
58. Srivastava A, Azad UP (2023) Nanobioengineered surface comprising carbon based materials for advanced biosensing and biomedical application. Int J Biol Macromol 253:126802. https://doi.org/10.1016/j.ijbiomac.2023.126802
59. Warriner K, Reddy SM, Namvar A, Neethirajan S (2014) Developments in nanoparticles for use in biosensors to assess food safety and quality. Trends Food Sci Technol 40:183–199. https://doi.org/10.1016/j.tifs.2014.07.008
60. Nam YH, Lee S, Lee SK, Kim JH, Park JH (2024) Microfluidic chip with integrated separation, mixing, and concentration operations for rapid and sensitive bacterial detection utilizing synthetic inorganic antibodies. Sens Actuat B Chem 404:135202. https://doi.org/10.1016/j.snb.2023.135202
61. Xu L, Lu Z, Cao L, Pang H, Zhang Q, Fu Y, Xiong Y, Li Y, Wang X, Wang J, Ying Y, Li Y (2017) In-field detection of multiple pathogenic bacteria in food products using a portable fluorescent biosensing system. Food Control 75:21–28. https://doi.org/10.1016/j.foodcont.2016.12.018
62. Yoo MS, Shin M, Kim Y, Jang M, Choi YE, Park SJ, Choi J, Lee J, Park C (2017) Development of electrochemical biosensor for detection of pathogenic microorganism in Asian dust events. Chemosphere 175:269–274. https://doi.org/10.1016/j.chemosphere.2017.02.060
63. Gambhir RP, Rohiwal SS, Tiwari AP (2022) Multifunctional surface functionalized magnetic iron oxide nanoparticles for biomedical applications: a review. Appl Surf Sci Adv 11:100303. https://doi.org/10.1016/j.apsadv.2022.100303
64. Madhusudan P, Chellukuri N, Shivakumar N (2018) Smart packaging of food for the 21st century—a review with futuristic trends, their feasibility and economics. Mater Today Proc 5:21018–21022. https://doi.org/10.1016/j.matpr.2018.06.494

65. Lin D, Harris KD, Chan NWC, Jemere AB (2018) Nanostructured indium tin oxide electrodes immobilized with toll-like receptor proteins for label-free electrochemical detection of pathogen markers. Sens Actuat B Chem 257:324–330. https://doi.org/10.1016/j.snb.2017.10.140
66. Bagdeli S, Abbasi Kajani A, Taheri-Kafrani A (2024) Bioinspired amino acid-functionalized cobalt ferrite nanocomposite: a nanozyme-based colorimetric sensor for sensitive and selective quantification of phenolic compounds and ascorbic acid antioxidant capacity. Food Chem 457:140144. https://doi.org/10.1016/j.foodchem.2024.140144
67. Saritha GNG, Anju T, Kumar A (2022) Nanotechnology—big impact: how nanotechnology is changing the future of agriculture? J Agric Food Res 10:100457. https://doi.org/10.1016/j.jafr.2022.100457
68. Shalileh F, Sabahi H, Golbashy M, Dadmehr M, Hosseini M (2023) A simple smartphone-assisted paper-based colorimetric biosensor for the detection of urea adulteration in milk based on an environment-friendly pH-sensitive nanocomposite. Anal Chim Acta 1284:341935. https://doi.org/10.1016/j.aca.2023.341935
69. Thakur M, Wang B, Verma ML (2022) Development and applications of nanobiosensors for sustainable agricultural and food industries: Recent developments, challenges and perspectives. Environ Technol Innov 26:102371. https://doi.org/10.1016/j.eti.2022.102371
70. Juanola-Feliu E, Colomer-Farrarons J, Miribel-Català P, Samitier J, Valls-Pasola J (2012) Market challenges facing academic research in commercializing nano-enabled implantable devices for in-vivo biomedical analysis. Technovation 32:193–204. https://doi.org/10.1016/j.technovation.2011.09.007
71. Kumar Ganechary P, Gupta T, Shekhar Yadav C (2023) Development of nanomaterial based biosensors for forensic applications. Mater Today Proc 95:88–100. https://doi.org/10.1016/j.matpr.2023.10.124
72. Mostajabodavati S, Mousavizadegan M, Hosseini M, Mohammadimasoudi M, Mohammadi J (2024) Machine learning-assisted liquid crystal-based aptasensor for the specific detection of whole-cell Escherichia coli in water and food. Food Chem 448:139113. https://doi.org/10.1016/j.foodchem.2024.139113
73. Caliskan-Aydogan O, Alocilja EC (2024) Parallel biosensor platform for the detection of carbapenemase-producing E. coli in spiked food and water samples. Food Control 163:110485. https://doi.org/10.1016/j.foodcont.2024.110485
74. Noah NM, Ndangili PM (2022) Green synthesis of nanomaterials from sustainable materials for biosensors and drug delivery. Sens Int 3:100166. https://doi.org/10.1016/j.sintl.2022.100166
75. Buckley JA, Thompson PB, Whyte KP (2017) Collingridge's dilemma and the early ethical assessment of emerging technology: the case of nanotechnology enabled biosensors Technol Soc 48:54–63. https://doi.org/10.1016/j.techsoc.2016.12.003
76. Raza SHA, Huang Z, Pang Y, Zhong R, Li X, Pant SD, Luo L, Lei H (2024) Rapid screening of prednisolone acetate adulterants in health foods using colloidal gold immunochromatographic assay. Chem Phys Lipids 105447. https://doi.org/10.1016/j.chemphyslip.2024.105447
77. Thakur A, Kumar A (2023) Exploring the potential of ionic liquid-based electrochemical biosensors for real-time biomolecule monitoring in pharmaceutical applications: from lab to life. Results Eng 20:101533. https://doi.org/10.1016/j.rineng.2023.101533
78. Bensalah J, Thakur A, Kumar A (2024) Investigating the adsorption processes of polymer resins for the removal of micropollutants: a comprehensive review in the field of environmental remediation. Environ Res 254:119128. https://doi.org/10.1016/j.envres.2024.119128
79. Thakur A, Kumar A (2022) Recent advances on rapid detection and remediation of environmental pollutants utilizing nanomaterials-based (bio)sensors. Sci Total Environ 834:155219. https://doi.org/10.1016/j.scitotenv.2022.155219
80. Verma C, Thakur A, Ganjoo R, Sharma S, Assad H, Kumar A, Quraishi MA, Alfantazi A (2023) Coordination bonding and corrosion inhibition potential of nitrogen-rich heterocycles: Azoles and triazines as specific examples. Coord Chem Rev 488:215177. https://doi.org/10.1016/j.ccr.2023.215177

81. Thakur A, Kumar A, Sharma S, Ganjoo R, Assad H (2022) Computational and experimental studies on the efficiency of Sonchus arvensis as green corrosion inhibitor for mild steel in 0.5 M HCl solution. Mater Today Proc 66:609–621. https://doi.org/10.1016/j.matpr.2022.06.479
82. Thakur A, Kaya S, Abousalem AS, Sharma S, Ganjoo R, Assad H, Kumar A (2022) Computational and experimental studies on the corrosion inhibition performance of an aerial extract of Cnicus Benedictus weed on the acidic corrosion of mild steel. Process Saf Environ Prot 161:801–818. https://doi.org/10.1016/j.psep.2022.03.082
83. Chausali N, Saxena J, Prasad R (2023) Nanotechnology as a sustainable approach for combating the environmental effects of climate change. J Agric Food Res 12:100541. https://doi.org/10.1016/j.jafr.2023.100541
84. Ahmadvand N, Mohammadi-Manesh E (2021) Introduction of recombinant fullerene-PbI2 core-shell and investigation of its use as surface modifier of biosensors. Mater Today Commun 29:102985. https://doi.org/10.1016/j.mtcomm.2021.102985
85. Manzano M, Vizzini P, Jia P, Adam PM, Ionescu RE (2016) Development of localized surface plasmon resonance biosensors for the detection of Brettanomyces bruxellensis in wine. Sens Actuat B Chem 223:295–300. https://doi.org/10.1016/j.snb.2015.09.099
86. Mousavi Nodoushan S, Nasirizadeh N, Amani J, Halabian R, Imani Fooladi AA (2019) An electrochemical aptasensor for staphylococcal enterotoxin B detection based on reduced graphene oxide and gold nano-urchins. Biosens Bioelectron 127:221–228. https://doi.org/10.1016/j.bios.2018.12.021
87. Yousef Elahi M, Bathaie S, kS.H. Kazemi, M.F. Mousavi, DNA immobilization on a polypyrrole nanofiber modified electrode and its interaction with salicylic acid/aspirin. Anal Biochem 411:176–184. https://doi.org/10.1016/j.ab.2011.01.006
88. Jin L, Li T, Yang T, Liang X, Wu B, Zou D, Hu L, Huang G, Zhang J (2020) NMR rapid detection of *Salmonella* in milk based on ultra-small iron oxide nanobiosensor. Int Dairy J 110:104807. https://doi.org/10.1016/j.idairyj.2020.104807
89. Wani MY, Ganie NA, Dar KA, Dar SQ, Khan AH, Khan NA, Zahmatkesh S, Manzar MS, Banerjee R (2023) Nanotechnology future in food using carbohydrate macromolecules: a state-of-the-art review. Int J Biol Macromol 239:124350. https://doi.org/10.1016/j.ijbiomac.2023.124350
90. Mohammadi L, Pal K, Bilal M, Rahdar A, Fytianos G, Kyzas GZ (2021) Green nanoparticles to treat patients with malaria disease: an overview. J Mol Struct 1229:129857. https://doi.org/10.1016/j.molstruc.2020.129857
91. Ma X, Xu X, Xia Y, Wang Z (2018) SERS aptasensor for *Salmonella* typhimurium detection based on spiny gold nanoparticles. Food Contr 84:232–237. https://doi.org/10.1016/j.foodcont.2017.07.016
92. Chauhan D, Kumar Y, Chandra R, Kumar S (2022) Nanostructured zirconia@reduced graphene oxide based ultraefficient nanobiosensing platform for food toxin detection††Electronic supplementary information (ESI) available. Sens Diagn 1:550–557. https://doi.org/10.1039/d2sd00012a
93. Goel A, Rastogi A, Pandey S, Kulshrestha S, Goel S (2023) An emergent biotechnology hierarchy: biosensors. Mater Today Proc https://doi.org/10.1016/j.matpr.2023.03.363
94. Sharma D, Thakur A, Sharma MK, Kumar A, Jakhar K, Kumar S, Sihmar A, Dahiya H, Kumar A, Sharma AK, Om H (2023) A convenient synthesis, electrochemical profiling, and morphological studies of a pyridine-based 1,3,4-oxadiazole hybrid: a promising study for corrosion mitigation of mild steel in strongly acidic environment. Inorg Chem Commun 158:111554. https://doi.org/10.1016/j.inoche.2023.111554
95. Sharma D, Thakur A, Sharma MK, Jakhar K, Kumar A, Sharma AK, Om H (2023) Synthesis, electrochemical, morphological, computational and corrosion inhibition studies of 3-(5-Naphthalen-2-yl-[1,3,4]oxadiazol-2-yl)-pyridine against Mild Steel in 1 M HCl. Asian J Chem 35:1079–1088
96. Savaş Kaya Hassane Lgaz ATAKDÖINK, Ben Ahmed S (2023) Molecular insights into the corrosion inhibition mechanism of omeprazole and tinidazole: a theoretical investigation. Mol Simul 49:1632–1646. https://doi.org/10.1080/08927022.2023.2256888

Efficient Strategies for Detecting Food Contaminants Using Nanobiosensors

Jithendar Pothuganti, Giridhar Babu Anam, Arkadeep Mukherjee, Sada Venkateswarulu, and Nadeem Akthar

Abstract Food production and distribution have expanded globally led to food safety as a top concern. Conventional approaches to detect contamination are frequently labor-intensive, time-consuming, and necessitate complex laboratory conditions. Nanobiosensors are a combination of biological and nanoscale materials used to detect various chemical residues associated with diseases, poisons, and pesticides with high sensitivity and specificity. It then discusses how different nanomaterials such as metal and metal oxide-based nanoparticles, nanowires, nanorods, carbon nanotubes, quantum dots, and dendrimers improve signals for detection capabilities and facilitate real-time monitoring. The effectiveness of key detection techniques, including electrochemical, optical, and piezoelectric sensing in various applications, is examined. The chapter places a strong emphasis on useful design factors for creating portable, reliable nanobiosensors that may be used in both lab and field environments. Future alternatives for producing low-cost, readily available nanobiosensors that ensure public health and trust among consumers in food systems are discussed, as well as challenges such as scalability and regulatory barriers.

Keywords Biosensors · Food contamination · Food safety · Nanomaterials · Nanobiosensors

---

J. Pothuganti · G. B. Anam (✉)
Department of Basic Sciences, School of Sciences and Humanities, SR University, Warangal, Telangana, India
e-mail: a.giridharbabu@sru.edu.in

A. Mukherjee
Department of Civil Engineering, Yeungnam University, Gyeongsan, Gyeongbuk, Korea

S. Venkateswarulu
Nanotechnology Centre, VSB-Technical University of Ostrava, Ostrava, Czech Republic

N. Akthar
CBS Bio Platforms Inc., Calgary, AB, Canada

D. Sillu et al. (eds.), *Nanobiosensors for the Food Industry*, Smart Nanomaterials Technology, https://doi.org/10.1007/978-981-95-0136-6_6

1 Introduction

As food production and distribution expanded worldwide, food safety emerged as an important global challenge. The globalization of the food system has brought many benefits, including increased food availability, lower prices, and more diverse diets. But it has also brought new food safety challenges [1]. The consolidation and globalization of food production and supply chains make foodborne disease outbreaks far more likely. People in developing countries are generally at greater risk, due to liberalized food safety laws [2]. Traditionally, foodborne organisms were identified by microbiological counts, where the target bacteria were inoculated on to nutrient agar slants, incubated and allowed to form colonies. After sufficient colonies grew, the colonies were enumerated and extrapolated to estimate pathogen prevalence [3]. For instance, culture-based methods, including the growth of *Escherichia coli* (*E. coli*) O157:H7 on Sorbitol MacConkey agar, and CHROMagar and Cefsulodin-Irgasan-Novobiocin, proved to have very high sensitivity [4, 5]. Other supplemental techniques, including gram staining following pre-enrichment and selective enrichment phases, contributed as well.

However, the basis of culture-based methods has a major limitation being that most pathogens can take 18–24 h to provide their results and therefore is usually labor-intensive and requires specialized laboratory materials and/or expertise [6–8]. They can be inaccessible or impractical especially within resource-limited settings. Food safety monitoring and detection has a wider area of approach from researchers and industries. Therefore, the technologies must be rapid, impactful, and of broad access, and should be deployable at an on-site or point-of-need setting [9–12]. In order to detect these contaminations, immunoassays approach that has been widely adopted for the detection of foodborne pathogens is enzyme-linked immunosorbent assay (ELISA) [13]. Since the development of the traditional ELISA, many improvements have been made, such as in the case of the double antibody sandwich ELISA which allows for more sensitive detection of antibodies. Antibodies (monoclonal or polyclonal) are used as well for immunological methods targeting antigens on bacterial cell membrane. Monoclonal antibodies are more specific for single clone [14, 15]. Another alternative to ELISA for pathogen detection is agglutination-based methods. These methods use antibodies to find specific antigens and see clumping of the antigen and antibody [16]. Some commercial agglutination tests utilize antibody-coated latex beads or colloidal gold particles for rapid serological identification of bacteria in food. Although agglutination offers semiquantitative information and rapid outcomes, the interpretation of borderline results or interference from other antigens on whole cells is a disadvantage of these methods. Despite these barriers, both immunological and culture-independent techniques are crucial to improving food safety [17]. Promising technologies, such as advanced biosensor technology, blockchain solutions for increasing food traceability, AI analysis of records and complaints, and whole genome sequencing (WGS) to identify and track foodborne pathogens, will hopefully increase the health burden caused by food contamination [18, 19]. The overview of the effective strategies were summarized in Fig. 1.

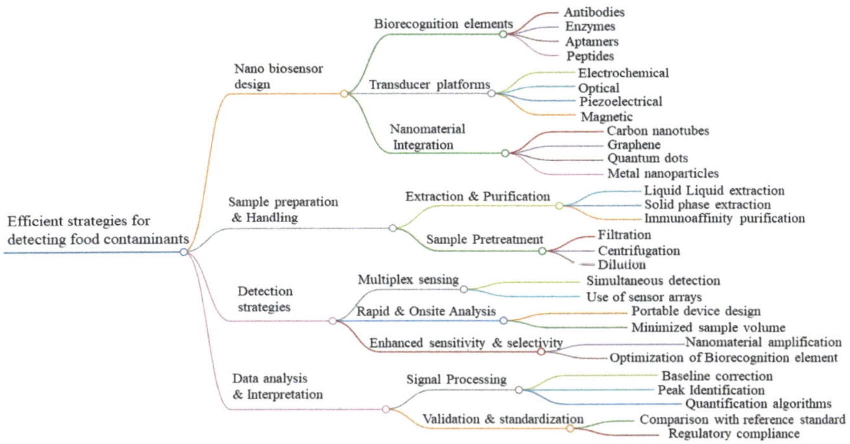

Fig. 1 Overview of strategies for detecting food contaminants by using nanobiosensors

2 Nanobiosensor Design

Nanosensor technology combines nanomaterials and biorecognition elements for the sensitive and specific detection of biological or chemical targets. Nanomaterials such as gold nanoparticles, graphene, when integrated with biorecognition components (e.g., antibodies, enzymes, DNA) and transducers (e.g., electrochemical, optical, etc.) can turn biological events into detectable signals. A well-designed nanobiosensor can lead to higher sensitivity to allow analysis of moderate concentrations of analytes which is an important part of the preliminary diagnosis as well as monitoring of basic diseases. For time-sensitive domains such as medical diagnostics and food safety monitoring, faster response times are critical and the focus on efficient design also plays a pivotal role here [20]. Biosensor could effectively be integrated to the other interactive technologies such as microfluidics and lab-on-a-chip systems through a well-established design for exploring the properties and functions [21]. The potential of nanobiosensors for the analysis of food, drinks, and the environment is myriad, a careful design of such tools makes it possible to exploit their appearance as an alternative solution for accurate and reliable food contaminant detection [22].

2.1 *Biorecognition Element*

The basic feature of a biosensor is its capacity to employ different biological molecules including enzymes, antibodies, nucleic acids, and cells as selectable elements for recognizing target substances. The biorecognition element functions as sensor sensitivity and selectivity and therefore controls the devices performance standards. However, the implementation of a biorecognition element on the sensor surface requires proper immobilization management because it affects biological activity stability and measurement repeatability. The biosensors detect and quantify target substances in actual samples is of their biorecognition elements include antibodies, enzymes, peptides, etc. [22, 23].

2.1.1 Antibodies

Antibodies show high affinity and specificity for target molecules, including proteins, hormones, and pathogens. They widely used in diagnostic applications, but their performance is critically dependent on a few parameters including environmental factors that influence their selectivity and sensitivity. For instance, the selectivity is affected by the interaction affinity of the antibody to the antigen, which overrides the structure of the antibody and the composition of the antigen-binding site with respect to the epitopes of the target molecule. Low detection limits require high-affinity antibodies from the outset, as they can outcompete other serological substances for their target. Sensitivity is critically dependent on antibody concentration, assay format, and detection method used. As an example, secondary antibodies which amplify the signal can greatly increase sensitivity. The selected detection methods such as ELISA (enzyme-linked immunosorbent assay) or fluorescence can impact the detection sensitivity of low-concentration analyte [24].

Additional parameters impacting the overall performance of antibody-based detection systems are the stability of the antibody, the specificity of the assay conditions, and matrix effects caused using biological samples. Therefore, the optimization of these sensors for structural, procedural, and environmental aspects is a prerequisite to establishing robust assays for clinical diagnostics, environmental monitoring, and food safety testing [25, 26].

2.1.2 Enzymes

Enzymes are important functional elements for the detection of substrates as an analyte, provide highly selective and sensitive techniques for identification and quantification. However, the enzyme detection ability toward target depends on multiple parameters which need to comprehend to create efficient analytical approaches. The specificity of enzyme-based detection methods is largely conferred by the active site of the enzyme, and the size, shape, and chemical properties of this site determines the ability of selective binding to the target substrate or analyte which enables discrimination between closely related molecules [27]. The detection system sensitivity can be influenced intrinsically by the enzyme and its affinity to substrate concentration, and other factors such as the presence of inhibitors or activators [28]. The performance of enzymes in these analytical applications is affected by a number of factors: the optimal temperature for a given enzyme; the pH of the environment in which the reaction is taking place (which alters the structure and, therefore, catalytic activity of the enzyme at play); the concentration of the substrate or analyte, the quantity of enzyme present and the presence of molecules that modulate the activity of the enzyme being studied (either inhibiting or enhancing enzyme activity) [16].

2.1.3 Aptamers

Aptamers are short single-stranded nucleic acid molecules which have the capacity to bind to diverse target analytes such as small molecules, proteins, cells, and even entire organisms with remarkable specificity and affinity. The selectivity and sensitivity of aptamer depend on multiple parameters such as the sequence, structure, and interactions serving as the binding of targets. The ability of an aptamer to specifically bind to a target analyte is strongly influenced by its sequence. The novel three-dimensional configurations of aptamers, like hairpins, stem-loops, or G-quadruplexes, can also form from specific sequences, which greatly promote affinity and selectivity toward the target, and several computational methods were produced recently [29]. Aptamers differ in their structural characteristics like the number and position of stem-loops, sequences of specific motifs, and overall folding, and they are known to affect the performance of aptamer in analyte detection and quantitation. Stable and well-defined aptamer structures tend to show higher selectivity and sensitivity than those with flexible or disordered conformations. In addition, the binding interactions between the target analyte and aptamer are also significant for the aptamer performances. However, the binding affinity and specificity can be affected by the strength of the hydrogen bonds, electrostatic interactions, and Van der Waals forces between the aptamer and the target. Multiple high-affinity interactions of aptamers with target analytes typically lead to higher selectivity and sensitivity. The optimal performance of the aptamer can depend on its inherent features as well as the experimental conditions including pH, temperature, and ionic strength. Tweaking these parameters may help to increase the selectivity and sensitivity of the aptamer-based detection system, and harbor unique properties to be engineered [30, 31].

2.1.4 Peptides

Peptides have been investigated as biomarkers and diagnostic tools. Peptidomic analysis, the study of peptide signatures, has become an attractive platform for biomarker discovery since it can be used to identify disease-associated proteins. pH governs peptide selectivity in analysis while the charge of the peptides can be changed through the pH of the medium which changes its interaction with the analytical platform. Optimizing this parameter can increase the selectivity and sensitivity of peptide-based assays. Moreover, peptide binding can also vary depending on its length and sequence so it's also important to consider its use as a biomarker or analytical tool [32–34].

2.2 Transducer Platforms

Transducer platforms translate the physical or chemical characteristics of the analyte into a detectable electric signal (i.e., integrating various molecular techniques, which allows accurate and sensitive quantification). An important descriptor used to choose the transducer platform is the limit of detection (LOD-the lowest concentration of an analyte that can be quantified) which indicates the sensitivity of the transducer, the noise level of the system, and the signal-to-noise ratio, collectively determine the achievable LOD [35]. The dynamic range is defined as the range of the analyte concentration that can be measured with acceptable accuracy by the transducer. A large dynamic range is desirable to allow the quantification of analytes over a wide range of concentrations. The selectivity of the transducer is also an important parameter, it specifies its ability to distinguish the target analyte from other interfering substances. The selection of the appropriate transducer platform is context-dependent, relying on the specific application and the target analyte as well as the desired performance characteristics [35].

2.2.1 Electrochemical Platforms

Electrochemical nanobiosensors are the potential for food contamination detection due to their rapid and sensitive detection characteristics. These sensors utilize nanomaterials to enhance the performance of the immobilized biological recognition element (e.g., enzyme, antibody, or nucleic acid) through the advantages of a large surface-to-volume ratio and modified electron transfer kinetics. The sensitivity and selectivity of electrochemical nanobiosensors depend upon several parameters, such as electrochemical transduction mechanisms such as amperometry, potentiometry, cyclic voltammetry, or impedance spectroscopy affect the sensitivity and selectivity of the sensor. The transduction mechanism selected is contingent on the analyte and the biological recognition element employed. The changes in parameters such as pH, temperature, buffer composition, and applied potential can affect the performance measurements of the sensor. Proficient optimization of these experimental conditions remains essential to produce sensors with accurate detection capabilities for food contaminants.

2.2.2 Optical Properties

Food contaminants' detection has been revolutionized through the development of optical nanobiosensors as an emerging diagnostic technology. The use of nanoparticles including gold and silver or quantum dots enhances the sensitivity of the optical nanobiosensors through biomolecular recognition elements such as antibodies or aptamers are attached to their surfaces. Surface plasmon resonance (SPR) is one such, which detect variations in its refractive index while performing label-free and real-time food contaminant monitoring. Quantum dots and fluorescent polymers which alter their emission properties upon binding analyte to it, which allows to detect food contaminants using selective fluorescence-based detection processes. Microfluidic systems can be coupled with optical nanobiosensors for improved functionality such as better handling of samples and reagents and multiplex detection [21]. Integrated platforms like these allow quick, automated, and miniaturized analysis of food samples. Piezoelectric nanobiosensors provide sensitive detection of analyte due to changes in sensor surface parameters through the piezoelectric effect which produces electrical signals from mechanical force applied to specific materials. All aspects of the piezoelectric nanobiosensor design, such as piezoelectric material, electrode configuration, and fluidic system, can be optimized to achieve sensitivity, selectivity, and response time improvements. Resonance frequency, temperature, and flow rate are some effects that need careful consideration [6, 21]

2.3 Nanomaterial Integration

Gold, silver, and quantum dot nanoparticles can be functionalized using specific molecules/antibodies to identify the presence of contaminants in food samples [36, 37]. Nanoscale particle sensors have enabled the rapid, sensitive, and selective detection of a few foodborne pathogens, toxins, and chemical residues. Nanotechnology can be employed for the improvement of spectroscopic techniques, e.g., surface-enhanced Raman spectroscopy (SERS) or surface plasmon resonance (SPR) for food contaminants detection.

2.3.1 Gold Nanoparticles (AuNPs)

The properties of gold nanoparticles (AuNP's) are markedly different from those of bulk gold. These nano-sized particles (usually in the range of 1 to 100 nm) exhibit extraordinary optical properties because of surface plasmon resonance (SPR). SPR is better understood as a collective oscillation of electrons on a particle's surface induced by light. The resulting phenomenon gives AuNPs and their distinctively rich colors, a 20 nm gold nanoparticle solution displays a wine-red hue, while larger particles transition toward purple/blue. These size-dependent colourimetric changes constitute a strong visual signal in biosensing applications [38]. Specific interaction of gold with thiol groups (-SH) provides a convenient way for surface modification via stable Au-S bonds. Many biological molecules can be clicked onto the surface of the Au nanoparticle. At the nanoscale, the contact area between the biomolecules and their surroundings becomes all the more intimate as the ratio between the volume and a surface area increases rapidly, facilitating an enormous surface area for biomolecule binding [39].

AuNPs are unique optical transducers due to their localized surface plasmon resonance (LSPR). The binding of target molecules on AuNP surfaces changes the surrounding local refractive index and thus leads to both up and downshifts of the LSPR peak. This principle allows for extraordinarily sensitive detection, even single-molecule detection in some instances. Where AuNPs have an extinction coefficient that increases significantly over that of organic dyes (roughly 2.7×10^8 M^{-1} cm^{-1} for 13 nm particles) [40]. Another advanced detection mechanism is provided by aggregation-based sensing. During the binding of target molecules to AuNPs, AuNPs aggregate which leads to plasmon coupling among AuNPs, thus changing color from red to blue. This forms a basic powerful colorimetric sensor that can at times be detected by the naked eye, which makes it extremely useful for point-of-care diagnostics [41].

Gold nanoparticles are also excellent electronic conductors in biosensors. Electrons transfer occurs rapidly between biomolecules and electrode surfaces due to their high surface area and excellent conductivity. As signal enhancers, AuNPs can improve the detection signals by several orders of magnitude upon implementing them in electrochemical biosensors. It is especially helpful for the detection of small molecules or low-concentration analytes [42]. In contrast to other nanomaterials, gold nanoparticles are resistant to oxidation and do not lose their properties over time, which makes them the first choice among other nanomaterials. Another major benefit is the biocompatibility of gold makes these particles safe for biological applications, including in-vivo sensing [37]. Despite the high cost of producing high-quality gold nanoparticles, it is important to ensure uniform size and shape distribution for reliable sensing. However, complex food matrices may interfere; thus, this must be addressed in sensor design.

2.3.2 Graphene and Graphene Oxide

Graphene is simply one layer of carbon atoms bonded together in a hexagonal lattice, making it unique compound. Its two-dimensional form generates a vast surface area where it can interact with biological molecules. Imagine it as a very thin sheet of homemade paper, but with every single atom capable of some interaction with your target molecule. Graphene's most valuable properties for biosensing are its very high electrical conductivity (64 mS/cm) and electron mobility (200,000 $cm^2/V.s$) [43, 44]. These properties mean that when a biological molecule binds to the graphene surface, it can rapidly and effectively transfer that interaction as an electrical signal.

Graphene has another important property is its mechanical strength. It has Young's modulus of 1 TPa, so extremely stable, and biosensors can perform over long ages. This material also has great thermal conductivity (5000 W/m·K) that can keep the sensor smoothly operational at various temperatures[45].

Graphene oxide starts with the same basic structure as graphene, but adds oxygen-containing functional groups to it, greatly altering its properties. These oxygen groups serve as anchor points for biological molecules. Due to the presence of oxygen-containing groups (carboxyl, hydroxyl, and epoxy), the contact angle of GO is less than 30°, thus making GO highly hydrophilic, similar to graphene, colloquially to be more hydrophobic [46]. The electrical conductivity of GO is less than that of pure graphene, yet this biostability becomes an advantage in some biosensing applications. This lowers conductivity (0.1–5 S/cm) providing a larger signal and enhancing the baseline for detection as biological molecules interact with it. In addition, the interlayer spacing of GO (0.6–1.2 nm) is larger than in graphene (0.34 nm) [47, 48], allowing larger biomolecules to be embedded between GO sheets.

Reduced graphene oxide is a well-designed compromise between GO and graphene. Controlled reduction processes remove some of the oxygen groups, which restores some electrical conductivity (up to 100 S/cm, depending on reduction level) while leaving sufficient functional groups for biomolecule attachment. The reduction process can be carefully tuned to adjust the surface chemistry, enabling researchers to achieve a desired sp2/ sp3 (graphene-like/ GO-like) ratio on rGO. The tunability offers rGO versatility in developing biosensors as it can fulfill special requirements in terms of conductivity and surface functionalization for diverse applications [49].

2.3.3 Carbon Nanotubes (CNTs)

Carbon-based cylindrical molecules with diameters in nanometers act as highly sensitive platforms to detect several food contaminants. Electrical properties of CNTs make them suited for sensing applications, and electrical conductivity enables them to quickly transport electrons upon the binding of target molecules on their surface. These are like super-sensitive electric wires that react to the smallest degree of external stimuli. The interaction of a contaminant molecule with the CNT surface leads to measurable changes in the electrical properties of the device, which can be detected and quantified [36, 50]. One of the key benefits of CNTs is their high surface area-to-volume ratio. Think of a sheet of graphene rolled up this tube has a very big surface area for molecule interaction, but a very small footprint. It can then bind many biological recognition elements (i.e., antibodies, enzymes, or DNA sequences) that can selectively bind to food contaminants [51].

Unlike many other materials that may suffer degradation in changing environments, CNTs offer stability over a wide range of pH, temperatures, and chemical conditions associated with food testing. Such stability allows for consistent and reproducible results over time [52]. Their specialized electronic composition can identify minute concentrations of contaminants often down to part-per-billion levels. This is especially relevant for dangerous chemicals such as pesticides, heavy metals, or bacterial toxins, where dim concentrations are needed to ensure food is safe to eat. It allows these sensors to be adapted to detect specific contaminants through different functionalization methods, which involve decorating the CNT surface with different molecular recognition elements. CNTs, for instance, can be functionalized with antibodies (for bacterial contamination detection) or specific proteins (for chemical adulteration detection) [53]. CNTs therefore hold promise for detecting food contamination and providing a low-cost alternative to conventional methods for food analysis and quality testing.

2.3.4 Metal and Metal Oxide Particles

Due to their unique properties, metal oxide nanoparticles provide significant improvements to biosensing applications for food contaminants in a variety of ways. Metal oxide nanoparticles, such as ZnO, TiO_2, and Fe_3O_4, with a high surface-to-volume ratio, can offer a larger number of active sites for interaction with analytes. This enhances the sensitivity of biosensors. Functionalization of metal oxide nanoparticles with specific receptors (e.g., antibodies or aptamers) enables highly selective capture of target contaminants, including pathogens (bacteria and viruses) or toxins. Due to their splendid electrical and optical properties, these nanoparticles could rectify different signal transduction mechanisms (e.g., electrochemical, optical) and ensure stronger and detectable signals. The innate ability of certain metal oxide nanoparticles to act as antimicrobials prohibits pathogen proliferation and enhances the durability and reliability of biosensors.

2.3.5 Nanowires

Nanowires have a high surface-to-volume ratio. This results in increased capture probabilities of the target analytes, thus enhancing the sensitivity of the biosensors. Several signal transduction mechanisms can be improved by using nanowires which have unique electrical and optical properties. Nanowires conduct, which means this can help increase current and signal. Fluorescence signals can be aggregated by the nanowires to improve the intensity and collection efficiency. This causes enhanced photon emissions which can create a much stronger optical signal. Nanowires may minimize background noise and non-specific binding, leading to more accurate and precise biosensing. This is especially useful for the detection of trace amounts of analytes [54–56].

2.3.6 Quantum Dots

Semiconductors are among the classes of materials that attract enormous attention to preparing optical and electronic applications and among this class, quantum dots (QDs) are great candidates to develop the signal detection in biosensors owing to their extraordinary properties. QDs have strong photoluminescence brightness and photostability, allowing bright and stable signals to be emitted over time, thus improving the sensitivity of biosensors. They have wide absorption spectra and narrow and size-tunable emission spectra, enabling simultaneous excitation of many QDs in a single-light source and weakening background interference for accurate detection. Another advantage of this being utilized for low-energy imaging is that the large Stoke's shift associated with QD's minimizes self-absorption, enhancing signal and resulting in improved detection accuracy. QD surfaces can further be immobilized with specific biomolecules, such as antibodies, peptides, or nucleic acids, allowing them to target specific analytes and thus improve selectivity and biosensor efficiency [57–59].

2.3.7 Dendrimers

Dendrimers are highly branched, tree-like macromolecules increase signal detection significantly and possess unique structural and chemical properties for biosensing applications. The numerous surface functional entities of dendrimers enable them to house a high amount of biorecognition agents which may enhance the binding event with the target analyte. It can be accounted that the tree-branched three-dimensional structure of dendrimers minimizes steric hindrance, which in turn reduces the non-specific binding and increases the signal-to-noise ratio. High structural uniformity and mono-dispersity guarantee the consistent and reproducible performance of biosensors, while a high density of functional groups allows biosensors to be functionalized with multiple recognition elements, which can significantly improve the biosensor's sensitivity and selectivity. In addition, dendrimer-based biosensors show good chemical stability, which is necessary for allowing sufficient service life in practice for permanent use. Dendrimers can increase signal transduction pathways, such as in electrochemical sensors, where dendrimers cause multiple electron transfer pathways that result in strengthened signals. All these properties make dendrimers ever improving in several biosensing applications [60, 61] (Fig. 2).

3 Sample Preparation and Handling

Food contaminants detection require proper sample management as an initial step to successfully detect them. The accuracy and reliability of test results depends on effective sample preparation however improper sample handling leads to misidentification of results (Fig. 3).

3.1 Extraction and Purification

3.1.1 Liquid–Liquid Extraction (LLE)

Liquid–liquid extraction has become a gold standard method to separate and purify analytes from matrices. The target analyte partitions between two incompatible liquid phases which usually combine organic solvent and aqueous solution. The extraction of non-polar or moderately polar contaminants including pesticides, mycotoxins, and heavy metals succeeds through this method [62].

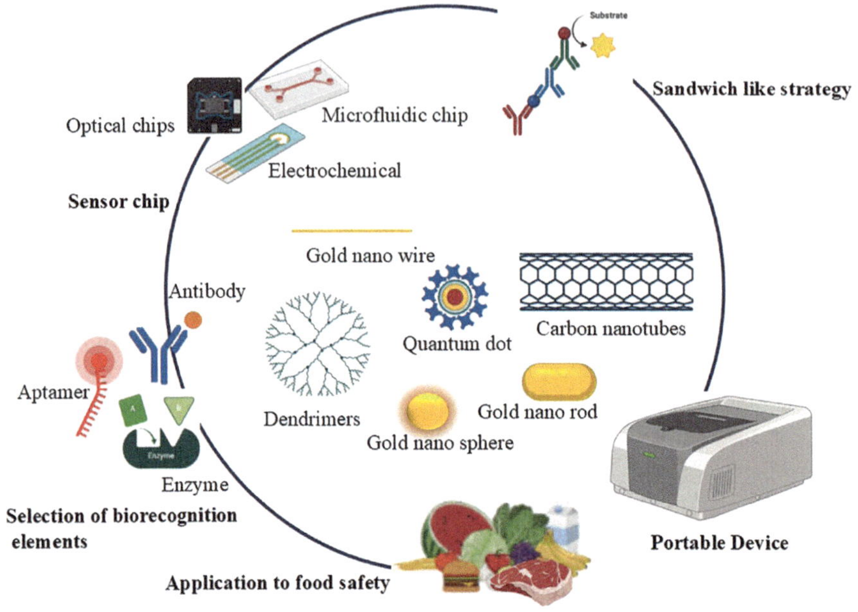

Fig. 2 Combination of different nanomaterials and sensors for the detection of pathogens effectively

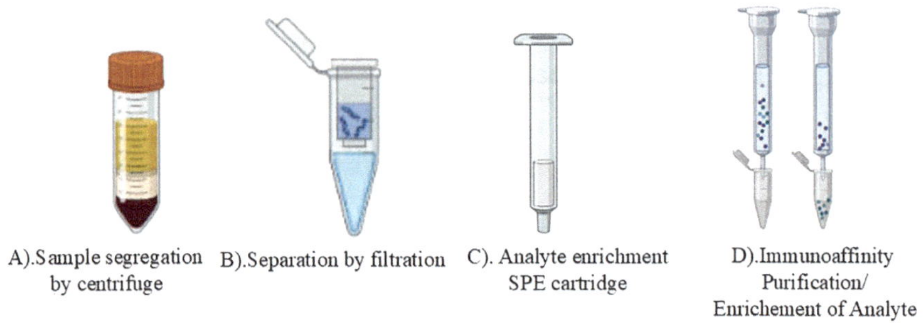

Fig. 3 Various sample preparation strategies

3.1.2 Solid-Phase Extraction (SPE)

Solid-phase extraction applies solid sorbent materials to selectively hold target analytes during their concentration process from various liquid samples. SPE provides better selectivity as well as enhanced sensitivity and cleaner samples than traditional LLE extraction. SPE offers the capability to extract both polar and non-polar compounds with its ability to obtain antibiotics, heavy metals [63], dioxins, and polychlorinated biphenyls (PCBs), etc.

3.1.3 Immunoaffinity Purification

The process of Immunoaffinity purification enables researchers to use solid support-bound antibodies which specifically isolate target compounds from multi-component food samples. The technique shows high specificity because it removes interfering compounds thus leading to enhanced detection limits with decreased matrix influences. The specific contamination analysis of mycotoxins and veterinary drug residues along with other eligible contaminants benefits greatly from immunoaffinity purification through antibody-based solid surface capture.

3.2 Sample Pretreatment

Detecting food contaminants entirely depends on successful sample pretreatment. The analysis preparation utilizes different techniques to modify the sample before measuring such as filtration alongside dilution and centrifugation. The analytical techniques function to attain precise and dependable results by performing isolation procedures while concentration actions and separation of interferences.

3.2.1 Filtration

The filtration process serves to eliminate big particulates together with debris and other unwanted components from the sample. Different filter types including membrane filters (Nylon and PVDF), paper filters, and glass fiber filters provide a solution for achieving this goal in elementary biochemical analysis. To determine a suitable filter, one needs to consider the size and type of present contaminants and the characteristics of the sample solution. The main function of filtration is to stop obstructing analytical equipment while safeguarding their operational integrity.

3.2.2 Dilution

Numerous samples are required to be diluted prior to analysis as their concentrations are above the detection limits of the method. A diluted sample permits detection techniques to operate within their linear range when measuring the target analyte concentration. The correct level of dilution enables consistent analytical results while preventing high-concentration elements from affecting the tests.

3.2.3 Centrifugation

Centrifugation allows to segregate the sample in a tube in to different layers with variable molecular mass due to applied centrifugal force. When the applied large mass molecules sediment at the bottom, while small mass molecules still be in the top layer. Whereas, this process can be enhanced by increasing the rate of centrifugal force.

4 Detection Strategies

4.1 Multiplex Sensing

Multiplex sensing involves the simultaneous determination of several target analytes with a single sensor or sensor array. This can be achieved by effective integration of respective recognition elements. These binding events are then transduced into detectable signals, e.g., optical, electrochemical, or mass-based properties, that are measurable and quantifiable.

Spatial multiplexing involves the use of an array of distinct sensing elements, whereas spectral multiplexing entails using distinct spectral signatures associated Scanning generates a more complex analytical data set (multiple analytes) requiring sophisticated data analysis approaches. That may include multivariate statistic methods, machine learning algorithms, or other data processing tools to extract developed features from multiplex data. Mass spectrometry hyphenated techniques like (1) liquid chromatology mass spectrometry (LC–MS) (2) gas chromatography mass spectrometry, and (3) matrix-assisted laser desorption/ionization (MALDI), (4) hyperspectral imaging, (5) near infrared (NIR) chromatography, (6) Fourier transform infrared spectroscopy (FTIR), (7) Raman spectroscopy, and (8) X-ray photoelectron spectroscopy that can provide simultaneous spatial and spectral information about multiple analytes or material properties [64–66]. Electrochemical sensor arrays or electrodes-based platforms allowing simultaneous detection of several redox-active analytes or electrochemical processes. Electrochemical sensor arrays and or electrode-based platforms allow the simultaneous detection of multiple (photo)redox-active analytes or electrochemical processes [66].

4.2 Rapid and Onsite Analysis

Portable devices for rapid and onsite analysis provide immediate results in fields such as environmental monitoring, medical diagnostics, and food safety. They work best when the right kind of samples can be analyzed quickly without using up a lot of sample volume for the test. These devices are generally portable and lightweight, making them easy to carry to remote or field locations. Reducing the amount of sample required reduces costs due to low reagent volumes consumed, which can be significant in the case of precious or limited resources and are usually much faster to process, allowing quick analysis [67–69].

4.3 Enhanced Sensitivity and Selectivity

Metal nanoparticles can enhance the fluorescence or surface-enhanced Raman scattering (SERS) signals once they are immobilized with biorecognition elements, ligand molecules SERS capable of detecting trace amounts of contaminants through enrichment or matrix elimination. Signal amplification by gold nanoparticles can enhance the surface plasmon resonance (SPR) signal of the surrounding medium thus enabling detection of target analytes in the lower concentration ranges. Likewise, carbon nanotubes and graphene can be used in electrochemical sensors, which can enhance the amplification of the signal due to their high surface area and good electrical conductivity [70–72].

Several mechanisms that account for signal amplification by nanomaterials include increased surface area, improved electron transfer to get better catalytic properties, signal transduction, etc. Nanomaterials have a huge surface-to-volume ratio, allowing for a larger contact between the target analyte and the detection platform, which enhances the sensitivity. Carbon nanotubes and graphene are excellent electrical conductors, which increase the efficiency of electron transfer during electrochemical detection and provide a superior signal amplification ability as quantum dots and plasmonic nanoparticles convert the interaction of the target analyte into a measurable signal, which can be optical or electrical and enhance the detection efficiency. Additionally, a combination of two or more types of nanoparticles has also been reported to improve signal intensities, for instance, the use of gold nanoparticles with graphene in various plasmonic sensors increased detection limits to femtomolar levels [72, 73]. Having appropriate biorecognition elements along with suitable nanoparticles would amplify the detection signal, for example, antibodies are used as biorecognition elements for monitoring trace analytes in sandwich type Assay to detect a single analyte. As aptamers are more sensitive than antibodies, two aptamers can be applied together as signal amplification strategies in the future [74, 75].

5 Data Analysis and Interpretation

Signal enhancements improve the quality and clarity of any signal by removing undesired components with loudness and removing or squelching the added noise. Many methods use techniques such as low-pass, high-pass, or band-pass filtering, where noise is interpreted in the time-frequency domain, e.g., as moving averages, median filtering, and wavelet smoothing. After preprocessing, relevant characteristics from the signal such as peaks or zero-crossings can be extracted. Fourier analysis, time-frequency analysis, and wavelet decomposition are methods for this purpose. Once features are identified, it is critical to understand the underlying physical or biological processes generating the signal. Answering this question often requires comparing the properties of the signal with known models or performing statistical analysis to find significant features in the signal, often assisted by various visualization techniques [76].

5.1 Baseline Correction

The use of various methods to eliminate the noise that can cause undesirable baseline shifts or signal differences will allow an analysis of the signal possible and provide more reliable and accurate signals. Environmental variables like noise and drift can sequester the raw signal and lead to false conclusions through baseline variations. Background Correction: There are various methods for background correction, all with their pros and cons. Common approaches are polynomial fitting, moving average, and Savitzky-Golay filtering, which can aid in to denoise the signal and estimating baselines.

5.2 Peak Identification

Peak identification is the detection and identification of significant or notable points in a signal or dataset. Some of the most common techniques for peak identification are threshold based approach, local maxima identification.

5.3 Quantification Algorithms

Quantitation algorithms take the output of the processor and convert continuous-time or continuous-amplitude input to discrete-time or discrete-amplitude output. Quantization is the process of mapping a sample value to a finite number of levels, and it helps with digital signal processing and storage. Therefore, quantitation algorithms try to cause as less information loss as possible for this conversion, where the signal in its digital form preserves the essential content of the original signal. Algorithms commonly used for quantitation are uniform quantization, non-uniform quantization, adaptive quantization, and sigma-delta modulation.

5.4 Validation and Standardization

Verification and standardization are essential parts of the development and application of food analytical methods for food regulatory compliance. Validation of analytical procedures establish the reliability, accuracy, and consistency of anthranilic acid for the compositional, safety, and quality determinations of food. The performance characteristics of the method (accuracy, precision, linearity, limit of detection, limit of quantification, and selectivity) are assessed during validation. This process generates a definition standard, which describes a common set of protocols, procedures, and guidelines regarding the analytical method that can be reproducibly executed in equivalent devices/manufacturers and different operators (Table 1).

6 Advantages and Limitations

6.1 Advantages

The quick identification can take anywhere from minutes to hours, greatly outspeeding the classical technique. Such systems also show high sensitivity with detection limits at nanomolar to femtomolar range. Nearly all are portable and easy to use for in-field testing the food processing industries.

6.2 Limitations

Nanomaterials are costly to produce, which may hinder them from achieving ubiquity. Moreover, scaling up the production of reproducible sensors on a larger scale poses another major obstacle. In addition, there are regulatory/validation challenges, as some testing methods require food safety approval and need to undergo validation testing. All of the factors together affect the wide integration and acceptance of nanomaterial-based sensors in the different applications.

7 Future Directions

They can be used to create smart sensors for supply chain real-time monitoring through IoT integration. Multiplexing, the simultaneous detection of multiple pathogens, is offered by these sensors. Moreover, they are synthesized through green approaches that make them sustainable and cost-effective nanomaterials. This is a very important way to improve the efficiency, sensitivity, and sustainability of detection systems.

Table 1 Summary of various contaminants detected by using nano-biosensors in food and other matrices

Target contaminant	Nanomaterial	Sensor type	Recognition element	LOD	References
Salmonella	Graphene oxide	Electrochemical	Aptamer	10 CFU/mL	Gupta et al. [77]
E. coli O157:H7	Gold nanoparticles	Optical	Antibody	100 CFU/mL	Hassan et al. [78]
Listeria monocytogenes	Carbon nanotubes	Electrochemical	Aptamer	5 CFU/mL	Feng et al. [79]
Aflatoxin B1	Quantum dots	Fluorescent	Antibody	0.02 ng/mL	Lu et al. [80]
Bisphenol A	Molecularly imprinted polymer-coated magnetic nanoparticles	Electrochemical	Molecular imprinting	0.05 ng/mL	Metwall et al. [81]
Melamine	Gold nanoparticles	Colorimetric	Antibody	0.1 mg/L	Jazayeri et al. [82]
Pesticide residues (organophosphates)	Carbon nanotubes	Electrochemical	Acetylcholinesterase	0.5 µg/L	Cheng et al. [83]
Heavy metals (lead, cadmium)	Graphene	Electrochemical	Metallic ion-binding ligands	0.1 µg/L, 0.05 µg/L	Meng et al. [84]
Mycotoxins (ochratoxin A)	Magnetic nanoparticles	Electrochemical	Aptamer	0.01 ng/mL	Arguobi et al. [85]
Chloramphenicol	Gold nanoparticles	Surface plasmon resonance	Antibody	32 pg/mL	Yuan et al. [86]
Phthalates	Molecularly imprinted polymer-coated magnetic nanoparticles	Electrochemical	Molecular imprinting	0.01 ng/mL	Li et al. [87]
Acrylamide	Gold nanoparticles	Colorimetric	Aptamer	0.038 nmol L^{-1}	Khoshibin et al. [88]
Histamine	Quantum dots	Fluorescent	Antibody	0.1 mg/L	Wu et al. [89]

(continued)

Table 1 (continued)

Target contaminant	Nanomaterial	Sensor type	Recognition element	LOD	References
Tetracycline	Molecularly imprinted polymer-coated magnetic nanoparticles	Electrochemical	Molecular imprinting	1.5×10^{-7} mol L^{-1}	[90]
Nitrite	Graphene oxide-gold nanoparticle hybrid	Electrochemical	Nitrite-reducing enzyme	0.1 µM	Jiang et al. [91]
Pesticide residues (organochlorines)	Graphene quantum dots	Fluorescent	Enzyme inhibition	0.01 µg/L	Gaviria-Arroyave et al. [92]
Allergens (peanut)	Gold nanoparticles	Colorimetric	Antibody	0.1 µg/mL	Gómez-Arribas et al. [93]
Gluten	Gold nanoparticles	Colorimetric	Antibody	8 mg/kg	Marra et al. [94]
Patulin	Molecularly imprinted polymer-coated magnetic nanoparticles	Electrochemical	Molecular imprinting	0.01 ng/mL	Huang et al. [95]
Fumonisins	Quantum dots	Fluorescent	Aptamer	0.05 ng/mL	Guo et al. [96]
Aflatoxin M1	Quantum dots	Fluorescent	Antibody	0.01 ng/mL	Kaur et al. [97]
Sulfonamides	Molecularly imprinted polymer-coated magnetic nanoparticles	Electrochemical	Molecular imprinting	2 ng/mL	Oshagi et al. [98]
Streptomycin	Gold nanoparticles	Colorimetric	Aptamer	0.1 ng/mL	Zhao et al. [99]
Enrofloxacin	Graphene oxide	Electrochemical	Aptamer	0.01 ng/mL	Guo et al. [100]
Quinolones	Gold nanoparticles	Surface plasmon resonance	Aptamer	0.01 ng/mL	Wang et al. [101]
Aminoglycosides	Gold nanoparticles	Colorimetric	Aptamer	0.1 ng/mL	Luan et al. [102]

8 Conclusion

In short, nanomaterials biosensors have a great potential in the fields of food safety through the fast and precise detection of food contaminants. Their high sensitivity and specificity make it possible to detect harmful elements in very small amounts, thus ensuring consumer safety. These biosensors have become highly efficient in a wide range of applications as the combination of various nanomaterials including gold nanoparticles, graphene, and quantum dots improves the performance of biosensors well. In conclusion, while many challenges remain in terms of scalability, cost, and regulatory approval, advances in nanotechnology and biosensor design pave a future where food safety monitoring can be achieved in a more efficient, reliable, and accessible manner. Such innovations will greatly assist public health and create further confidence in global food systems. Continued R&D in this area are critical for addressing these limitations and fully embracing the potential of nanobiosensors to protect the safety and quality of our food supply.

References

1. Saravanan A et al (2021) Methods of detection of food-borne pathogens: a review. Environ Chem Lett 19:189–207
2. Benard O, Lanoi D, Ruth OO (2018) Food safety, E-S Yehia (ed). IntechOpen, Rijeka
3. Leblanc-Maridor M, Beaudeau F, Seegers H, Denis M, Belloc C (2011) Rapid identification and quantification of Campylobacter coli and Campylobacter jejuni by real-time PCR in pure cultures and in complex samples. BMC Microbiol 11(1):113
4. Church DL, Emshey D, Semeniuk H, Lloyd T, Pitout JD (2007) Evaluation of BBL CHROMagar O157 versus sorbitol-MacConkey medium for routine detection of Escherichia coli O157 in a centralized regional clinical microbiology laboratory. J Clin Microbiol 45(9):3098–3100
5. March SB, Ratnam S (1986) Sorbitol-MacConkey medium for detection of Escherichia coli O157: H7 associated with hemorrhagic colitis. J Clin Microbiol 23(5):869–872
6. Franco-Duarte R et al (2019) Advances in chemical and biological methods to identify microorganisms—from past to present. Microorganisms 7(5):130. https://doi.org/10.3390/microorganisms7050130
7. Mancini N et al (2010) The era of molecular and other non-culture-based methods in diagnosis of sepsis. Clin Microbiol Rev 23(1):235–251
8. Hassall J, Coxon C, Patel VC, Goldenberg SD, Sergaki C (2024) Limitations of current techniques in clinical antimicrobial resistance diagnosis: examples and future prospects. npj Antimicrob Resist 2(1):16
9. Patel A, Wolfram A, Desin TS (2024) Advancements in detection methods for salmonella in food: a comprehensive review. Pathogens. https://doi.org/10.3390/pathogens13121075
10. Priyanka B, Patil RK, Dwarakanath S (2016) A review on detection methods used for foodborne pathogens. Indian J Med Res 144(3):327–338
11. Chiou J, Leung AHH, Lee HW, Wong W-T (2015) Rapid testing methods for food contaminants and toxicants. J Integr Agric 14:2243–2264
12. Priyanka B, Patil RK, Dwarakanath S (2016) A review on detection methods used for foodborne pathogens. Indian J Med Res 144:327–338

13. Law JWF, Ab Mutalib NS, Chan KG, Lee LH (2015) Rapid methods for the detection of foodborne bacterial pathogens: principles, applications, advantages and limitations. Front Microbiol 5:770
14. Berry JD, Gaudet RG (2011) Antibodies in infectious diseases: polyclonals, monoclonals and niche biotechnology. New Biotechnol 28(5):489–501
15. Chiu ML, Goulet DR, Teplyakov A, Gilliland GL (2019) Antibody structure and function: the basis for engineering therapeutics. Antibodies 8(4):55. https://doi.org/10.3390/antib8040055
16. Özdemir M et al (2011) A comparison of immuncapture agglutination and ELISA methods in serological diagnosis of brucellosis. Int J Med Sci 8(5):428
17. Kinns H, Pitkin S, Housley D, Freedman DB (2013) Internal quality control: best practice. J Clin Pathol 66(12):1027–1032
18. Meliana C, Liu J, Show PL, Low SS (2024) Biosensor in smart food traceability system for food safety and security. Bioengineered 15(1):2310908
19. Ellahi RM, Wood LC, Bekhit AE-DA (2023) Blockchain-based frameworks for food traceability: a systematic review. Foods 12:3026
20. Wityk P et al (2025) Reusable biosensor for easy RNA detection from unfiltered saliva. Sensors. https://doi.org/10.3390/s25020360
21. Sekhwama M, Mpofu K, Sudesh S, Mthunzi-Kufa P (2024) Integration of microfluidic chips with biosensors. Discov Appl Sci 6:458
22. Mishra GK, Barfidokht A, Tehrani F, Mishra RK (2018) Food safety analysis using electrochemical biosensors. Foods. https://doi.org/10.3390/foods7090141
23. Mehrotra P (2016) Biosensors and their applications—a review. J Oral Biol Craniofacial Res 6:153–159
24. Yeung D et al (2016) Evaluation of highly sensitive immunoassay technologies for quantitative measurements of sub-pg/mL levels of cytokines in human serum. J Immunol Methods 437:53–63
25. Ma H, Ó'Fágáin C, O'Kennedy R (2020) Antibody stability: a key to performance—analysis, influences and improvement. Biochimie 177:213–225
26. Byrne B, Stack E, Gilmartin N, O'Kennedy R (2009) Antibody-based sensors: principles, problems and potential for detection of pathogens and associated toxins. Sensors 9(6):4407–4445
27. Lee I (2008) "Enzyme kinetics and mechanism" by Paul F. Cook and W. W. Cleland. J Enzym Inhib Med Chem 23:296–296
28. Kilic NM et al (2023) Novel approaches to enzyme-based electrochemical nanobiosensors. Biosensors. https://doi.org/10.3390/bios13060622
29. Wong F et al (2024) Deep generative design of RNA aptamers using structural predictions. Nat Comput Sci 4:829–839
30. Ilgu M, Nilsen-Hamilton M (2016) Aptamers in analytics. Analyst 141:1551–1568
31. Yoo H, Jo H, Oh SS (2020) Detection and beyond: challenges and advances in aptamer-based biosensors. Mater Adv 1:2663–2687
32. Tertis M, Hosu O, Feier B, Cernat A, Florea A, Cristea C (2021) Electrochemical peptide-based sensors for foodborne pathogens detection. Molecules 26(11):3200. https://doi.org/10.3390/molecules26113200
33. Cunningham R, Ma D, Li L (2012) Mass spectrometry-based proteomics and peptidomics for systems biology and biomarker discovery. Front Biol 7(4):313–335
34. Li L, Wu J, Lyon CJ, Jiang L, Hu TY (2023) Clinical peptidomics: advances in instrumentation, analyses, and applications. BME Front 4:0019
35. Polat EO et al (2022) Transducer technologies for biosensors and their wearable applications. Biosensors. https://doi.org/10.3390/bios12060385
36. Norizan MN et al (2020) Carbon nanotubes: functionalisation and their application in chemical sensors. RSC Adv 10:43704–43732
37. Saha K, Agasti SS, Kim C, Li X, Rotello VM (2012) Gold nanoparticles in chemical and biological sensing. Chem Rev 112:2739–2779

38. Ghosh SK, Pal T (2007) Interparticle coupling effect on the surface plasmon resonance of gold nanoparticles: from theory to applications. Chem Rev 107:4797–4862
39. Mahato K et al (2019) Gold nanoparticle surface engineering strategies and their applications in biomedicine and diagnostics. 3 Biotech 9(2):57
40. Guo L et al (2015) Strategies for enhancing the sensitivity of plasmonic nanosensors. Nano Today 10:213–239
41. Su H, Sun B, Chen L, Xu Z, Ai S (2012) Colorimetric sensing of dopamine based on the aggregation of gold nanoparticles induced by copper ions. Anal Methods 4:3981–3986
42. Malathi S, Pakrudheen I, Kalkura SN, Webster TJ, Balasubramanian S (2022) Disposable biosensors based on metal nanoparticles. Sens Int 3:100169
43. Ghislandi M, Tkalya E, Alekseev A, Koning C, de With G (2015) Electrical conductive behavior of polymer composites prepared with aqueous graphene dispersions. Appl Mater Today 1:88–94
44. Martín A, Escarpa A (2014) Graphene: the cutting–edge interaction between chemistry and electrochemistry. TrAC Trends Anal Chem 56:13–26
45. Pop E, Varshney V, Roy AK (2012) Thermal properties of graphene: fundamentals and applications. MRS Bull 37:1273–1281
46. Liu L, Zhang J, Zhao J, Liu F (2012) Mechanical properties of graphene oxides. Nanoscale 4:5910–5916
47. Hu Y-X et al (2022) Expanded interlayer spacing of graphene oxide achieved by electrostatic cation intercalation towards superior sodium ion storage. Ionics 28:3833–3842
48. Hamidah NL et al (2019) Improving the proton conductivity of graphene oxide membranes by intercalating cations. SN Appl Sci 1:630
49. Das P, Ibrahim S, Chakraborty K, Ghosh S, Pal T (2024) Stepwise reduction of graphene oxide and studies on defect-controlled physical properties. Sci Rep 14:294
50. Sánchez-Romate XF, Jiménez-Suárez A, Ureña A (2020) Handbook of carbon nanotubes, Abraham J, Thomas S, Kalarikkal N (eds). Springer International Publishing, Cham, pp 1–35
51. Ferrier DC, Honeychurch KC (2021) Carbon nanotube (CNT)-based biosensors. Biosensors 11(12):486. https://doi.org/10.3390/bios11120486
52. Nurazzi NA-O et al (2021) Mechanical performance and applications of CNTs reinforced polymer composites—a review. Nanomaterials 11(9):2186. https://doi.org/10.3390/nano11092186
53. Yang N, Chen X, Ren T, Zhang P, Yang D (2015) Carbon nanotube based biosensors. Sens Actuators B Chem 207:690–715
54. Ramesh M, Janani R, Deepa C, Rajeshkumar L (2022) Nanotechnology-enabled biosensors: a review of fundamentals, design principles, materials, and applications. Biosensors 13(1):40. https://doi.org/10.3390/bios13010040
55. Hsu C-Y et al (2023) Nanowires properties and applications: a review study. S Afr J Chem Eng 46:286–311
56. Ambhorkar P, Wang Z, Ko H, Lee S, Koo KI, Kim K, Cho DI (2018) Nanowire-based biosensors: from growth to applications. Micromachines 9(12):679. https://doi.org/10.3390/mi9120679
57. Hu X et al (2020) Photoluminescence of InAs/GaAs quantum dots under direct two-photon excitation. Sci Rep 10:10930
58. Xu Q et al (2018) Two-dimensional quantum dots: fundamentals, photoluminescence mechanism and their energy and environmental applications. Mater Today Energy 10:222–240
59. El-Zeiny ME, Ehab AO (2024) Advances in semiconductor physics, devices and quantum dots—nanotechnology and future challenges, Jean-Luc A, Tingting Y, Daniela M (eds). IntechOpen, Rijeka
60. Satija J, Sai VVR, Mukherji S (2011) Dendrimers in biosensors: concept and applications. J Mater Chem 21:14367–14386
61. Kuswandi B, Futra D, Heng LY (2017) Nanotechnology applications in food, Oprea AE, Grumezescu AM (eds). Academic Press, pp 307–333

62. Jiang H, Yang S, Tian H, Sun B (2023) Research progress in the use of liquid-liquid extraction for food flavour analysis. Trends Food Sci Technol 132:138–149
63. Khan WA, Arain MB, Soylak M (2020) Nanomaterials-based solid phase extraction and solid phase microextraction for heavy metals food toxicity. Food Chem Toxicol 145:111704
64. Zupančič U et al (2021) Modern techniques in biosensors: detection methods and commercial aspects, Dutta G, Biswas A, Chakrabarti A (eds). Springer Singapore, Singapore, pp 63–93
65. Wu Y et al (2020) Emerging design strategies for constructing multiplex lateral flow test strip sensors. Biosens Bioelectron 157:112168
66. Grabowska I, Hepel M, Kurzątkowska-Adaszyńska K (2022) Advances in design strategies of multiplex electrochemical aptasensors. Sensors. https://doi.org/10.3390/s22010161
67. Rasheed S et al (2024) Advances and challenges in portable optical biosensors for onsite detection and point-of-care diagnostics. TrAC Trends Anal Chem 173:117640
68. Dragone R, Grasso G, Muccini M, Toffanin S (2017) Portable bio/chemosensoristic devices: innovative systems for environmental health and food safety diagnostics. Front Public Health 5:80
69. Duan C et al (2022) Portable instruments for on-site analysis of environmental samples. TrAC Trends Anal Chem 154:116653
70. Jo NR, Shin YB (2020) Enhancing biosensing sensitivity of metal nanostructures through site-selective binding. Sci Rep 10(1):1024
71. Liu Z et al (2015) Enhancing refractive index sensing capability with hybrid plasmonic–photonic absorbers. J Mater Chem C 3:4222–4226
72. Chu CH et al (2023) Enhanced optical sensing with metal-insulator-metal nanohole arrays integrated with silver nanoparticles. Results Phys 55:107203
73. Wu L, Xiong E, Zhang X, Zhang X, Chen J (2014) Nanomaterials as signal amplification elements in DNA-based electrochemical sensing. Nano Today 9:197–211
74. Morales MA, Halpern JM (2018) Guide to selecting a biorecognition element for biosensors. Bioconjugate Chem 29(10):3231–3239
75. Kumar Kulabhusan P, Hussain B, Yüce M (2020) Current perspectives on aptamers as diagnostic tools and therapeutic agents. Pharmaceutics 12(7):646. https://doi.org/10.3390/pharmaceutics12070646
76. Maharana K, Mondal S, Nemade B (2022) A review: data pre-processing and data augmentation techniques. Glob Transit Proc 3:91–99
77. Gupta R et al (2023) Salmonella typhimurium detection and ablation using OmpD specific aptamer with non-magnetic and magnetic graphene oxide. Biosens Bioelectron 234:115354
78. Hassan A-RHA-A, de la Escosura-Muñiz A, Merkoçi A (2015) Highly sensitive and rapid determination of Escherichia coli O157:H7 in minced beef and water using electrocatalytic gold nanoparticle tags. Biosens Bioelectron 67:511–515
79. Feng J, Dai Z, Tian X, Jiang X (2018) Detection of Listeria monocytogenes based on combined aptamers magnetic capture and loop-mediated isothermal amplification. Food Control 85:443–452
80. Lu Y-N et al (2022) Quantum dot microsphere-based immunochromatography test strip enabled sensitive and quantitative on-site detections for multiple mycotoxins in grains. Food Chem 376:131868
81. Metwally MG, Shehab OR, Ibrahim H, El Nashar RM (2022) Electrochemical detection of Bisphenol A in plastic bottled drinking waters and soft drinks based on molecularly imprinted polymer. J Environ Chem Eng 10:107699
82. Jazayeri MH, Aghaie T, Avan A, Vatankhah A, Ghaffari MRS (2018) Colorimetric detection based on gold nano particles (GNPs): an easy, fast, inexpensive, low-cost and short time method in detection of analytes (protein, DNA, and ion). Sens Bio Sens Res 20:1–8
83. Chen D et al (2017) Electrochemical acetylcholinesterase biosensor based on multi-walled carbon nanotubes/dicyclohexyl phthalate modified screen-printed electrode for detection of chlorpyrifos. J Electroanal Chem 801:185–191
84. Meng R et al (2023) The innovative and accurate detection of heavy metals in foods: a critical review on electrochemical sensors. Food Control 150:109743

85. Argoubi W, Algethami FK, Raouafi N (2024) Enhanced sensitivity in electrochemical detection of ochratoxin A within food samples using ferrocene- and aptamer-tethered gold nanoparticles on disposable electrodes. RSC Adv 14:8007–8015
86. Yuan J, Addo J, Aguilar M-I, Wu Y (2009) Surface plasmon resonance assay for chloramphenicol without surface regeneration. Anal Biochem 390:97–99
87. Li L, Lu Y, Wang C, Cheng L (2024) Fabrication of magnetic molecularly imprinted polymers for selective extraction of dibutyl phthalates in food matrices. Foods 13(9):1397. https://doi.org/10.3390/foods13091397
88. Khoshbin Z, Moeenfard M, Abnous K, Taghdisi SM (2023) Nano-gold mediated aptasensor for colorimetric monitoring of acrylamide: Smartphone readout strategy for on-site food control. Food Chem 399:133983
89. Wu G, Dou X, Li D, Xu S, Zhang J, Ding Z, Xie J (2022) Recent progress of fluorescence sensors for histamine in foods. Biosensors 12(3):161. https://doi.org/10.3390/bios12030161
90. Zeb S, Wong A, Khan S, Hussain S, Sotomayor MDPT (2021) Using magnetic nanoparticles/MIP-based electrochemical sensor for quantification of tetracycline in milk samples. J Electroanal Chem 900:115713
91. Jiang J, Fan W, Du X (2014) Nitrite electrochemical biosensing based on coupled graphene and gold nanoparticles. Biosens Bioelectron 51:343–348
92. Gaviria-Arroyave MI, Arango JP, Barrientos Urdinola K, Cano JB, Peñuela Mesa GA (2023) Fluorescent nanostructured carbon dot-aptasensor for chlorpyrifos detection: Elucidating the interaction mechanism for an environmentally hazardous pollutant. Anal Chim Acta 1278:341711
93. Gómez-Arribas LN, Benito-Peña E, Hurtado-Sanchez MDC, Moreno-Bondi MC (2018) Biosensing based on nanoparticles for food allergens detection. Sensors 18(4):1087. https://doi.org/10.3390/s18041087
94. Marra D et al (2024) Advances in gluten detection: a rapid colorimetric approach using core-satellite magnetic particles. Biosens Bioelectron X 21:100545
95. Huang Q et al (2019) Molecularly imprinted poly(thionine)-based electrochemical sensing platform for fast and selective ultratrace determination of patulin. Anal Chem 91:4116–4123
96. Guo X, Qiao Q, Zhang M, Fauconnier ML (2022) Nuclease triggered "signal-on" and amplified fluorescent sensing of fumonisin B1 incorporating graphene oxide and specific aptamer. Int J Mol Sci 23(16):9024. https://doi.org/10.3390/ijms23169024
97. Kaur G, Sharma S, Singh S, Bhardwaj N, Deep A (2022) Selective and sensitive electrochemical sensor for aflatoxin M1 with a molybdenum disulfide quantum dot/metal–organic framework nanocomposite. ACS Omega 7:17600–17608
98. Oshaghi S (2024) Nano-sized magnetic molecularly imprinted polymer solid-phase microextraction for highly selective recognition and enrichment of sulfamethoxazole from spiked water samples. J Chromatogr A 1729:465016
99. Zhao J et al (2017) Colorimetric detection of streptomycin in milk based on peroxidase-mimicking catalytic activity of gold nanoparticles. RSC Adv 7:38471–38478
100. Guo X et al (2024) Magnetic nanoparticles-based electrochemical aptasensor for the detection of enrofloxacin in chicken. Int J Electrochem Sci 19:100537
101. Wang P, Ding L, Zhang Y, Jiang X (2023) A novel aptamer biosensor based on a localized surface plasmon resonance sensing chip for high-sensitivity and rapid enrofloxacin detection. Biosensors 13(12):1027. https://doi.org/10.3390/bios13121027
102. Luan Y, Wang N, Li C, Guo X, Lu A (2020) Advances in the application of aptamer biosensors to the detection of aminoglycoside antibiotics. Antibiotics 9(11):787. https://doi.org/10.3390/antibiotics9110787

Nanobiosensors for the Detection of Pesticides in Raw and Processed Food

Kritika Saini, Anita Sharma, Parul Dhar, Shyna Bhalla, Harinderjeet Kaur, and Saddat Saba

Abstract The intake of pesticides in raw and processed food poses severe health issues. Therefore, there is a need for the accurate detection and identification of pesticides to ensure consumer food safety. Thus, understanding the concept of biosensing has enabled researchers to develop nanobiosensors, an indispensable tool with various nanomaterials and composites to increase the sensitivity as well as the specificity of pesticide detection in raw and processed food. The different applications of nanomaterials have enabled researchers to use advanced technologies in biosensors for the transfer of signals to enhance their efficiency and sensitivity. Nanomaterials like carbon nanotubes, graphene, gold and silver nanoparticles, dendrimers, nanohybrides, and quantum dots are commonly employed to produce biosensors with improved sensitivity and specificity of detection due to their exclusive magnetic, optical, thermal, and physical properties. Regarding the types of transducers, the most commonly used were amperometric, impedimetric, conductometric, optical, colorimetric, and whole-cell biosensors. All nanoparticles and new composites used in biosensors need to be classified for their enhanced performance, quick detection, and unobtrusive and effective use in pesticide analysis. Hence, we intend to summarize the different sensing techniques used in pesticide detection, their design, working principles, and advances in sensing systems.

Keywords Nanobiosensors · Nanomaterials · Food safety · Pesticide contamination

Abbreviations

%	Percentage
Ng	Nanogram
L	Liter

K. Saini (✉) · A. Sharma · P. Dhar · S. Bhalla · H. Kaur · S. Saba
Department of Life Sciences, University School of Sciences, Rayat Bahra University Kharar, Mohali, India
e-mail: kritika.19210@rayatbahrauniversity.edu.in

D. Sillu et al. (eds.), *Nanobiosensors for the Food Industry*, Smart Nanomaterials Technology, https://doi.org/10.1007/978-981-95-0136-6_7

μg	Microgram
Nm	Nanometer
mAb	Monoclonal antibody
OPs	Organophosphate
CB	Carbamate
OPPs	Organophosphorus pesticides
AChE	Acetylcholinesterase
ATC	Acetylthiocholine chloride
HPLC	High-performance liquid chromatography
GC	Gas chromatography
LCMS	Liquid chromatography with mass spectrometry
ELISA	Enzyme-linked immunosorbent-based techniques
AuNPs	Gold nanoparticles
AgNPs	Silver nanoparticles
CNTs	Carbon nanotubes
MWCNT	Multiwalled carbon nanotubes
QDs	Quantum dots
SPR	Surface plasmon resonance
rGO	Graphene oxide
CQDs	Carbon quantum dots
GQDs	Graphene quantum dots
GeQDs	Germanium quantum dots
SiQDs	Silicon quantum dots
CNTs	Carbon nanotubes
GR	Graphene
CB	Carbon black
CNHs	Carbon nanohorns
Cds	Carbon dots
NPs	Nanoparticles
DNA	Deoxyribonucleic acid
RNA	Ribonucleic acid
LOD	Limit of detection

1 Introduction

Due to the rapid increase in population, there is an urgent demand to increase the agricultural and food products, which has led to the usage of various agrochemicals [1]. Among them, the pesticides are indispensable compounds which contain toxic compounds and are used in the fields to protect the crops from the pests. The word "pesticide" refers to the class of chemicals or biological agents used to control or eradicate hazardous pests, insects, weeds, and plant diseases at a level that is acceptable. These pesticides can work by attracting and reducing the pests [2].

This has significantly increased food output and improved the quality of produce. For example, the average yield of maize, wheat, and rice worldwide more than quadrupled when pesticide use increased by 15–20 times [3]. According to an estimate, the world's consumption of pesticides has skyrocketed in recent years, reaching 719.51thousand tons in 2021 [4]. The worldwide consumption of pesticides has increased by 96% from the year 1996 to 2021. The countries with the highest consumption of pesticides globally in 2021 were China, the US, France, Australia, Argentina, Ukraine, Malaysia, Brazil, Canada, and Spain. India ranked 13th in this category [5]. The primary classification of pesticides worldwide is based on their chemical structure; these include organochlorines, malathion organophosphorus, carbamates, chlorophenols, and synthetic pyrethroids [6]. Some pesticides are categorized based on the target pests, which include herbicides, insecticides, acaricides, miticides, rodenticides, algaecides, and fungicides.

Organophosphate (OPs) is widely used to improve agricultural products, household applications, and human and pet medications; however, improper use of OPs can result in a number of health issues, as well as environmental pollution of the land, soil, and water, which in turn affects food safety. OPs include the well-known compounds parathion, abamectin, diazinon, malathion, and organochlorines [7]. Human health issues are caused by the extensive use of these pesticides in the creation of crop leaf residues in soil, water, fruits, and vegetables. Since the 1950s, malathion, a well-known and often used organophosphate, which has been used extensively in agriculture to control fruit flies on crops [8]. The different pesticides on the basis of their chemical names, classification and mode of action are described in (Table 1).

The most harmful pesticides to human health are carbamate (CB) and organophosphorus pesticides (OPPs), which block acetylcholinesterase (AChE) and cause a variety of metabolic responses. As a result, OPs are now the preferred option for farmers and agricultural workers. The OPs toxicity is directly correlated with their capacity to suppress acetylcholinesterase enzyme activity. An essential component of the neurological system, AchE is crucial to the breakdown of acetylcholine. Acetylcholine buildup results in organ failure. OPs also prevent acetylthiocholine chloride (ATCl) from hydrolyzing. ATCl deposition causes brain dysfunction, which in turn causes the SLUDGE syndrome.

Table 1 Categories of pesticides on the basis of their chemical names, target pests, and mode of action

Pesticides	Classification	Mode of action	References
Organophosphates	Herbicides	Acetyl cholinesterase inhibitor	[9]
Carbamates	Herbicides/insecticides	ATP synthase inhibitor	[10]
Abamectin	Insecticides/miticides/ acaricide	Neurotransmitter inhibitor	[11]
Organochlorines	Insecticides	Sodium channel inhibitors	[12]
Parathions	Insecticides	Acetyl cholinesterase inhibitor	[13]
Malathion	Insecticides	Acetyl cholinesterase inhibitor	[14]

As the use of pesticides (herbicides, fungicides, and insecticides) in farming increases, concerns about food safety grow. Only 0.1% of the pesticides sprayed actually hit their targets; the rest end up in the soil and can affect future crops. Pesticide residues are linked to over 3.5% of total deaths, which shows the need for careful application and better management. The World Health Organization estimated that every year, 1.5 children are suffering from diarrhea as a result of taking contaminated food, which directly causes over 3 million fatalities. However, excessive exposure to OPs can have negative health effects, including headaches, dizziness, nausea, increased perspiration, hyperventilation, and other indications of illness [15, 16]. Every year, phosphate poisoning from the use of OPs insecticides causes over 200,000 deaths throughout the world [17]. To ensure safety, we need quick and effective ways to measure pesticide levels in food.

Consequently, there is need to develop the sensitive method for the detection in raw food, fruits, vegetables, agricultural crops, milk, eggs, and meat. Various analytical methods such as High-Performance Liquid Chromatography (HPLC) [18], Gas Chromatography (GC) [19], Liquid chromatography with Mass Spectrometry (LCMS) [20], and Enzyme-linked immunosorbent-based techniques (ELISA) [21] are generally being used to determine the pesticide residues in food products and in the food sector. On the other hand, these approaches have some limitations, are highly expensive and time consuming and require sophisticated laboratory, and specialized workers to operate. Therefore, the development of fast, reliable, and cost-effective screening methodologies like use of nanobiosensors along with various nanoparticles for pesticide residues in raw and processed food samples owing to their accurate sensitivity in an affordable manner, which gives more hope to the food industry on their applications [22]. The biosensors appear as an alternative method for the detection of the pesticides in raw and processed food because of their sensitivity and specificity.

2 Classification of Pesticide Biosensors Based on Various Technologies

Biosensors are the promising devices consist of bioreceptors to determine the analytes and which can convert the biological signals into the electrical signals. These are the simple and affordable methods used for the detection of pesticides in raw and processed foods. Biosensors can be distinguished by their biorecognition components, which can include cells, DNA, enzymes, aptamers, RNA, or antibodies. The transducer and biorecognition component of the biosensor convert biological signals into digital signals (Fig. 1).

The biosensors are based on the use of different nanomaterials like gold nanoparticles, (AuNPs), silver nanoparticles (AgNPs), carbon nanotubes (CNTs), Multiwalled Carbon nanotubes (MWCNTs), Quantum dots (QDs), for the detection of pesticides in food commodities. Nanomaterials are frequently used in the biosensors as they can

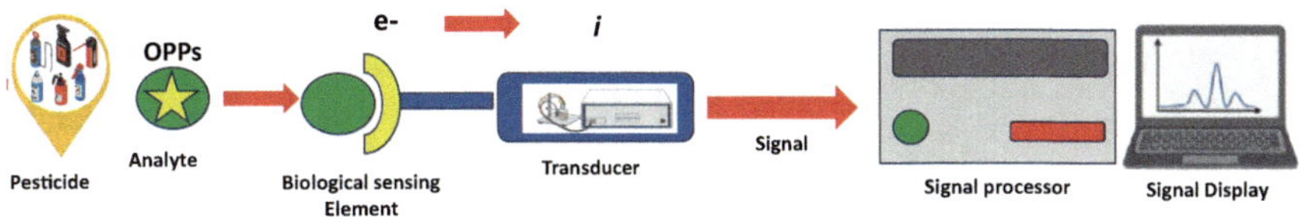

Fig. 1 The schematic working mechanism of the biosensor

increase biosensors efficacy and enable the cost effective detection of the pesticides. These nanomaterials have various properties like significant performance improvement, faster, efficient and affordable detection of substances, etc. because of which they are commonly used as biosensors [23, 24]. They improve performance as they have unique electrical and optical properties which help in generating high surface-to-volume contact ratio, electrical conductivity, biocompatibility, catalytic activity, absorption ability and can be modified easily with different functional groups like gold nanoparticles, silver nanoparticles, gold nanorods, carbon nanotubes, silver nanowires, gold nanostars, and copper nanowire [23, 24]. Although there are many different kinds of transducers, the widely used biosensors for detecting pesticides in raw and processed food are Optical (colorimetric and fluorescence), electrochemical, and whole cell biosensors [25]. The benefit of using colorimetric biosensor is that it can be confirmed without the need of the optical equipment. But this transduction's drawback is its lower sensitivity compared to others, such electrochemical and fluorescence. Although these biosensors show low selectivity and limited detection range have prompted research into new detection methods based on nanoparticles. Therefore, to confirm food safety, there is an urgent requirement to develop new convenient and accurate methods to determine the pesticide residues in food. Figure 2 shows a schematic of the transducers needed to build a biosensor, along with the choice of nanomaterials to increase sensitivity.

This book chapter covers the different transducers and nanomaterials used in detecting pesticide residues in food commodities adopted by different detection methods.

3 Type of Transducers Used for the Detection of Pesticides

The primary purpose of the transducers is to convert analytical signal into measurable electrical signals. There are number of different approaches: Electrochemical (amperometric, impedimetric, and conductometric), Optical (colorimetric and fluorescence) transducers have been studied for the detection of the pesticides in food samples. The types of transducers employed and their analytical performance in detecting pesticides in food are described below.

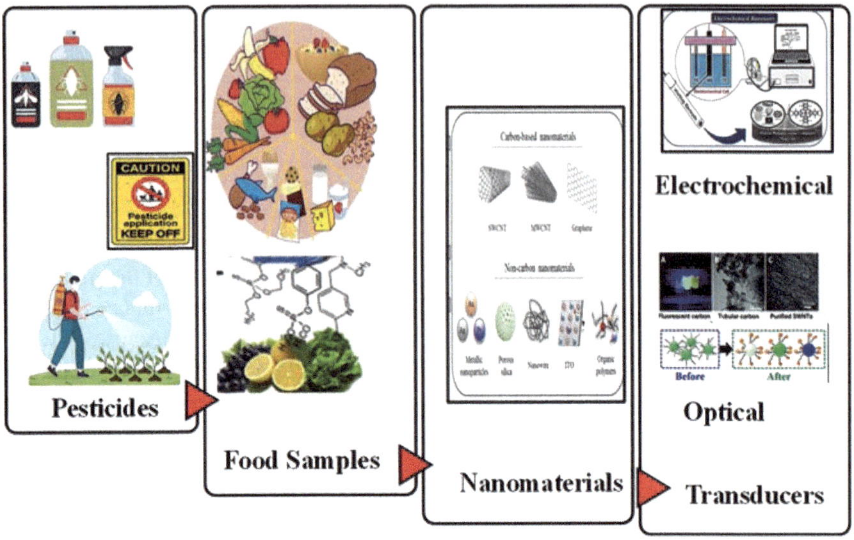

Fig. 2 A diagram outlining the steps required to build a biosensor

3.1 Electrochemical Transducer

Electrochemical biosensors are devices that detect and measure specific molecules by converting chemical information into electrical signals. These biosensors use a working electrode to identify substances called analytes, which include various biological materials like enzymes, nucleic acids, proteins, viruses, cells, tissues, antibodies, and pesticides. When these substances are detected, the sensor generates an electrical signal that shows their concentration, which can be measured in voltage, current, or impedance [26]. The most common tool for detecting pesticides is the electrochemical transducer. Electrochemical sensors are easy to make and use simple, low-cost materials and tools. They are portable and can work with different types of materials and biological components, providing good and stable results. Many researchers combine this transducer with optical and infrared methods. The most frequently used nanomaterials are metal nanoparticle probes, like gold and iron, which provide good stability during detection. The target detection method most often used for biosensor production is AChE.

Several studies have highlighted the production of biosensors using AChE, for example, Cesarino et al. [27] developed an electrochemical acetylcholinesterase (AChE) biosensor using a polyaniline (PANI) and multiwalled carbon nanotubes (MWCNTs) modified glassy carbon electrode (GC) for detecting carbamate pesticides in fruits and vegetables like apples, broccoli, and cabbage. The GC/MWCNT/PANI/AChE biosensor had detection limit of 1.4 $\mu mol\ L^{-1}$ for carbaryl and 0.95 $\mu mol\ L^{-1}$ for methomyl. Another researcher, Wang et al. [28] developed an electrochemical biosensor utilizing a graphene membrane enhanced with magnetic nanoparticles

(Fe_3O_4) for the detection of chlorpyrifos through the inhibition of the enzyme AChE. This membrane features a large specific surface area and facilitates high electron transfer, thereby improving detection capabilities and ensuring effective immobilization of AChE. The abundant active sites provided by the nanocomposite promote catalytic reactions, creating an optimal environment for the AChE reaction. The biosensor demonstrates outstanding sensitivity, achieving a limit of detection (LOD) of 20 ng L^{-1}. Additionally, it offers the capability for multiple uses, as it can be restored to approximately 90% of its original activity by immersion in a pralidoxime chloride solution [28].

Dong et al. [29] developed an electrochemical biosensor to detect malathion and methyl parathion in Chinese cabbage by inhibiting AChE. They used a three-dimensional (3D) reduced graphene oxide (rGO) film combined with nickel foam and gold nanoparticles (AuNPs/rGO/NF). This design enhances AChE adsorption and electron transfer, offering an environmentally friendly solution. The biosensor achieved a low detection limit (LOD) of 2.17×10^{-2} ng L^{-1}. Li et al. [30] created a sensor using rGO modified with tetraethylenepentamine (TEPA) and copper nanowires (CuNWs) to boost the electrode's conductivity and loading capacity for AChE. The combination of copper's catalytic properties with other nanomaterials in the rGO-TEPA/Cu NWs nanocomposite improved the sensor's surface area and detection sensitivity.

Recently, Hou et al. [31] introduced an innovative, sensitive, and reliable electrochemical biosensor designed to detect chlorpyrifos in samples of orange and spinach. This biosensor integrates MXene, carbon nanotubes (CNTs), and AuPd-acetylcholinesterase (AChE) and is based on a glassy carbon electrode (GCE).It exhibited outstanding repeatability, reproducibility, and strong anti-interference capabilities. In a related study, Bilge et al. [32] demonstrated that modifications in the electrode production process and the inclusion of biomass can significantly improve the detection of pesticides in real samples. Their research highlighted the effectiveness of carbon-based materials for pesticide detection through electrochemical methods. Furthermore, Miglione et al. [33] developed a novel biosensor specifically capable of detecting the pesticide dichlorvos on fruit peels.

3.2 *Optical Transducers*

Optical biosensors offer a potential that puts them at the forefront, compared to conventional analytical techniques, for real-time and label-free detection of biological and chemical substances in a very sensitive, specific, and cost-effective way. These biosensors are secured with various advantages like high specificity, high sensitivity, small size, and cost-effectiveness.

Due to high sensitivity, this transducer enables the detection of very small concentrations of an analyte even in the attomolar and femtomolar values. Such sensors considered very unique as they have the ability to provide rich information about the sample without interference from either an electric or a magnetic field [34].

Different types of optical transducers like Fluorescence-based biosensors which (measure changes in fluorescence intensity or wavelength upon pesticide binding), Surface Plasmon Resonance (SPR) transducer (detect changes in refractive index caused by pesticide binding), Colorimetric transducer (measure changes in color or absorbance upon pesticide binding) are used for the detection of pesticides in the food commodities. Liu et al. [35], developed an optical biosensor for detection of isocarbophos in samples of peaches, tea, and cabbage with a LOD of 2.48×10^3 ng L^{-1}. In another study, Bala et al. [36] constructed an optical biosensor using AgNPs to detect malathion in apples with a LOD of 0.5 pM by using aptamers that react with AgNPs to produce a colorimetric response. Since the AgNPs interact with the aptamer and peptide link, they appear yellow when the pesticide is not present and orange when malathion is present.

Compared to conventional techniques, colorimetric transducers are more affordable, portable, and lighter, have a smaller sample size, and require less equipment. For qualitative or quantitative study, these transducers can be customized to a particular analyte or a variety of analytes. One advantage of colorimetric detection is that it does not require optical instruments for visualization or confirmation. The disadvantage of this transduction is that it is less sensitive than other types, such as fluorescence or electrochemical [37]. The colorimetric transducers, which used AuNPs as the primary nanomaterial for the detection of malathion, in vegetable samples with detection time of 5 min and a low LODs of 4.03 ng/mL^{-1} [38]. Liu et al. [39] developed a colorimetric immunochromatographic strip (ICA) to detect dicofol, which is an organo-insecticide widely found on fruits, vegetables, teas, ornamental plants, and field crops by using a monoclonal antibody (mAb). The ICA with both apple and cucumber gave good sensitivity at a LOD of 50 μg L^{-1} and cross-reactivity test was strong for specificity of dicofol. These results were in support of LC-MS and immunocapture ELISA, which ascertain that the ICA approach is possible and reliable enough in searching for dicofol in fruits and vegetables [40].

3.3 *Whole Cell Transducers*

Whole cell transducers are the advanced technique which uses the genetic system of living organisms for the detection and quantification of pesticides. These are the biosensors based on the various transduction schemes amperometric, potentiometric and conductometric. Specific genes from the whole genomes of organism viz. bacteria, yeast, fungi, algae, cultured plants or animal cells are used as recognition elements in whole cell biosensors through production of measurable signal generated because of the presence of the targeted analyte [41]. The genes used could be naturally present in the applied organism or could be introduced for such purpose. These biosensors detect signals analytically by measuring the variations in metabolic parameters of organisms such as cell viability, bioluminescence, growth inhibition, cellular respiration, etc. [42, 43]. The degree of variation in such processes proportionate with the percentage amount of the specific analyte [41]. Detection of an insecticide chlorpyrifos using *Sinorhizobiummeliloti* endogenous plasmid pSymB, cloned in *E.coli* has been reported by Whangsuk et al. [44]. Whole cell biosensors proves advantageous over enzyme-based biosensors because of its high stability, low preparation cost, reduced requirements for purification and efficient recognition of cofactors [43, 45]. Based on the selection of studies and transducers described above, various transducers are summarized in the form of Table 2.

4 Type of Nanomaterials Used in Biosensors for the Detection of Pesticides

Researchers are currently looking into different nanomaterials for electrodes. The use of nanomaterials in biosensors provides high sensitivity and selectivity. One of the methods used for the immobilization of nanomaterial on the surface of the electrode is nanofabrication. The use of nanomaterials in electrochemical and optical biosensors is high because they possess unique optical and electrical properties. The nanomaterials can be easily modified with the functional groups and which can generate high surface-to-volume ratio, high electrical conductivity, catalytic activity, redox properties and biocompatibility. Due to redox property the use of electrochemical biosensors along with the use of nanomaterials is high which enhances the sensitivity of the biosensors for the detection of pesticides. Table 3 shows the most often used nanomaterials used for the detection of pesticides in raw and processed food samples.

These nanomaterials-based biosensors are divided into two main groups:

Carbon-based nano-biosensors, which include carbon nanotubes (CNT), graphene (GR), carbon black (CB), carbon nanohorns (CNHs), and carbon dots (CDs) (Fig. 3) [64].

Non-carbon-based nano-biosensors, which include metal nanoparticles (Metal NPs), chitin and chitosan, and metal oxide nanoparticles (Fig. 3) [65].

4.1 Carbon-Based Nano-biosensors Used in Pesticide Detection

4.1.1 Carbon Nanotubes (CNTs)

CNTs are the carbon allotropes with cylindrical lattice of carbon atoms in one or more layers used to create nanohybrids, in conjucation with other nanomaterials. The carbon nanotubes are mostly used in the nanobiosensors for the detection of pesticides, pathogens, and heavy metals because of its high surface area, exceptional electrical conductivity, mechanical strength, electrochemical stability in both aqueous and non-aqueous solutions, and high thermal conductivity [66, 67].

Table 2 Different types of transducer used to detect pesticide residues in various food samples

Transducer type	Food samples (raw)	Pesticide type	LOD	References
Electrochemical transducer	Fruits: grape, apple, guava Vegetables: Lettuce, cabbage, onion	MP	5 pm	[46]
	Rice and pulse	Paraoxon-ethyl	96 pM	[47]
	Grains samples	Carbaryl Carbofuran Fenobucarb Isoprocarb	Carbaryl 4.8×10^{-8} mol L^{-1} Carbofuran 4.9×10^{-8} mol L^{-1} Fenobucarb 8.0×10^{-8} mol L^{-1} Isoprocarb 7.9×10^{-8} mol L^{-1}	[48]
	Rice and Apple	Profenophos, Carbendazim and Acetamiprid	Profenofos: 2.1 ng L^{-1} Acetamiprid: 4.6 ng L^{-1} Carbendazim: 6.1 ng L^{-1}	[49]
	Wheat flour	Phosmet	3.6 nM	[50]
	Sugarcane bagasse	Paraquat	1 μM	[51]
	Lettuce and cabbage	Carbofuran	0.025 ng/mL	[52]
	Orange and spinach	Chlorpyrifos	1.98 pM	[31]
	Cabbage, spinach	Carbaryl	0.15 ng/mL	[53]
	Oilve oil Olive fruits and oil samples	Fenthion Malathion	0.05 mg/Kg 0.06 pg mL^{-1}	[54] [55]
	Vegetable and fruit	Chlorpyrifos	50.3 nM 10 fM	[56] [57]
Optical transducers	Fruits: orange, papaya, apple Vegetables: potato, cabbage, tomato,	Pendimethalin	0.930 nM	[58]
	Apple	Malathion	4 pM	[59]
	Chinese cabbage, cucumber, apple	Triazophos	0.096 ng/ml	[60]

(continued)

Table 2 (continued)

Transducer type	Food samples (raw)	Pesticide type	LOD	References
Colorimetric transducers	Fruits: apple, orange Vegetables: spinach, tomato, cucumber	Chlorpyrifos Profenofos Cypermethrin	Chlorpyrifos 0.235 $\times$ 10^{-6} M Profenofos 4.891 $\times$ 10^{-6} M Cypermethrin 4.053 $\times$ 10^{-6} M	[61]
	Garlic chives	Paraxon	0.82 mgL^{-1}	[62]
	Rice, pear, apple, cabbage	Parathion	2.13 ng kg^{-1}	[63]
	Vegetable samples	Malathion	4.03 ng/mL^{-1}	[38]

Graphene and its derivatives are among the various carbon structures that can be produced [68]. The benefits of these nanomaterials have led to the widespread use of biosensors for the detection of various pathogens, heavy metals, pesticides and particularly in the biomedical field [69, 70]. Kaur et al. developed an electrochemical AChE-based biosensor for the detection of chlorpyrifos-methyl in spiked lettuce with a detection limit of 1 mg L^{-1} [71]. At the same time, Kaur et al. developed a different biosensor that detected malathion in lettuce with a LOD of 1 $\times$ 10^{-6} nM using electrochemical detection and AChE enzymes [72]. Another biosensor was developed for the detection of chlorpyrifos using AChE with a LOD of 0.05 μg/L [73], CNTs demonstrated exceptional sensitivity and detection time as biosensor constituents for various pesticide classes. A high electron transfer rate, strong mechanical and chemical stability, and a large surface-to-volume ratio are the main benefits of employing CNTs.

Researchers have developed various electrochemical biosensors for the detection of pesticides in food samples using CNTs. In contrast, an electrochemical biosensor for the detection of paraoxon in spinach juice samples was developed with LOD of 3 $\times$ 10^{-6} nmol L^{-1}. This biosensor was featured with CNT with amino acid ionic liquid, which enhances the electrochemical activity of the mineralized cell (M-Cell), which was then fused with organophosphatehydrolase as the biorecognition material [74]. An author, developed an another electrochemical biosensor based on AChE which can detect organophosphorus pesticides malathion and methyl parathion in Chinese cabbage samples, had a limit of detection of 1.88 $\times$ 10^{-4} ng L^{-1}, respectively.

4.1.2 Graphene and Graphene Oxide (rGO)

Graphene is a two-dimensional hexagonal network of covalently connected sp2 hybridized carbon atoms, which possess large surface area, higher mobility, and good thermal conductivity.

Table 3 Different nanomaterials used for detection of pesticides in different food samples

Nanomaterial	LOD	Biorecognition material	Pesticide	Food material	References
Gold nanoparticles	Organophosphorus: 19–77 ng L^{-1} Methomyl: 81 ng L^{-1}	AChE	11 organophosphorus pesticides and methomyl	Cabbage and apple	[85]
Gold nanoparticles	36 ng L^{-1}	Aptamer	Chlorpyrifos	Pak choi and apple	[81]
Silver nanoparticles	17×10^3 ng L^{-1}	AChE	Paraoxon	Carrot and apple	[30]
Silver nanoparticles	17×10^3 ng L^{-1}	AChE	Paraoxon	Cabbage and chives	[92]
Silver nanowires	212 nM	BChE	Paraoxon	Milk	[93]
Gold nanoparticles and UCNPs	0.36 nM	ABA	Acetamiprid	Chinese green tea	[94]
PtPd@NCS	$8.6 \times 10^{-6} - 7.1 \times 10^{-5}$ nM	AChE	Malathion, Methyl parathion and Chlorpyrifos	Corn grains and potato	[95]
Au@PtNPs MNPs	2.13 ng kg^{-1}	ssDNAs and mAbs	Parathion	Apple, rice, cabbage and pear	[96]
PDA-AuNPs	5×10^{-1} ng L^{-1}	Aptamer	Malathion	Cabbage and cauliflower	[97]
MWCNTs	0.3 nmol L^{-1}	Ds-DNA	Diazinon	Lettuce and tomato juice	[98]
MWCNTs	4×10^{-3} nM	AChE	Organophosphate	Cabbage and spinach	[99]
Fe_3O_4 and graphene	20 ng L^{-1}	AChE	Chlorpyrifos	Spinach and cabbage	[28]
Reduced graphene oxide	1.9 nmol L^{-1}	AChE	Carbamate	Tomatoes	[100]
Reduced graphene oxide	7.12×10^{-5} nM	Aptamer	Acetamiprid	Tea	[101]

(continued)

Table 3 (continued)

Nanomaterial	LOD	Biorecognition material	Pesticide	Food material	References
CdTe-quantum dots	Pirimicarb: 5×10^4 ng L^{-1} Dichlorvos: 1×10^4 ng L^{-1} Carbaryl: 1×10^4 ng L^{-1}	AChE	Carbamate and organophosphorus	Rice, lettuce and choy	[102]
Nanocarriers	26 nM	Candida rugosa lipase	Nitrofen	Apricot	[103]

Legends LOD: Limit of Detection, ABA: Acetamiprid-binding aptamer, AChE: Acetylcholinesterase, BChE: Butyrylcholinesterase, NCS: N-doped carbon shells, NPs: Nanoparticles, mAbs: Monoclonal antibodies, MNPs: Magnetic nanoparticles, MWCNTs: Multi-walled carbon nanotubes, PDA: Polydopamine, UCNPs: Upconversionnanoparticles, PtPd@NCS: Platinum Palliumnanocubes, CdTe-Quantum dots: Cadmium telluride(CdTe) quantum dots (QDs), MWCNTs: Multi-walled carbon nanotubes, Fe_3O_4 and graphene: Graphene oxide and iron oxide, PDA-AuNPs: Polydopamine and gold nanoparticles, ABA: Acetamiprid-binding aptamer, BChE: Butyrylcholinesterase, ssDNAs: Single-stranded DNA (ssDNA) and Ds-DNA: Double-stranded DNA

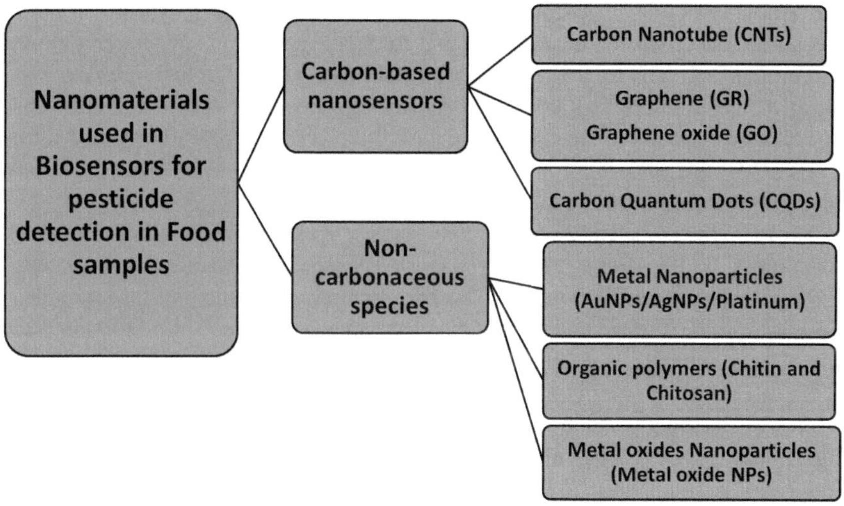

Fig. 3 Nanomaterial used in biosensors for pesticide detection in raw and processed food samples

The treatment of graphene with strong oxidants which adds carboxyl, hydroxyl, and epoxy groups to its structure, and resulting in graphene oxide, which is then reduced to produce graphene oxide (rGO) [75]. The graphene oxides result in better conductivity which improves the sensitivity and selectivity, when graphene oxides used as nanomaterials in the construction of the biosensors. The better conductivity in relation to graphene oxide, helps in improving the solvent dispersion because of the presence of functional groups graphene and reduced graphene oxide have outstanding qualities, including very high mechanical strength and stiffness, attractiveness for flexible device construction, high optical transparency, excellent electrical conductivity, and good biocompatibility; as a result, they may find use in portable electronic devices [76]. Nanoparticles of graphene and rGO were proven to be an excellent alternative for the detection of pesticides in food samples. The biosensors LOD's ranged from 2.17×10^{-2} [29] to 3.9×10^{-2} ng L^{-1} [30], which is an order of 10^4 difference. The more sensitive biosensor also uses AuNPs, which have specific properties that make them ideal for biosensors, which accounts for the discrepancy.

4.1.3 Quantum Dots (QDs)

Quantum dots (QDs) are nanoparticles having size ranges from 1 to 10 nm and remain restricted to three dimensional with quantized energy states as their size is smaller than Bohr radius [77]. QDs are generally classified into two classes: Group IV elements and Group II-I, III-V and IV-VI elements. The Group IV elements include Carbon Quantum dots (CQDs), Graphene Quantum dots (GQDs), Germanium Quantum dots (Ge QDs), and Silicon Quantum dots (Si QDs) while Group II-I, III-V and IV-VI elements include ZnS, ZnSe, ZnO, and CdS [78, 79]. The special optical characteristics of QDs that set them apart from conventional fluorescent markers, namely their excellent photostability and adjustable fluorescence [80]. Lin et al. [81] used CdTe QDs with pH-sensitive fluorescence to create an OPs sensor. AChE protonated CdTe QDs, reducing their fluorescence, and hydrolyzed ACh into acetic acid. OP can be detected because it prevents the hydrolysis mentioned above, which also prevents changes in fluorescence [79].

Yan et al. developed a hybrid-based biosensor in combination of Ir nanorods with cadmium sulfide quantum Dots (Ir NRs@CdSQDs) used to identify paraoxon in cabbage pakchoi, and lettuce. The electrochemiluminescence system includes Ir NRs as anodic emitter and CdSQDs as cathode emitter. In the immobilization of biocomposite of AChE and ChOx in the presence of organophosphate insecticides, the production of H_2O_2 during the enzyme reaction raises the cathode signal and lowers the anodic. This biosensor exhibits long-term acceptable stability with a high LOD of 1.67×10^{-3} nM, and RSD < 5% [82]. Korram et al. used Cadmium Telluride Quantum Dots (CdTeQDs) to create another device for sensitive trace detection without the spectral interference characteristics. In many optical detections of organophosphate pesticides such as paraoxon, dichlorvos, malathion, and triazophos in apple and tomato juice, this fluorescent biosensor is based on the mechanism of inhibition of the enzyme AChE. Mixing the sample with AChE and the solution containing AChE, CHOx, and QDs is all that this straightforward detection method requires as a single step. Showing greater sensitivity towards paraoxon but less sensitivity to malathion, with an LOD of 1.62×10^{-6} to 0.23 nM [83].

4.2 Nobel Metal Nanoparticles Used in Pesticide Detection

4.2.1 Gold Nanoparticles (AuNPs)

Gold Nanoparticles have unique optical and electrical properties, they are frequently employed to build biosensors. These nanoparticles considered as the best transducers for optical sensing system. One of the most significant optical characteristics of AuNPs is localized surface plasmon resonance (LSPR). The size and form of AuNPs have a significant impact on LSPR. Sharp nanoparticles (NP) corners cause charge polarization to increase, which in turn causes LSPR spectra to redshift, whereas highly symmetric structures intensify dipole resonances, which in turn causes LSPR intensities to increase [84]. Zhao et al. [85] showed with his studies that AuNP-based optical biosensors have low LOD as demonstrated in the detection of methomyl having 81 ngL^{-1} LOD value. Another study reported the detection of eleven organophosphate pesticides, LODs found were between 19 and 77 ng L^{-1}. These LOD values are a result of the extraordinary conductivity of AuNPs and the increased surface area resulting from the nanoscale effect. Due to the powerful bond between Acetylcholinesterase (AChE) and mercaptomethamidophos, the immobilizing mercaptomethamidophos was able to remain on the electrode surface through Au–S bonds, ensuring that high concentrations of AChE could attach to the surface [37].

Lin et al. [81] developed an electrochemical biosensor with a limit of detection of 36 ng L^{-1} that uses Au–S bonds to detect chlorpyrifos in apple and pakchoi samples. In this study, ferrocyanide probe was used, which was immobilized on the surface of the gold nanoparticle electrode and the signal was amplified. Due to the presence of AuNPs on the surface of electrode the surface area of the electrode was increased which shows, high electrical conductivity for the detection of the pesticides [81]. An optical biosensor was developed with a limit of detection of $2.48 \times 10^{-3} ngL^{-1}$ for the detection of isocarbophos in samples of peaches, tea, and cabbage. The AuNPs have been used for the detection of isocarbophos, chlorpyrifos, carbamate, and methomyl samples with high sensitivity and specificity ranging with LOD from $70 \times 10^{-3} ngL^{-1}$ [86] to $2.48 \times 10^{-3} ngL^{-1}$ [87].

4.2.2 Silver Nanoparticles (AgNPs)

Silver Nanoparticles are also used in biosensors for their high sensitivity, nontoxic, environmentally benign, and selectivity towards pesticides. AgNPs have special optical characteristics allowing them to respond to changes in the surrounding medium and offer color response such as the SPR phenomenon. The development of a visual biosensor for pesticide detection makes use of this optical characteristic. Acetylcholinesterase (AChE) activity is inhibited by organophosphate insecticides, which is the basis of biosensors' sensing concept. Acetylcholine (ACh) substrates are hydrolyzed by the enzyme AChE to produce choline and acetic acid. It is insufficient to depend only on AChE activity when using AgNPs as indicators in AChE inhibition-based biosensors. The conversion of chloride to betaine and hydrogen peroxide (H_2O_2) requires the catalysis of choline oxidase, a second enzyme [88, 89]. Thereby, AgNP and H_2O_2 spontaneously undergo redox, changing color as a result of Ag^0 in AgNP changing to Ag^+ ions, which clearly shows the existence of OP. A biosensor was created to detect organophosphate pesticide using acetylcholinesterase (AChE) and silver nanoparticle (AgNP) immobilized on the alginate-chitosan film matrix. The basis of the sensing principle is the inhibition of AChE activity by organophosphate insecticides [90]. Ag_3PO_4 nanoparticles' ability to mimic oxidases was demonstrated in one study as a means of accurately and swiftly detecting chlorpyrifos. Chlorpyrifos is oxidized to create sulfide ions and chlorpyrifos oxons when Ag_3PO_4 NPs nanozymes are present. Sulfide ions in the reaction system can detect chlorpyrifos pesticides and use feedback inhibition to prevent Ag_3PO_4 catalysis for TMB [91]. Zheng et al. [92] developed an electrochemical biosensor based on AChE immobilization to identify paraoxon residues in Chinese cabbage and chives. The AgNPs were used on the electrode and here chitosan was used as binder, and the electrode achieved a low limit of detection of 4×10^3 ng L^{-1}.

5 Future Perspectives

Nowadays the use of single biosensors for the detection of pesticides is limited. As for the detection of the pesticide parathion, the single biosensor known as bioimmunoassay, which is based on the catalysis of bimetallic nanoenzymes (Au@Pt), is not commonly used. Biosensors for multiple residues of pesticides that can simultaneously detect multiple pesticides are the focus of excellent studies. To detect 12 pesticides in such situations, Zhao et al. [85] created an electrochemical biosensor utilizing nanogold/mercaptomethamidophos belonging to the Au-S family of organophosphates. By binding firmly to AChE, the mercaptomethamidophos functionalized in AuNPs enables the indirect competitive approach to detect pesticides. Excellent electrochemical properties were demonstrated by the investigation, with a broad linear range of 0.1 ~ 1500 ng/mL^{-1} and a LOD of 0.019 ~ 0.077 ng/mL^{-1}. This biosensor was used to identify the herbicides dichlorvos and trichlorfon in apple and cabbage food samples. This biosensor has shown great promise for usage in food safety and environmental sensing.

Recently, in order to overcome the drawbacks of using natural enzymes, sensors based on nanozymes have been developed that combine the properties of a nanomaterial and a bioenzyme. However, because they are primarily based on a colorimetric approach, they are susceptible to severe color interference caused by the nanozymes and samples themselves. Several attempts have been made in recent years to overcome color interference, and Wu et al. [104] recently proposed a promising electrochemical sensor based on nanozymes with manganese dioxide nanosheets

(MnNS) to detect OPs. The authors claim that the MnNS eliminated color interference and made it possible to detect paraoxon in the pak choi sample with great precision and accuracy. Future research will increasingly concentrate on fusing traits to produce biosensors that are more selective and sensitive. Using various biorecognition materials, including aptamers, antibodies, DNA, and nanoenzymes, provides another avenue for further study. Since most research uses AChE, it is less selective than manufactured oligonucleotides, especially for the target analytes. Last but not least, potential future developments include the creation of portable, field-use multi-residue detection biosensors that get beyond the limitations of conventional detection techniques.

6 Conclusion

Pesticides are utilized in agriculture and/or at different stages of the production of food crops. As a result, they may be harmful to both people and the environment. The maximum residue limits for any pesticide residue were set by the European Commission at 0.01 mg/kg. This chapter covered a number of biosensors created for the analysis of pesticides in food samples. For the detection of pesticides in foods, biosensors may offer a faster, more reliable, and more affordable option than conventional chromatographic techniques. The results show that detection may vary depending on the nanomaterial, biorecognition material, transducer used to produce response signals and the relevant analyte. Electrochemical transducers, which are notable for their high sensitivity and user-friendliness, were the most often used. In addition, the primary advantage of colorimetric transducers is their high sensitivity and visual detection capability. Although they were employed less frequently, fluorescence, amperometric, impedimetric transducers were also employed and showed good sensitivity. One potential solution to increase the sensitivity of biosensors is the use of nanomaterials. The most widely used nanomaterials include AuNPs, AgNPs, CNTs, and nanohybrids. One of the highlights was the use of magnetic nanoparticles to prepare samples for a magnetic detection system. In addition to helping or influencing a biosensor's analytical signals, biological recognition materials are crucial to these devices because they are in charge of the chemical interaction between pesticides. AChE and aptamers were the most often used biological components among these biomaterials, per the investigations.

References

1. Bolat G, Yaman YT, Abacı S, Seyyar S (2023) Poly-arginine/graphene oxide functionalized disposable sensor for monitoring fenitrothion pesticide residues in water and cucumber samples. Mater Today Chem 30:101517

2. Abubakar Y, Tijjani H, Egbuna C, Adetunji CO, Kala S, Kryeziu TL, Ifemeje JC, Patrick-Iwuanyanwu KC (2020) Pesticides, history, and classification. In: Natural remedies for pest, disease and weed control. Academic Press, pp 29–42
3. Tiwari B, Kharwar S, Tiwari DN (2019) Pesticides and rice agriculture. In: Cyanobacteria. Academic Press, pp 303–325
4. Azeem S, Sohail M, Azeem A, Haq MZU, Hassan Z, Khan AA, Shahzad U (2024) Guardian crops cultivating resilience against pests. In: Revolutionizing pest management for sustainable agriculture. IGI Global, pp 181–202
5. Worldometers (2023) Pesticide use by country. https://www.worldometers.info/food-agriculture/pesticides-by-country/. Accessed 08 Aug 2023.
6. Dhuldhaj UP, Singh R, Singh VK (2023) Pesticide contamination in agro-ecosystems: toxicity, impacts, and bio-based management strategies. Environ Sci Pollut Res 30(4):9243–9270
7. SharmaA SA, Attri K, Kumar M, Kumar P, Suttee A, Singla N (2020) Global trends in pesticides: a looming threat and viable alternatives. Ecotoxicol Environ Saf 201:110812
8. Newhart K (2006) Environmental fate of malathion. Calif Environ Prot Agency 11:1–20
9. Colovic MB, Krstic DZ, Lazarevic-Pasti TD, Bondzic AM, Vasic VM (2013) Acetylcholinesterase inhibitors: pharmacology and toxicology. Curr Neuropharmacol 11(3):315–335
10. Mahmood Q, Bilal M, Jan S (2014) Herbicides, pesticides, and plant tolerance: an overview. In: Emerging technologies and management of crop stress tolerance, pp 423–448
11. Zia R, Taj A, Younis S, Bukhari SZ, Latif F, Feroz Y, Fatima K, Imran A, Bajwa SZ (2022) Application of nanosensors for pesticide detection. In: Nanosensors for smart agriculture. Elsevier, pp 259–302
12. Van Dyk JS, Pletschke B (2011) Review on the use of enzymes for the detection of organochlorine, organophosphate and carbamate pesticides in the environment. Chemosphere 82(3):291–307
13. Deo RP, Wang J, Block I, Mulchandani A, Joshi KA, Trojanowicz M, Scholz F, Chen W, Lin Y (2005) Determination of organophosphate pesticides at a carbon nanotube/organophosphorus hydrolase electrochemical biosensor. AnalyticaChimicaActa 530(2):185–189
14. Azizan MAI, Taufik S, Norizan MN, Rashid JIA (2023) A review on surface modification in the development of electrochemical biosensor for malathion. BiosensS Bioelectron X 13:100291
15. James T, Wyke S, Marczylo T, Collins S, Gaulton T, Foxall K, AmlôtR, Duarte-Davison R (2018) Chemical warfare agent simulants for human volunteer trials of emergency decontamination: a systematic review. J Appl Toxicol 38(1):113–121
16. Tchounwou PB, Patlolla AK, Yedjou CG, Moore PD (2015) Environmental exposure and health effects associated with malathion toxicity. Toxic Hazard Agrochem 51:2145–2149
17. Phopin K, Tantimongcolwat T (2020) Pesticide aptasensors-state of the art and perspectives. Sensors 20(23):6809
18. Hossain A, Islam MT, Islam MS, Ahmed S, Sarker KK, Gathala MK (2024) Chemical weed management in maize (Zeamays L.) under conservation agricultural systems: an outlook of the Eastern Gangetic Plains in South-Asia. IntechOpen
19. Chowdhury MA, Fakhruddin AN, Islam MN, Moniruzzaman M, Gan SH, Alam MK (2013) Detection of the residues of nineteen pesticides in fresh vegetable samples using gas chromatography–mass spectrometry. Food Control 34(2):457–465
20. Gao B, Poma G, Malarvannan G, Dumitrascu C, Bastiaensen M, Wang M, Covaci A (2022) Development of an analytical method based on solid-phase extraction and LC-MS/MS for the monitoring of current-use pesticides and their metabolites in human urine. J Environ Sci 111:153–163
21. Tang J, Zhang M, Cheng G, Li A, Lu Y (2008) Development of IC-ELISA for detection of organophosphorus pesticides in water. J Environ Sci Health B 43(8):707–712
22. Guruprasath N, Sankarganesh P, AdeyeyeS A, Babu AS, Parthasarathy V (2024) Review on emerging applications of nanobiosensor in food safety. J Food Sci 7(6):401–419
23. Wang W, Wang X, Cheng N, Luo Y, Lin Y, Xu W, Du D (2020) Recent advances in nanomaterials-based electrochemical (bio) sensors for pesticides detection. TrAC Trends Anal Chem 132:116041

24. Baig N, Kammakakam I, Falath W (2021) Nanomaterials: a review of synthesis methods, properties, recent progress, and challenges. Mater Adv 2(6):1821–1871
25. Zhao F, Wu J, Ying Y, She Y, Wang J, Ping J (2018) Carbon nanomaterial-enabled pesticide biosensors: design strategy, biosensing mechanism, and practical application. TrAC Trends Anal Chem 106:62–83
26. Silvestrini M, Fruk L, Moretto LM, Ugo P (2015) Detection of DNA hybridization by methylene blue electrochemistry at activated nanoelectrode ensembles. J Nanosci Nanotechnol 15(5):3437–3442
27. Cesarino I, Moraes FC, Lanza MR, Machado SA (2012) Electrochemical detection of carbamate pesticides in fruit and vegetables with a biosensor based on acetylcholinesterase immobilised on a composite of polyaniline–carbon nanotubes. Food Chem 135(3):873–879
28. Wang H, Zhao G, Chen D, Wang Z, Liu G (2016) A sensitive acetylcholinesterase biosensor based on screen printed electrode modified with Fe_3O_4 nanoparticle and graphene for chlorpyrifos determination. Int J Electrochem Sci 11(12):10906–10918
29. Dong P, Jiang B, Zheng J (2019) A novel acetylcholinesterase biosensor based on gold nanoparticles obtained by electroless plating on three-dimensional graphene for detecting organophosphorus pesticides in water and vegetable samples. Anal Methods 11(18):2428–2434
30. Li S, Qu LM, Wang JF, Ran XQ, Niu X (2020) Acetylcholinesterase based rGO-TEPA-Copper nanowires biosensor for detecting malathion. Int J Electrochem Sci 15(1):505–514
31. Hou X, Ding R, Jiang W, Yang Q, Mao X (2023) Detection of chlorpyrifos via carbon nanotubes and AuPd nanoparticles-enhanced MXene electrode-based acetylcholinesterase biosensor. Microchem J 195:109319
32. Bilge S, Dogan-Topal B, Gürbüz MM, Ozkan SA, Sınağ A (2024) Recent trends in core/shell nanoparticles: their enzyme-based electrochemical biosensor applications. Microchim Acta 191(5):240
33. Miglione A, Raucci A, Mancini M, Gioia V, Frugis A, Cinti S (2024) An electrochemical biosensor for on-glove application: organophosphorus pesticide detection directly on fruit peels. Talanta 283:127093
34. Herrera-Domínguez M, Morales-Luna G, Mahlknecht J, Cheng Q, Aguilar-Hernández I, Ornelas-Soto N (2023) Optical biosensors and their applications for the detection of water pollutants. Biosensors 13(3):370
35. Liu B, Tang Y, Yang Y, Wu Y (2021) Design an aptamer-based sensitive lateral flow biosensor for rapid determination of isocarbophos pesticide in foods. Food Control 129:108208
36. Bala R, Mittal S, Sharma RK, Wangoo N (2018) A supersensitive silver nanoprobe based aptasensor for lowcost detection of malathion residues in water and food samples. Spectrochim Acta Part A Mol Biomol Spectrosc 196:268–273
37. MirresACM SBEPMD, Tessaro L, GalvanD AJC, AquinoA JN, Conte-Junior CA (2022) Recent advances in nanomaterial-based biosensors for pesticide detection in foods. Biosensors 12(8):572
38. Sahu B, Kurrey R, Deb MK, Khalkho BR, Manikpuri S (2023) Recognition of malathion pesticides in agricultural samples by using α-CD functionalized gold nanoparticles as a colorimetric sensor. Talanta 259:124526
39. Liu J, Song S, Wu A, Kuang H, Liu L, Xiao J, Xu C (2021) Development of immunochromatographic strips for the detection of dicofol. Analyst 146:2240–2247
40. Berkal MA, Nardin C (2023) Pesticide biosensors: trends and progresses. Anal Bioanal Chem 415:5899–5924
41. Matter IA, Darwesh OM, Matter HA (2021) Nanosensors for herbicides monitoring in soil. In: Nanomaterials for soil remediation. Elsevier, pp 221–237
42. Wang X, Lu X, Chen J (2014) Development of biosensor technologies for analysis of environmental contaminants. Trends Environ Anal Chem 2:25–32
43. Saini RK, Bagri LP, Bajpai AK (2017) Smart nanosensors for pesticide detection. In: New pesticides and soil sensors. Academic Press, pp 519–559

44. Whangsuk W, Dubbs JM, Sallabhan R, Somsongkul K, Mongkolsuk S, Loprasert S (2010) ChpR is a chlorpyrifos-responsive transcription regulator in Sinorhizobiummeliloti. J Mol Microbiol Biotechnol 18(3):141–147
45. De Carvalho CC (2011) Enzymatic and whole cell catalysis: finding new strategies for old processes. Biotechnol Adv 29(1):75–83
46. Gissawong N, Srijaranai S, Nanan S, Mukdasai K, Uppachai P, Teshima N, Mukdasai S (2022) Electrochemical detection of methyl parathion using calix [6] arene/bismuth ferrite/multiwall carbon nanotube-modified fluorine-doped tin oxide electrode. Microchim Acta 189(12):461
47. Paneru S, Kumar D (2023) A novel electrochemical biosensor based on polyaniline-embedded copper oxide nanoparticles for high-sensitive paraoxon-ethyl (PE) detection. Appl Biochem Biotechnol 195(7):4485–4502
48. Della Pelle F, Angelini C, Sergi M, Del Carlo M, Pepe A, Compagnone D (2018) Nano carbon black-based screen printed sensor for carbofuran, isoprocarb, carbaryl and fenobucarb detection: application to grain samples. Talanta 186:389–396
49. Lu Y, Tan Y, Xiao Y, Li Z, Sheng E, Dai Z (2021) A silver @ gold nanoparticle tetrahedron biosensor for multiple pesticides detection based on surface-enhanced Raman scattering. Talanta 234:122585
50. Bilal S, Hassan MM, Ur Rehman MF, Nasir M, Sami AJ, Hayat A (2021) An insect acetylcholinesterase biosensor utilizing WO_3/g-C_3N_4 nanocomposite modified pencil graphite electrode for phosmet detection in stored grains. Food Chem 346:128894
51. Thongsomboon W, Sonjai J, Jakmunee J, Lerdsri J, Reanpang P, Upan J (2024) Manganese oxide@ nanocellulose modified poster paper-based electrode as a novel electrochemical sensor for sensitive determination of paraquat. J Appl Electrochem 54(5):1085–1094
52. Li Y, Zhao R, Wang M, Han G (2023) Efficient immobilization of acetylcholinesterase onto amino functionalized nitrogen-doped reduced graphene oxide for highly sensitive pesticides biosensors. ChemistrySelect 8(32):e202300917
53. Li Y, Zhao R, Shi L, Han G, Xiao Y (2017) Acetylcholinesterase biosensor based on electrochemically inducing 3D graphene oxide network/multi-walled carbon nanotube composites for detection of pesticides. RSC Adv 7(84):53570–53577
54. Aghoutane Y, Bari NE, Laghrari Z, Bouchikhi B (2021) Electrochemical detection of fenthion insecticide in olive oils by a sensitive non-enzymatic biomimetic sensor enhanced with metal nanoparticles. Chem Proc 5(1):64
55. Aghoutane Y, Diouf A, Österlund L, Bouchikhi B, El Bari N (2020) Development of a molecularly imprinted polymer electrochemical sensor and its application for sensitive detection and determination of malathion in olive fruits and oils. Bioelectrochemistry 132:107404
56. Dey B, Ahmad MW, Syed A, Bahkali AH, Verma M, Choudhury A (2024) Iron metal organic framework decorated carbon nanofiber-based electrode for electrochemical sensing platform of chlorpyrifos in fruits and vegetables. Mater Sci Semicond Process 181:108669
57. Talan A, Mishra A, Eremin SA, Narang J, Kumar A, Gandhi S (2018) Ultrasensitive electrochemical immuno-sensing platform based on gold nanoparticles triggering chlorpyrifos detection in fruits and vegetables. Biosens Bioelectron 105:14–21
58. Pratibha Kapoor A, Rajput JK, Kumar A (2022) Dualistic fluorescence as well as portable smartphone-assisted RGB-relied sensing assay for the ultra-sensitive determination of pendimethalin in food and water samples by AIEE active organic probes. Anal Chem 94(50):17685–17691
59. Bala R, Swami A, Tabujew I, Peneva K, Wangoo N, Sharma RK (2018) Ultra-sensitive detection of malathion using quantum dots-polymer based fluorescence aptasensor. Biosens Bioelectron 104:45–49
60. Guo Y, Liu R, Liu Y, Xiang D, Liu Y, Gui W, Li M, Zhu G (2018) A non-competitive surface plasmon resonance immunosensor for rapid detection of triazophos residue in environmental and agricultural samples. Sci Total Environ 613:783–791
61. Zhu J, Yin X, Zhang W, Chen M, Feng D, Zhao Y, Zhu Y (2023) Simultaneous and sensitive detection of three pesticides using a functional poly (sulfobetaine methacrylate)-coated paper-based colorimetric sensor. Biosensors 13(3):309

62. Gai P, Pu L, Wang C, Zhu D, Li F (2023) CeO_2@NC nanozyme with robust dephosphorylation ability of phosphotriester: a simple colorimetric assay for rapid and selective detection of paraoxon. Biosens Bioelectron 220:114841
63. Chen G, Jin M, Ma J, Yan M, Cui X, Wang Y, Zhang X, Li H, Zheng W, Zhang Y (2020) Competitive bio-barcode immunoassay for highly sensitive detection of parathion based on bimetallic nanozyme catalysis. J Agric Food Chem 68:660–668
64. Hegde V, Bhat MP, Lee JH, Kurkuri MD, Kim CS, Lee KH (2024) Carbon-based nanomaterials: multifaceted role in agrochemical recognition, remediation, and release. Nano Today 57:102388
65. Yang M, Xiao L, Chen WT, Deng X, Hu G (2024) Recent advances on metal-organic framework-based electrochemical sensors for determination of organic small molecules. Talanta 280:126744
66. Sireesha M, Jagadeesh Babu V, Kranthi Kiran AS, Ramakrishna S (2018) A review on carbon nanotubes in biosensor devices and their applications in medicine. Nanocomposites 4:36–57
67. Dhull V (2018) Fabrication of AChE/SnO_2-cMWCNTs/Cu nanocomposite-based sensor electrode for detection of methyl parathion in water. Int J Anal Chem 2018:2874059
68. Baruah A, Newar R, Das S, Kalita N, Nath M, Ghosh P, Chinnam S, Sarma H, Narayan M (2024) Biomedical applications of graphene-based nanomaterials: recent progress, challenges, and prospects in highly sensitive biosensors. Discover Nano 19(1):103
69. ChakravartyM VA (2021) Nanotechnology-based antiviral therapeutics. Drug Deliv Transl Res 11:748–787
70. ArduiniF CS, Scognamiglio V, Moscone D (2016) Nanomaterials in electrochemical biosensors for pesticide detection: advances and challenges in food analysis. Microchim Acta 183:2063–2083
71. Kaur N, Thakur H, Kumar R, Prabhakar N (2016) An electrochemical sensor modified with poly(3,4-ethylenedioxythiophene)-wrapped multi-walled carbon nanotubes for enzyme inhibition-based determination of organophosphates. Microchim Acta 183:2307–2315
72. Kaur N, Bhatnagar A, Bhalla A, Prabhakar N (2021) Determination of an organophosphate pesticide using antibody immobilised hybrid nanocomposites. Int J Environ Anal Chem 101(11):1485–1498
73. ChenD FuJ, Liu Z, Guo Y, Sun X, Wang X, Wang Z (2017) A Simple acetylcholinesterase biosensor based on ionic liquid/multiwalled carbon nanotubes-modified screen-printed electrode for rapid detecting chlorpyrifos. Int J Electrochem Sci 12:9465–9477
74. Han L, Chen D, Li F (2018) Rational integration of biomineralization, microbial surface display, and carbon nanocomposites: ultrasensitive and selective biosensor for traces of pesticides. Adv Mater Interfaces 5:1801332
75. LiaoC LY, Tjong SC (2018) Graphene nanomaterials: synthesis, biocompatibility, and cytotoxicity. Int J Mol Sci 19:3564
76. Razaq A, Bibi F, ZhengX PR, Hassan S, Jafri M (2022) Oxide-based flexible composites: from fabrication to applications. Materials 15:1012
77. Kazemifard N, Ensafi AA, Dehkordi ZS (2021) A review of the incorporation of QDs and imprinting technology in optical sensors-imprinting methods and sensing responses. New J Chem 45:10170–10198
78. Farzin MA, Abdoos H (2021) A critical review on quantum dots: from synthesis toward applications in electrochemical biosensors for determination of disease-related biomolecules. Talanta 224:121828
79. Wang Z, Yao B, Xiao Y, Tian X, Wang Y (2023) Fluorescent quantum dots and its composites for highly sensitive detection of heavy metal ions and pesticide residues: a Review. Chemosensors 11(7):405
80. Peihua M, Xiaoxue J, Yiyang H, Junheng T, Wang Q, Wei C-I (2024) Recent progress of quantum dots for food safety assessment: a review. Trends Food Sci Technol 143:104310
81. Lin Z, Liu X, Li Y, Li C, Yang L, Ma K, Zhang Z, Huang H (2021) Electrochemical aptasensor based on Mo_2C/Mo_2N and gold nanoparticles for determination of chlorpyrifos. Microchim Acta 188:170

82. Yang G, He Y, Zhao J, Chen S, Yuan R (2021) Ratiometric electrochemiluminescence biosensor based on Ir nanorods and CdS quantum dots for the detection of organophosphorus pesticides. Sens Actuators B Chem 341:130008
83. Korram J, Dewangan L, Karbhal I, Nagwanshi R, Vaishanav SK, Ghosh KK, Satnami ML (2020) CdTe QD-based inhibition and reactivation assay of acetylcholinesterase for the detection of organophosphorus pesticides. RSC Adv 10:24190–24202.
84. Nguyen HA, Mai QD, Ngo XD, Tran QH, Doan QT, Le TNL, Bui VH, Anh-Tuan L (2022) Gold nanoparticle-based optical nanosensors for food and health safety monitoring: recent advances and future perspectives. R Soc Chem 12:10950–10988
85. Zhao G, Zhou B, Wang X, Shen J, Zhao B (2021) Detection of organophosphorus pesticides by nanogold/mercaptomethamidophos multi-residue electrochemical biosensor. Food Chem 354:129511
86. HouL ZX, KongM JG, Sun Y, MoW LT, Ye F, Zhao S (2020) A competitive immunoassay for electrochemical impedimetric determination of chlorpyrifos using a nanogold-modified glassy carbon electrode based on enzymatic biocatalytic precipitation. Microchim Acta 187:204
87. LiuB TY, YangY WuY (2021) Design an aptamer-based sensitive lateral flow biosensor for rapid determination of isocarbophos pesticide in foods. Food Control 129:108208
88. Sequeira CA (2021) Silver nanoparticles for hydrogen peroxide sensors. Biomed J Sci Tech Res 40(4):32438–32443
89. Hermanto D, Ismillayli N, Muliasari H, Wirawan R, Kamali SR (2024) Silver-based plasmonic nanoparticles for biosensor organophosphate pesticides using a single film containing acetylcholinesterase/choline oxidase. Glob J Environ Sci Manag 10(1):1–12
90. Dhony H, Nurul I, Saprini H, Siti RK, Rahadi W, Handa M, Rochmad KS (2024) Inhibitive determination of organophosphate pesticides using acetylcholinesterase and silver nanoparticle as colorimetric probe. Environ Eng Res 29(4):230503
91. ZhangC QM, Wang J, Liu Y (2023) Recent advances in nanoparticle-based optical sensors for detection of pesticide residues in soil. Biosensors 13(4):415
92. Zheng Q, Yu Y, FanK JF, WuJ YingY (2016) A nano-silver enzyme electrode for organophosphorus pesticide detection. Anal Bioanal Chem 408:5819–5827
93. Turan J, Kesik M, Soylemez S, Goker S, Coskun S, Unalan HE, Toppare L (2016) An effective surface design based on a conjugated polymer and silver nanowires for the detection of paraoxon in tap water and milk. Sens Actuators B Chem 228:278–286
94. Yang L, Sun H, Wang X, Yao W, Zhang W, Jiang L (2019) An aptamer based aggregation assay for the neonicotinoid insecticide acetamiprid using fluorescent upconversion nanoparticles and DNA functionalized gold nanoparticles. Microchim Acta 186:1–11
95. Ma L, Zhou L, He Y, Wang L, Huang Z, Jiang Y, Gao J (2018) Hierarchical nanocomposites with an N-doped carbon shell and bimetal core: novel enzyme nanocarriers for electrochemical pesticide detection. Biosens Bioelectron 121:166–173
96. Chen G, Jin M, Ma J, Yan M, Cui X, Wang Y, Wang J (2019) Competitive bio-barcode immunoassay for highly sensitive detection of parathion based on bimetallic nanozyme catalysis. J Agric Food Chem 68(2):660–668
97. Xu G, Hou J, Zhao Y, Bao J, Yang M, Fa H, Hou C (2019) Dual-signal aptamer sensor based on polydopamine-gold nanoparticles and exonuclease I for ultrasensitive malathion detection. Sens Actuators B Chem 287:428–436
98. Zare AR, Ensafi AA, Rezaei B (2019) Animpedimetric biosensor based on poly (l-lysine)-decorated multiwall carbon nanotubes for the determination of diazinon in water and fruits. J Iran Chem Soc 16(12):2777–2785
99. Mahmoudi E, Fakhri H, Hajian A, Afkhami A, Bagheri H (2019) High-performance electrochemical enzyme sensor for organophosphate pesticide detection using modified metal-organic framework sensing platforms. Bioelectrochemistry 130:107348
100. da Silva MK, Vanzela HC, Defavari LM, Cesarino I (2018) Determination of carbamate pesticide in food using a biosensor based on reduced graphene oxide and acetylcholinesterase enzyme. Sens Actuators B Chem 277:555–561

101. Yi J, Liu Z, Liu J, Liu H, Xia F, Tian D, Zhou C (2020) A label-free electrochemical aptasensor based on 3D porous CS/rGO/GCE for acetamiprid residue detection. Biosens Bioelectron 148:111827
102. Apilux A, Siangproh W, Insin N, Chailapakul O, Prachayasittikul V (2017) Paper based thioglycolic acid (TGA)-capped CdTe QD device for rapid screening of organophosphorus and carbamate insecticides. Anal Methods 9(3):519–527
103. Cheng Y, Lai OM, Tan CP, Panpipat W, Cheong LZ, Shen C (2021) Proline-modified UIO-66 as nanocarriers to enhance Candida rugosa lipase catalytic activity and stability for electrochemical detection of nitrofen. ACS Appl Mater Interfaces 13(3):4146–4155
104. Wu J, Yang Q, Li Q, Li H, Li F (2021) Two-dimensional MnO_2 nanozyme-mediated homogeneous electrochemical detection of organophosphate pesticides without the interference of H_2O_2 and color. Anal Chem 93:4084–4091

Nanobiosensors for Detection of Heavy Metals in Food

Sahar Akhavan-Mahdavi and Danial Dehnad

Abstract Heavy metal contamination in food is a serious health concern due to the toxic nature of metals such as lead (Pb), mercury (Hg), cadmium (Cd), and arsenic (As) and their tendency to accumulate in the human body over time. Meanwhile, nanobiosensors, known for their high sensitivity, selectivity, and fast response, offer an advanced solution for the detection of small levels of these contaminants in food. This chapter provides a comprehensive overview of the latest developments in nanobiosensor technology for the detection of heavy metals in various food matrices. It begins by examining the sources and health effects of heavy metal contamination in the food supply chain. Then, the principles and mechanisms underlying nanobiosensors are reviewed, emphasizing their advantages over conventional detection methods. This chapter highlights key advances in the functionalization of nanomaterials that enhance the sensitivity and selectivity of these sensors. Practical applications are demonstrated through case studies of heavy metal detection in real food samples. Finally, this chapter addresses current challenges and future perspectives, including regulatory considerations and the integration of nanobiosensors into food safety monitoring systems.

Keywords Nanobiosensors · Heavy metal detection · Food safety · Toxicity monitoring · Nanomaterials functionalization

---

S. Akhavan-Mahdavi (✉)
Department of Food Materials and Process Design Engineering, Gorgan University of Agricultural Science and Natural Resources, Gorgan, Iran
e-mail: akhavan.sahar@gmail.com

D. Dehnad
Department of Food Nanotechnology, Research Institute of Food Science and Technology (RIFST), Mashhad, Iran

D. Sillu et al. (eds.), *Nanobiosensors for the Food Industry*, Smart Nanomaterials Technology, https://doi.org/10.1007/978-981-95-0136-6_8

1 Introduction

Heavy metal contamination in food has become a pressing global concern due to its potential risks to human health and the environment. Heavy metals, including lead (Pb), mercury (Hg), cadmium (Cd), and arsenic (As), can enter the food chain through various environmental pathways such as contaminated water, soil, and air, as well as through food processing and packaging [1]. These metals are non-biodegradable and can accumulate in living organisms, leading to toxic effects even at low exposure levels. The impact of heavy metal contamination in food is far-reaching, as it affects not only human health but also food security and safety regulations worldwide [2]. The presence of heavy metals in food has been linked to numerous adverse health outcomes, ranging from neurodevelopmental issues and cognitive impairment to organ damage and carcinogenicity. Vulnerable populations, such as children, pregnant women, and individuals with weakened immune systems, are particularly susceptible to the toxic effects of heavy metals. As a result, regulatory agencies such as the World Health Organization (WHO), the U.S. Food and Drug Administration (FDA), and the European Union (EU) have established stringent limits on the permissible levels of these contaminants in food products [3].

Recent advancements in nanotechnology have paved the way for the development of nanobiosensors, which offer a promising solution for the rapid, sensitive, and accurate detection of heavy metals in food. Nanobiosensors utilize nanoscale materials to enhance the sensitivity of traditional biosensors, allowing for the detection of even trace amounts of heavy metals in complex food matrices. These innovative devices hold great potential for ensuring food safety and protecting public health by providing real-time monitoring of contaminants [4].

This chapter focuses on the application of nanobiosensors in detecting heavy metals in food. It aims to explore the principles behind these sensors, their advantages over conventional methods, and their practical applications in ensuring food safety. By integrating nanotechnology with biosensor design, these tools offer enhanced performance in terms of sensitivity and accuracy, which are crucial for the effective monitoring of contaminants in various food products.

2 Heavy Metals and Associated Concerns

Heavy metals are defined as elements with an atomic weight exceeding 63.5 and a specific gravity greater than 5.0, which pose significant risks to both human health and the environment. Major heavy metals include lead (Pb), cadmium (Cd), cobalt (Co), chromium (Cr), copper (Cu), iron (Fe), arsenic (As), nickel (Ni), zinc (Zn), and mercury (Hg). The expansion of industrial activities, excessive use of chemicals in agriculture, and increased vehicular traffic have contributed to substantial ecological imbalances which impact food safety, food resources, and consumer well-being [2].

Heavy metals can originate from both natural sources, such as rocks, and anthropogenic activities, including metal mining, smelting, waste disposal, and pesticide use. The industrial revolution and economic globalization have led to a marked increase in environmental contaminants, which persist in the environment without dissolving, resulting in long-term accumulation. Consequently, food security issues related to these contaminants have become a global apprehensiveness [5].

Soil serves as the foundation for food crops and can be significantly contaminated by heavy metals from both point sources (e.g., energy-intensive industries like thermal power plants and coal mines, and chemical industries such as gold-mines, smelting, electroplating, textiles, leather, and e-waste processing) and non-point sources (e.g., soil erosion, agricultural runoff, and open freight storage). Heavy metals not only impact human health but also adversely affect soil biota, including beneficial insects, invertebrates, and mammals [1, 6]. Many medicinal plants and greenhouse vegetables have been found to accumulate metals when grown near industrial areas [7, 8].

Heavy metals exert their toxicity primarily through their ability to interfere with biological processes by forming stable bonds with cellular macromolecules, such as proteins and enzymes. Upon entering the body, these metals can undergo oxidation and accumulate in tissues, disrupting normal cellular functions [9]. For instance, they may inhibit enzyme activity by binding to active sites, impairing metabolic pathways and biochemical reactions. Heavy metals like lead and cadmium can mimic essential metal ions, interfering with the function of vital enzymes and proteins, and causing oxidative stress through the generation of reactive oxygen species (ROS) [10]. This oxidative stress can lead to cellular damage, inflammation, and apoptosis. Additionally, heavy metals can accumulate in organs such as the liver, kidneys, and brain, where they contribute to chronic health issues including neurotoxicity, renal dysfunction, and carcinogenicity. Their persistence in the body and their potential to disrupt numerous physiological processes underscore the severity of their toxic effects [3].

The health concerns associated with heavy metals arise from their potential toxicity and ability to accumulate in biological systems. Table 1 summarizes the health concerns associated with different heavy metals commonly found in food.

3 Heavy Metals in Food

In many regions, the levels of heavy metals in food and beverages do not pose significant risks to human health. However, increasing environmental contamination necessitates continuous monitoring to protect public health. Table 2 reviews existing studies on heavy metal contamination in common dietary staples, including fruits and vegetables, milk and dairy products, meat and derivatives, edible oils, and alcoholic beverages. These food products are key components of daily diets in many regions, and their heavy metal concentrations can vary significantly due to various factors [2].

Table 1 Health concerns associated with various heavy metals

Heavy metal	Health concerns	References
Lead (Pb)	• Carcinogenesis, mutagenesis, teratogenesis • Anemia, central nervous system disorders • Renal failure, convulsions, death	[11]
Cadmium (Cd)	• Osteomalacia (soft bones) • Hypertension • Renal dysfunction • Slow excretion, long-term accumulation effects	[12]
Copper (Cu)	• Digestive disturbances • Hemolysis • Neurological issues • Conditions like Menkes and Wilson's disease	[13]
Zinc (Zn)	• Digestive and metabolic disorders • Paralysis of the central nervous system • Anemia, growth stagnation, comatose states	[14]
Arsenic (As)	• Cardiovascular diseases • Reproductive problems • Cancer, diabetes • Neurological effects, especially in children	[15]
Mercury (Hg)	• Neurological damage • Tremors, sleep disturbances, irritability • Cognitive impairments, developmental problems in children	[16]
Chromium (Cr)	• Allergic reactions • Respiratory and gastrointestinal cancers • Disruption of trace element absorption • Organ damage	[17]
Cobalt (Co)	• Allergic contact dermatitis • Respiratory issues • Cardiovascular diseases • Severe respiratory conditions in industrial settings	[18]
Nickel (Ni)	• Headaches, vertigo • Respiratory disorders • Carcinogenic effects • Skin allergies, cardiovascular diseases	[19]

Fruits and vegetables play a vital role in health maintenance and disease prevention but can accumulate both essential and toxic heavy metals due to factors like soil contamination, agricultural practices, and environmental conditions [20]. Studies have shown varying levels of heavy metals in these foods, particularly in areas with high industrial activity. Commonly tested foods include spinach, lettuce, apples, bananas, grapes, and berries, with some findings exceeding Codex Alimentarius standards. For instance, lettuce from the USA and Romania showed high levels of arsenic and zinc while Egyptian and Chinese samples had significant concentrations of nickel and copper, respectively [21, 22]. Milk and dairy products, essential in many diets, also show variable heavy metal concentrations, influenced by factors such as animal

Table 2 Summary of heavy metal concentrations in various food products

Analyzed product	Identified heavy metal	Concentrations	References
Raw cow milk	Hg^{2+}	0.0014 mg/kg	[30]
	As^{3+}	0.012 mg/kg	
	Pb^{2+}	0.1016 mg/kg	
	Cd^{2+}	0.07 mg/kg	
	Cr^{2+}	0.1044 mg/kg	
	Cu^{2+}	0.0656 mg/kg	
Milk	As^{3+}	1.01 μg/kg	[31]
	Al^{2+}	71.89 μg/kg	
	Ni^{2+}	98.53 μg/kg	
	Cd^{2+}	nd*	
	Pb^{2+}	0.85 μg/kg	
Full fat UHT milk	As^{3+}	2.33 μg/kg	[32]
	Cd^{2+}	5.00 μg/kg	
	Pb^{2+}	2.66 μg/kg	
	Hg^{2+}	3.66 μg/kg	
Halloumi cheese	As^{3+}	12.33 μg/kg	[32]
	Cd^{2+}	44.33 μg/kg	
	Pb^{2+}	35.33 μg/kg	
	Cu^{2+}	591.33 μg/kg	
	Hg^{2+}	15.00 μg/kg	
Full-fat yogurt	As^{3+}	6.00 μg/kg	[32]
	Cd^{2+}	36.33 μg/kg	
	Pb^{2+}	3.00 μg/kg	
	Cu^{2+}	15.66 μg/kg	
	Hg^{2+}	6.00 μg/kg	
Cream with 10% milk fat	As^{3+}	0.019 mg/kg	[33]
		0.029 mg/kg (over 70% milk fat)	
Pork meat products	Cr^{2+}	0.15–0.23 mg/kg	[34]
	Cd^{2+}	0.01–0.03 mg/kg	
	Hg^{2+}	0.01–0.02 mg/kg	
	Cu^{2+}	1.08–1.21 mg/kg	
	Pb^{2+}	0.22–0.38 mg/kg	
	Hg^{2+}	0.003 mg/kg	
Bacon	Pb^{2+}	0.58 mg/kg	[35]
	Cd^{2+}	0.11 mg/kg	
	Cu^{2+}	1.02 mg/kg	

(continued)

Table 2 (continued)

Analyzed product	Identified heavy metal	Concentrations	References
	Zn^{2+}	42.1 mg/kg	
Ham	Pb^{2+}	0.65 mg/kg	[35]
	Cd^{2+}	0.13 mg/kg	
	Cu^{2+}	0.73 mg/kg	
	Zn^{2+}	33.5 mg/kg	
Salami	Pb^{2+}	0.96 mg/kg	[35]
	Cd^{2+}	0.21 mg/kg	
	Cu^{2+}	1.32 mg/kg	
	Zn^{2+}	32.19 mg/kg	
Sausage	Pb^{2+}	0.82 mg/kg	[35]
Beer	Cu^{2+}	0.0368 mg/L	[36]
	Cd^{2+}	0.0014 mg/L	
	Mn^{2+}	0.0954 mg/L	
	Pb^{2+}	0.006 mg/L	
	Zn^{2+}	1.5206 mg/L	
	Cd^{2+}	nd–0.00058 mg/L	[37]
	Pb^{2+}	nd–0.0133 mg/L	
	Cd^{2+}	nd–0.0143 mg/L	[38]
	Pb^{2+}	0.013–0.0329 mg/L	[39]
	Zn^{2+}	0.0527–0.226 mg/L	
White wine	V^{2+}	1 μg/L	[40]
	Cr^{2+}	3 μg/L	
	Mn^{2+}	5 μg/L	
	Fe^{2+}	61 μg/L	
	Ni^{2+}	6 μg/L	
	Cu^{2+}	6 μg/L	
	Zn^{2+}	59 μg/L	
	As^{3+}	nd	
	Pb^{2+}	30 μg/L	
	Pb^{2+}	0.85 μg/kg	

* nd: Not determined

feed, water quality, and cow breed. Studies have found elevated levels of metals like lead, cadmium, and copper, with Holstein Friesian cows exhibiting higher concentrations compared to Simmental cows [23]. Seasonal variations and environmental pollutants further affect these levels, with some findings surpassing Codex Alimentarius limits. Despite these concerns, milk and dairy products are generally not major sources of copper and zinc in the diet [24, 25]. Heavy metals in meat products often originate from contaminated feeds, particularly in industrialized regions. Research indicates higher metal concentrations in processed meats like salami and sausages compared to less processed options [26]. Although metal levels are usually within permissible limits, there are concerns about cumulative health effects. For example, lead and cadmium levels have exceeded limits in red meat from Asia and Africa, and varying metal concentrations have been observed in Chinese meats [27]. Vegetable oils can also contain trace metals that may accelerate oxidation, with contamination from soil and manufacturing processes being key contributors. While edible oils generally have high metal content in seeds, most metals are not transferred to the final oils [28]. Lastly, heavy metals in alcoholic beverages, including beer and wine, often stem from agricultural pesticides and production processes. Research shows that while some beverages may have detectable metal levels, they generally do not pose significant health risks, though methods like ion exchange and reverse osmosis can help reduce metal content [2, 29].

4 Current Detection Methods for Heavy Metals

The detection of heavy metals in food has traditionally relied on several sophisticated techniques, each offering a combination of strengths and limitations. Conventional methods such as atomic absorption spectroscopy (AAS), inductively coupled plasma mass spectrometry (ICP-MS), and X-ray fluorescence (XRF) are commonly used due to their sensitivity and reliability in measuring metal concentrations. However, they are often expensive, require skilled operators, and involve complex sample preparation steps, limiting their accessibility in resource-limited settings [41].

Inductively Coupled Plasma Atomic Emission Spectrometry (ICP-AES) is a widely used technique for detecting heavy metals, known for its swift analysis speed, high sensitivity, and accuracy. It allows for multi-element detection across a wide measurement range, making it an excellent tool for heavy metal analysis. However, it requires significant investment in equipment, and its accuracy may be compromised when analyzing solid samples without converting them to a liquid form [42]. Flame Atomic Absorption Spectrometry (FAAS) and Graphite Furnace Atomic Absorption Spectrometry (GF-AAS) are frequently employed in analytical laboratories due to their cost-effectiveness and ability to detect trace levels of metals such as cobalt (Co), cadmium (Cd), copper (Cu), iron (Fe), manganese (Mn), nickel (Ni), lead (Pb), and zinc (Zn). FAAS is especially noted for its low operational costs, but it is limited by its single-element detection capability and narrow linear response range. While FAAS and GF-AAS offer economic benefits, their detection capabilities are restricted to individual elements, require larger sample volumes, and enforce the use of flammable gases [43]. Inductively Coupled Plasma Mass Spectrometry (ICP-MS) is regarded as one of the most powerful techniques for analyzing trace heavy metals, with advantages such as multi-element detection, high sensitivity, and straightforward sample preparation. However, its high setup and maintenance costs, coupled with the need for specialized operators and multiple high-purity gases, present significant operational challenges [44]. These factors make ICP-MS less accessible for routine employment in smaller laboratories or low-resource settings [45]. Other techniques like Microwave Induced Plasma Optical Emission Spectrometry (MIP OES) and Electrochemical Methods are also employed, though their utility is often determined by the specific requirements of the analysis. For example, X-ray fluorescence (XRF) is useful for simultaneously analyzing multiple metals without sample digestion but may lack the sensitivity needed for trace-level detection. Chromatographic methods, such as Liquid Chromatography Coupled with Mass Spectrometry (LC-MS/MS) and Gas Chromatography-Tandem Mass Spectrometry (GC-MS), offer high precision but face limitations like high operational costs and low sample throughput. These methods are valuable for their specificity and accuracy, yet they require sophisticated infrastructure, limiting their widespread adoption [5].

Despite the trustworthiness and delicacy of traditional methods, they come with notable drawbacks such as high costs, complexity, and the need for specialized equipment and expertise. Furthermore, sample throughput is often slow, and these techniques are generally unsuitable for real-time, on-site detection. The emergence of biosensors and other novel technologies has gained momentum due to their compact size, affordability, and ease of use [46]. Biosensors, in particular, offer rapid detection and hold promise for widespread application in heavy metal analysis, providing a balance between cost and efficiency that conventional methods struggle to achieve. The limitations of these conventional methods underscore the growing need for innovative, cost-effective, and portable alternatives like nanobiosensors, which have the potential to revolutionize heavy metal detection by offering real-time, on-site monitoring with high sensitivity, even in complex food systems [47].

5 Application of Nanobiosensors for Heavy Metal Detection

Nanobiosensors represent a groundbreaking advancement in the detection of heavy metals, particularly for ensuring food safety. These devices integrate nanomaterials with biological recognition elements to create highly sensitive systems capable of identifying trace levels of contaminants. Compared to conventional techniques, nanobiosensors offer numerous advantages, such as lower detection limits, faster response times, and portability, making them ideal for on-site testing in food production and supply chains [48]. A nanobiosensor comprises two primary components: nanomaterials and biorecognition elements. The nanomaterials, such as gold nanoparticles (AuNPs), carbon nanotubes (CNTs), and quantum dots (QDs), provide a high surface-to-volume ratio, significantly enhancing sensitivity and enabling real-time detection at the nanoscale. On the other hand, biorecognition elements, including enzymes, antibodies, or DNA sequences, interact selectively with specific heavy metals, triggering a detectable signal. This combination allows nanobiosensors to detect minute concentrations of heavy metals with remarkable precision [49].

Nanobiosensors employ various detection mechanisms tailored to specific measurements. Electrochemical detection, for instance, monitors changes in electrical properties like current or resistance when heavy metals such as lead and mercury bind to the sensor, offering high sensitivity at trace levels [50]. Optical detection measures alterations in light properties, such as fluorescence or absorbance, often utilizing quantum dots for their fluorescent capabilities to detect metal interactions. Piezoelectric detection, on the other hand, gauges changes in mass or mechanical stress caused by metal binding, providing highly accurate results without the need for labels or secondary reactions [51]. Nanobiosensors generally exhibit greater sensitivity and specificity compared to traditional techniques like ICP-MS and AAS. While conventional methods provide accurate results, they require expensive instrumentation, skilled operators, and extensive sample preparation. Nanobiosensors, by contrast, can achieve detection limits as low as parts per billion (ppb) or parts per trillion (ppt) in a more user-friendly and accessible format. Furthermore, nanobiosensors offer quicker detection times, often in minutes, and are suitable for continuous monitoring, making them ideal for real-time applications in food safety [52].

Nanobiosensors are increasingly being used in real-world food safety applications to monitor contamination levels in various food products, including fish, vegetables, dairy, and grains. They provide a rapid, cost-effective, and portable method for detecting heavy metals, thus ensuring compliance with regulatory standards. In agricultural settings, these sensors have been deployed to monitor soil and water contamination, helping farmers reduce the risk of heavy metal uptake in crops. Additionally, food manufacturers are exploring the use of nanobiosensors in packaging systems to offer consumers real-time information about potential contaminants, enhancing food safety along the supply chain [53, 54].

6 Principle of Electrochemical Sensing of Heavy Metals

Electrochemical sensing typically employs a three-electrode setup consisting of a working electrode (WE), a reference electrode (RE), and a counter electrode (CE) (Fig. 1). The working electrode can be modified with various materials to enhance its sensitivity and detection limits [55]. The fundamental operation of electrochemical sensors involves the conversion of chemical signals into electrical signals through a potentiostat, which measures the potential difference [56]. During the measurement process, heavy metals undergo oxidation or reduction by either gaining or losing electrons at the working electrode surface. This electrochemical reaction generates a current proportional to the concentration of the heavy metals, allowing the electrochemical workstation to determine their concentration [57]. The performance of the sensing electrode/working electrode is critical to the overall detection capability. Consequently, researchers have developed various modified electrodes using a range of interface materials, including inorganic substances (such as metal and metal oxide nanoparticles, graphene, and carbon nanotubes), organic compounds (such as

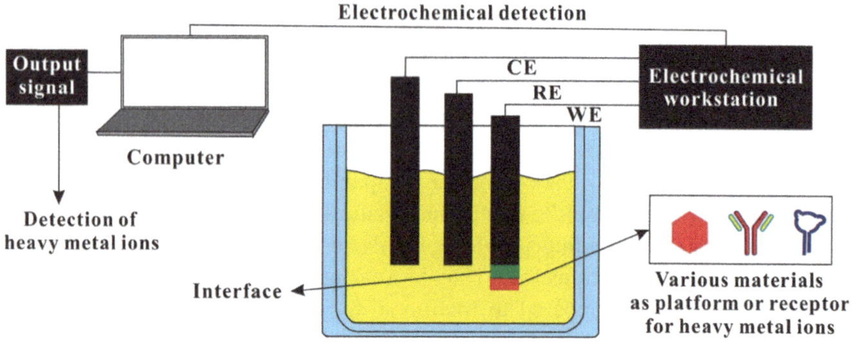

Fig. 1 Schematic of the general principle of electrochemical sensing of heavy metals [55]

metal-organic frameworks), and biomaterials. These modifications aim to enhance the sensitivity, reliability, and detection limits of heavy metal sensors [58].

6.1 Inorganic Nanomaterials

Inorganic nanomaterials are commonly applied to modify sensing electrodes owing to their advantageous properties such as good thermal stability, large specific surface area, strong adsorption ability, and high catalytic efficiency. These nanomaterials typically include metals, metal oxides, and carbon-based materials [55]. Metal nanomaterials exhibit unique physical and chemical properties including large surface areas, uniform particle sizes, and excellent electrical, optical, and catalytic characteristics. These properties enhance the sensitivity, linear range, and selectivity of electrochemical sensors for heavy metal detection. Gold (Au), silver (Ag), bismuth (Bi), and platinum (Pt) are frequently used metals in electrode modification, with Au nanoparticles (AuNPs) being particularly prominent thanks to their outstanding electrical conductivity, heat conductivity, chemical stability, and multifunctionality [59]. AuNPs are effective in detecting a range of heavy metals, such as Cd^{2+}, Pb^{2+}, Hg^{2+}, Cu^{2+}, and Zn^{2+}, with sensitivity levels that surpass recommended detection limits by WHO and national standards. For instance, a fluorinated graphene/gold nanocage (FGP/AuNC) composite has been used to detect multiple heavy metals in food with high sensitivity. Additionally, AuNPs have been employed in sensors to detect Pb^{2+} with a detection limit as low as 312 pmol L^{-1} in water [60].

Metal oxide nanomaterials are favored for their electronic properties, strong electron transfer capacity, and semiconducting nature. Common metal oxides include iron oxides (Fe_3O_4, Fe_2O_3), titanium dioxide (TiO_2), zinc oxide (ZnO), manganese dioxide (MnO_2), magnesium oxide (MgO), and cobalt oxides (Co_3O_4, Co_2O_3). Among these, Fe_3O_4 nanoparticles are particularly notable for their effectiveness in sensing applications. For example, Fe_3O_4 nanoparticles combined with room temperature ionic liquids (RTIL) have been used to detect As^{3+} in drinking water with a low detection limit. Furthermore, Fe_3O_4-AuNPs composites have demonstrated excellent sensitivity for As^{3+} detection. Other metal oxides, such as TiO_2, have been utilized to enhance the sensitivity of Pb^{2+} and Cd^{2+} detection because of their favorable adsorption properties [61]. Gold nanoparticle-based electrochemical sensors (AuNPs) have gained attention by virtue of the unique properties of gold nanoparticles, making them highly effective for detecting various analytes, including heavy metals and biological molecules. Functionalized AuNPs, often modified with ligands or biomolecules, selectively bind to heavy metal ions like mercury (Hg^{2+}), lead (Pb^{2+}), and cadmium (Cd^{2+}). When these metals interact with the nanoparticles, changes in electrical signals occur, allowing the sensor to detect them with high sensitivity and selectivity [62]. AuNPs offer distinct advantages due to their small size (1–100 nm), high surface area, and tunable shape and size, which enhance their sensitivity and selectivity in various sensing applications. Their stability and biocompatibility make them suitable for environmental and biological systems. Additionally, AuNPs possess unparalleled optical properties, such as surface plasmon resonance (SPR), enabling their use in colorimetric sensing for detecting heavy metals and biological molecules. Their versatility is further enhanced by the ability to functionalize AuNPs with specific ligands or biomolecules, ensuring targeted interactions with analytes while maintaining stability. Moreover, AuNPs can be synthesized through various methods, including physical, chemical, and environmentally-friendly biological approaches, expanding their applicability across fields like medicine and environmental monitoring.

Recent advances have led to the development of AuNP-based sensors for both electrochemical and colorimetric detection of heavy metals in food and water [62]. These sensors operate by utilizing the interaction between heavy metals and ligands on the nanoparticle surface. For instance, thymine-rich oligonucleotides form T-Hg^{2+}-T complexes in the presence of Hg^{2+}, enabling the detection of mercury. This mechanism has been used in fluorescence resonance energy transfer (FRET) aptasensors, where the presence of Hg^{2+} triggers fluorescence recovery, achieving an impressive detection limit of 7.5×10^{-13} M [63]. Similarly, Pb^{2+} can be detected via DNAzyme-induced cleavage of target sequences, leading to fluorescence recovery or colorimetric changes.

AuNP-based optical nanosensors are increasingly used in food safety monitoring, focusing on detecting contaminants such as heavy metals, pesticides, and food additives. Heavy metals like mercury (Hg), lead (Pb), copper (Cu), iron (Fe), nickel (Ni), cobalt (Co), and cadmium (Cd) are naturally present in the environment but become harmful at high concentrations due to industrial and agricultural activities. These metals accumulate in the food chain, posing significant health risks by damaging organs such as the liver, kidneys, bones, and the central nervous system. To mitigate these risks, AuNP-based sensors have been developed for sensitive monitoring of heavy metals in food products [64]. Besides oligonucleotide interactions, peptides rich in reactive groups, such as glutathione (GSH), have shown promise in detecting heavy metals. GSH, with its multiple binding sites, exhibits a high affinity for heavy metals, facilitating their detection through AuNP aggregation. This approach has been used for detecting Pb^{2+} and As^{3+} with remarkable sensitivity. For example, incubating GSH with Cd^{2+} before adding AuNPs prevented their binding to GSH, causing nanoparticle aggregation and achieving a limit of detection as low as 4.3 pM [65].

Niu et al. [66] evaluated the development of a novel "turn-on" fluorescence sensor for the detection of lead ions (Pb^{2+}), utilizing graphene quantum dots (GQDs) and gold nanoparticles (AuNPs). The detection mechanism is based on the fluorescence recovery resulting from the separation of GQDs and AuNPs upon the introduction of Pb^{2+}. The sensor demonstrates a wide detection range, from 50 to 4 μM, with a detection limit of 16.7 nM. It also exhibits high selectivity for Pb^{2+} over other metal ions, including mercury (Hg^{2+}). The system offers a simple, low-cost, sensitive, and selective approach for real-time Pb^{2+} detection, potentially applicable to other heavy metals, antigens, or DNA by modifying the sensing molecules. This makes it particularly useful for environmental monitoring and health protection [66].

Carbon-based nanomaterials, including carbon nanotubes (CNTs), graphene (GR), and graphene oxide, are extensively used in electrochemical sensors due to their high conductivity, surface area, and stability. CNTs, particularly multiwall carbon nanotubes (MWCNTs), have been used in sensors for simultaneous detection of Cd^{2+} and Pb^{2+} with high sensitivity. Graphene, a two-dimensional carbon material, is also popular for its excellent electrochemical properties. Reduced graphene oxide (rGO) has been used to enhance sensor performance for Pb^{2+} detection in various food samples. Additionally, metal-organic frameworks (MOFs) embedded in graphene aerogels have been employed to simultaneously detect multiple heavy metals with impressive sensitivity and selectivity [67, 68].

Cheng et al. [69] developed two metal-enhanced fluorescence (MEF) chemosensors, Ag-HMS and Au-HMS, by loading hexagonal mesoporous silica (HMS) materials with silver (Ag) and gold (Au) nanoparticles functionalized with a rhodamine derivative (R) for detecting Hg^{2+} in aqueous solutions. These nanocomposites were synthesized using an impregnation-reduction method and characterized by techniques such as TEM, XRD, and UV-Vis. Both sensors exhibited high selectivity for Hg^{2+}, with Ag-HMS demonstrating a stronger MEF effect (enhancement factor of 4.58) compared to Au-HMS (EF of 2.93). The detection limits for Hg^{2+} were 0.9 ppb for Ag-HMS and 1.4 ppb for Au-HMS. The fluorescence lifetime of R slightly decreased after introducing the nanoparticles, with Ag-HMS showing superior sensitivity over Au-HMS [69].

Zhu et al. [70] presented the fabrication of a graphene nanodots-encaged porous gold electrode for electrochemical detection of heavy metal ions, specifically Cu^{2+} and Pb^{2+}, using ion beam sputtering deposition (IBSD). The electrode was created by simultaneously sputtering a gold target and a composite of aluminum and graphene onto glass substrates, followed by acid corrosion to form a porous 3D structure. The graphene nanodots, trapped within the porous gold, enhance the detection range for metal ions. The electrode exhibited a linear detection range of 0.009–4 μM for Cu^{2+} and 0.006–2.5 μM for Pb^{2+}, with good sensitivity, reusability, and repeatability. Structural analysis was performed using SEM, TEM, and XPS, revealing the potential of this 3D porous electrode for future sensing applications [70].

Carbon nanotube-based biosensors have been employed to monitor mercury (Hg) contamination in fish, offering rapid detection with minimal sample preparation. Quantum dot-based optical sensors have successfully identified cadmium (Cd) in agricultural products, providing a portable and cost-effective solution for in-field testing. These case studies highlight the versatility of nanobiosensors, capable of detecting multiple heavy metals simultaneously in both liquid and solid samples [71]. Compared to conventional methods like ICP-MS and AAS, nanobiosensors exhibit greater sensitivity and specificity, achieving detection limits as low as parts per billion (ppb) or parts per trillion (ppt). Additionally, nanobiosensors are faster, easier to use, and more economical, making them ideal for real-time applications in food safety [71, 72].

6.2 Organic Polymers

Organic polymers enhance electrode performance in electrochemical sensors through direct coating or copolymerization with monomers. They are often used in conjunction with inorganic nanomaterials to create highly sensitive and selective sensors for detecting heavy metals. For example, a sensor designed by Yi et al. used poly(L-glutamic acid) (PGA) combined with graphene oxide (GO) to modify a glassy carbon electrode (GCE) for the simultaneous detection of Cd^{2+}, Hg^{2+}, and Cu^{2+} [73]. This modified electrode demonstrated excellent performance with a linear detection range of 0.25–5.5 μmol L^{-1} and low detection limits (LOD) of 0.015, 0.032, and 0.024 μmol L^{-1}, respectively. This sensor also achieved high recovery rates of 99.3–100.4% in water samples. Other organic polymers, such as 4-tert-butylcalix [4]arene-tetrakis and poly(2,5-dimercapto-1,3,4-thiadiazole), have been utilized to detect Pb^{2+} and Cd^{2+} in various water sources, highlighting the versatility of these materials in sensor development [74, 75].

Recent developments have integrated metal-organic frameworks (MOFs) with organic polymers for enhanced electrochemical sensing. For instance, Ye et al. created a sensor using lanthanide MOF ZJU-27 modified on a GCE to detect Cd^{2+} and Pb^{2+} in drinking water. The ZJU-27/GCE demonstrated a linear detection range of 0.5–2.0 μmol L^{-1} with LODs of 1.66 and 1.10 nmol L^{-1} for Cd^{2+} and Pb^{2+}, respectively. Another example involves the use of polypyrrole (PPy) combined with NH2-MIL-53(Al) MOF to modify a gold electrode for detecting Pb^{2+} and Cu^{2+} in tap water, achieving LODs of 0.315 and 0.244 μg L^{-1} [5]. Zhang et al. developed a highly sensitive electrochemical sensor using a porphyrin-functionalized MOF (porph@MOF) and Pb^{2+}-dependent DNAzyme for detecting Pb^{2+} in leaf vegetables. This sensor achieved a remarkable LOD of 5 pmol L^{-1} for Pb^{2+}, fitting it for monitoring Pb contamination in food. The sensor benefited from the high stability and peroxidase-mimicking activity of porph@MOF, along with the specificity provided by the DNAzyme [76].

Among the diverse range of organic nanomaterials employed in electrochemical sensing, electrospun nanofibers have emerged as a promising platform for the development of highly sensitive and selective nanosensors for heavy metal detection. Electrospinning is a versatile and cost-effective technique that produces nanofibers with high surface area-to-volume ratios, tunable diameters, and porous structures, which are advantageous for enhancing sensor performance.

Electrospun nanofibers can be tailored from a variety of polymers, such as polyacrylonitrile (PAN), polylactic acid (PLA), and polyvinyl alcohol (PVA), and can be functionalized with various sensing elements, including conductive polymers, nanoparticles, and biomolecules. This customization enables the fabrication of nanofibers that are specifically designed to interact with target heavy metals, facilitating improved detection sensitivity and selectivity [4]. One of the key advantages of electrospun nanofiber-based nanosensors is their ability to offer high surface-to-volume ratios, which enhances the contact area between the sensing material and the analyte. This characteristic significantly improves the sensor's response time and detection limits. For example, the incorporation of metallic nanoparticles, such as gold or silver, into electrospun nanofibers has been shown to amplify electrochemical signals, resulting in highly sensitive detection of heavy metals like lead, cadmium, and mercury [4, 77]. Moreover, electrospun nanofibers can be engineered to form flexible and lightweight sensor devices, making them suitable for a wide range of applications, including wearable sensors and portable detection systems. Their scalability and ease of fabrication further contribute to their practical utility in real-world scenarios, such as monitoring food safety and environmental pollution [78].

Zhang et al. [79] evaluated a reusable colorimetric sensor based on Poly(aspartic acid) (PASP) electrospun nanofiber hydrogel membranes (ENHM) for the detection of copper (Cu^{2+}) and iron (Fe^{3+}) ions in aqueous solutions. The sensor benefits from the high metal ion adsorption capacity of PASP and the large surface area of the nanofibers. Upon exposure to Cu^{2+} and Fe^{3+}, the sensor undergoes a visible color change, turning from white to blue for Cu^{2+} (with a detection limit of 0.3 mg/L) and from white to yellow for Fe^{3+} (with a detection limit of 0.1 mg/L), detectable by the naked eye. The sensor is highly sensitive, selective, and reusable, as the adsorbed ions can be removed through a desorption process, allowing for repeated use. This simple filtration-based detection system offers a practical approach for detecting Cu^{2+} and Fe^{3+} in water samples (Fig. 2).

Song et al. [80] evaluated a novel biosensor for detecting arsenic (III) (As (III)) using Surface-Enhanced Raman Scattering (SERS) with Au@Ag core–shell nanoparticles. This biosensor features high SERS efficiency and excellent selectivity, achieving a detection limit of 0.1 ppb, which is lower than EPA and WHO standards. The approach combines As (III) aptamers with Raman-labeled Au@Ag nanoparticles, enabling precise detection even in the presence of other ions. The biosensor has been successfully applied to lake water samples, demonstrating its effectiveness and potential for environmental and biomedical applications [80].

Bagheri et al. [81] developed a novel electrode composed of a triphenylphosphine-modified multi-walled carbon nanotube (MWCNT) composite with an ionic liquid binder for the sensitive, simultaneous electrochemical detection of Cd^{2+}, Pb^{2+}, and Hg^{2+}. The electrode's surface morphology and properties were characterized using SEM and electrochemical impedance spectroscopy. The detection was performed

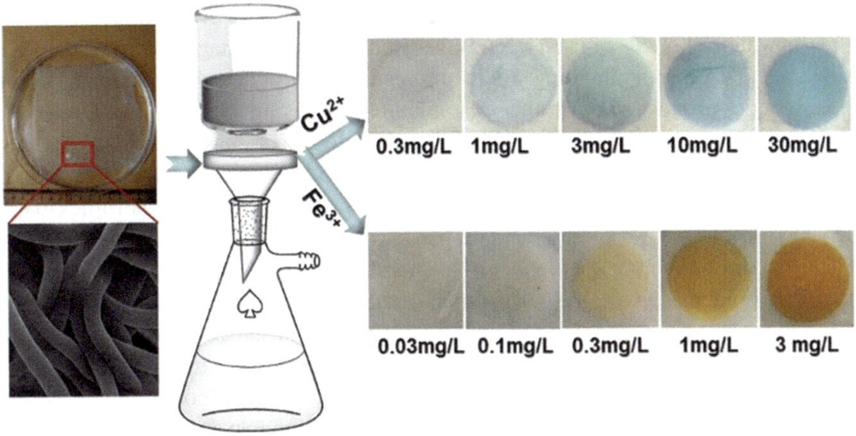

Fig. 2 Poly(aspartic acid) electrospun nanofiber hydrogel membrane-based reusable colorimetric sensor for Cu(II) and Fe(III) detection [79] (ACS Omega 2019, 4, 11, 14633–14639)

using square-wave anodic stripping voltammetry (SWASV), showing significantly improved responses compared to conventional carbon paste electrodes. The electrode could detect these heavy metals in the range of 0.1–150.0 nM, with detection limits of 6.0×10^{-5}, 9.2×10^{-5}, and 7.4×10^{-5} μM for Pb^{2+}, Hg^{2+}, and Cd^{2+}, respectively. The method was successfully applied to analyze heavy metal content in soil, gasoline, fish, tap water, and wastewater samples [81].

6.3 Biomaterials

Biomaterials are specialized substances with biological activity that own specific sites capable of binding to particular heavy metals. When used to modify electrodes, these biomaterials enhance the performance of electrochemical sensors by providing strong specificity, high sensitivity, and ease of preparation [82]. These sensors integrate sensitive biomaterials and electrochemical transducers, where biomaterials identify metal ions, and the transducer converts biochemical signals into electronic signals [83, 84]. Currently, key biomaterials for heavy metal detection include nucleic acids, enzymes, antibodies, and whole cells, with nucleic acids and enzymes being the most researched for electrode modification in electrochemical sensing of heavy metals in food [85]. Nucleic acids can bind with heavy metals to form metal-ion-guided base pairs, such as thymine (T)-Hg^{2+}-T and cytosine (C)-Ag+-C pairs. This binding ability has led to increased use of functional nucleic acids in electrode modification for heavy metal detection. Two prominent types of nucleic acids utilized are DNAzymes and aptamers [86].

DNAzymes are highly stable and exhibit strong specificity and binding affinity, making them suitable for heavy metal detection. For instance, a DNAzyme-based sensor combined with rolling circle amplification (RCA) was developed to detect Pb^{2+} in water. This sensor used DNAzymes immobilized on magnetic beads to cleave substrate DNA, producing long single-stranded DNA products that were detected via quantum dot probes [87]. Another example involved using a hairpin-like DNAzyme to enhance Pb^{2+} detection sensitivity through cascade reactions, achieving a detection limit of 0.043 nmol L^{-1} [88]. DNAzymes have also been integrated with nanoparticles for enhanced detection, such as a sensor employing hemin/G-quadruplex-based DNAzyme and Au–Pd bimetallic nanoparticles for Pb^{2+} detection [89].

Aptamers are short, synthetic single-stranded oligonucleotides (DNA or RNA) that can bind to a wide range of organic and inorganic targets with high specificity. They are produced through an in vitro Darwinian selection process known as Systematic Evolution of Ligands by Exponential Enrichment (SELEX), first introduced in 1990 [90]. Aptamers offer several advantages over antibodies, including low immunogenicity, ease of large-scale synthesis at a reasonable cost, physical stability, extended shelf life, and the ability to undergo simple chemical modifications. Initially, aptamers were extensively used in therapeutic and diagnostic research, particularly in drug development. Over time, their applications expanded to analytical chemistry, where they have been employed as immobilized ligands and in homogeneous assays [91].

In recent decades, numerous DNA-based chemical sensors have been developed, though many of them are incompatible with strong acidic or oxidative digestion reagents and are prone to non-specific absorption in real-world samples [92]. Several reviews have detailed the use of aptamers in food analysis, and Mao et al. recently discussed nanomaterial-based aptamer sensors for rapid arsenic detection, primarily for environmental applications such as water analysis [93]. Similarly, Sharma et al. reviewed nanoparticle-based aptasensors for the detection of Hg(II) and As(III) up to 2015, with many applications relying on gold nanoparticles (AuNPs) and various biosensing techniques like FRET (fluorescence resonance energy transfer), CRET (chemiluminescence resonance energy transfer), SPR (surface plasmon resonance), RRS (resonance Rayleigh scattering), SERRS (surface-enhanced resonance Raman scattering), and electrochemical aptasensors [94]. The most sensitive detection limit (10 fM) was achieved by Pelossof et al., who developed amplified SPR aptasensors using hemin/G-quadruplexes and Au nanoparticles [95]. This system included thiolated nucleic acid hairpin nanostructures specific to different analytes (DNA, adenosine monophosphate (AMP), or Hg(II)), and upon detection, the hairpins opened, forming hemin/G-quadruplexes. The high sensitivity was attributed to electronic coupling between the localized plasmon of the AuNPs and the surface plasmon wave, amplifying minute changes in the refractive index at the surface [91].

Aptamers, which are single-stranded DNA, RNA, or peptide sequences, offer high affinity and specificity for target metal ions. They are considered frugal alternatives to antibodies. Aptamer-based sensors have been developed for detecting various heavy metals, including Hg^{2+} and Ag^{+} in drinking water and red wine [96]. Another sensor used aptamer (GT)21-ssDNA to detect As^{3+} with high sensitivity and a detection limit of 0.058 ppb [97]. Enzymes possess active sites that interact specifically with substrates, and their activity can be affected by heavy metals. This interaction allows for the development of electrochemical sensors that measure changes in enzyme activity in response to metal ion presence.

Urease Inhibition-Based Sensors use urease immobilized on an electrode to measure heavy metal concentrations. For example, a sensor designed to detect Pb^{2+} and Hg^{2+} used urease and CeO_2 nanoparticles, demonstrating a detection range of 0.5–2.2 μmol L^{-1} for Pb^{2+} and 0.02–0.8 μmol L^{-1} for Hg^{2+} [47]. Another sensor utilizing choline oxidase (ChOx) showed high sensitivity for Pb^{2+} detection, with a detection limit of 0.04 nmol L^{-1} [83]. Additionally, catalase-based sensors have been developed for Hg^{2+} detection, providing reliable electrochemical performance [98].

Shen et al. [99] evaluated a label-free biosensor for detecting mercury(II) in milk using an aptamer-gated graphene field-effect transistor (Apt-SGGT). This sensor features a gate electrode modified with a structure-switching aptamer specific to Hg(II), and is further blocked with bovine serum albumin (BSA). The aptamer binds Hg(II) and undergoes a conformational change, redistributing surface charges and amplifying the channel current in the graphene transistor. The Apt-SGGT sensor demonstrates high sensitivity with a detection limit of 5 nM and a linear range from 5 to 200 nM for Hg(II). It shows excellent selectivity against other metal ions and a high recovery rate of 96.71% when testing real milk samples, indicating its practical potential for monitoring heavy metal residues in complex food matrices [99] (Fig. 3).

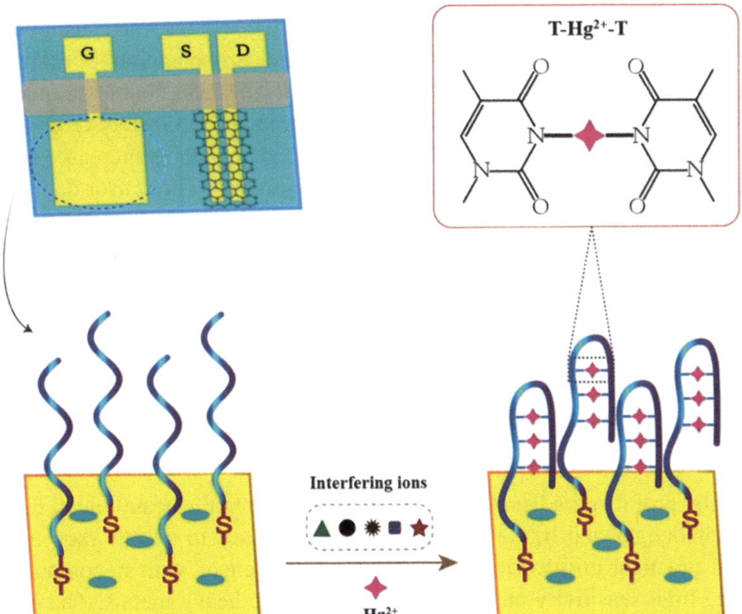

Fig. 3 Schematic of label-free biosensing of mercury(II) in milk using an aptamer-gated graphene field-effect transistor [99] (reproduced with permission)

Shen et al. [99] introduced a reusable and sensitive fluorescence polarization (FP) biosensor for detecting total mercury (Hg) in canned fish. The sensor utilizes single-stranded DNA (ssDNA) with thymine bases, which binds to magnetic nanoparticles (MNPs). In the presence of Hg^{2+}, the ssDNA forms thymine–Hg^{2+}–thymine complexes, causing an increase in the molar mass of the fluorophore and enhancing the FP signal. This method achieves a wide dynamic range (2.0 to 1.0 mM) and a low detection limit of 0.49 nM, with high selectivity for Hg^{2+} over other metal ions. The biosensor can be reused for up to six cycles and has been successfully applied to monitor Hg^{2+} in real canned fish samples, demonstrating its practical utility and accuracy [99].

Sun et al. [100] introduced a fluorescent biosensor for detecting mercury (II) (Hg^{2+}) using aptamer-functionalized magnetic beads. This method depends on magnetic separation to avoid the direct binding of Hg^{2+} with the fluorescent signal probe, which typically causes fluorescence quenching. Aptamers attached to magnetic beads form complexes with Hg^{2+}, and upon separation, the remaining fluorescent signal probe in solution emits a strong fluorescence, indicating the presence of Hg^{2+}. The assay is highly sensitive, with a detection limit of 0.2 nM, and demonstrates excellent accuracy and selectivity, successfully applied to river water and ribbon fish samples. This approach effectively overcomes the challenges of heavy metal interference and offers a rapid, reliable method for monitoring mercury contamination in environmental and food samples [100].

Sreekanth et al. [101] presented the development of an electrochemical DNA-based nanobiosensor for detecting cadmium (Cd(II)) ions in water, using a glassy carbon electrode (GCE) modified with multi-walled carbon nanotubes (MWCNTs). The sensor leverages the interaction between Cd(II) ions and double-stranded DNA (dsDNA), causing DNA unwinding into single-stranded DNA (ssDNA), which binds to the indicator dye ethyl green (EG). This interaction results in increased reduction peak current, allowing for the quantification of cadmium. The sensor demonstrated a linear detection range of 2–10 nM, with a sensitivity of 5 nA nM^{-1} and a detection limit of 2 nM, which is below the permissible cadmium exposure limit set by WHO. The study highlights the potential for MWCNT-modified electrodes in environmental monitoring for cadmium and other heavy metals [101].

7 Challenges and Limitations

Despite the promising advancements in nanobiosensors for heavy metal detection, several challenges and limitations need to be addressed to elevate their effectiveness and practical utility. One of the primary challenges with nanobiosensors is achieving high sensitivity and selectivity for specific heavy metals. The effectiveness of a nanobiosensor relies on its ability to detect low concentrations of target analytes amidst complex matrices. However, the sensitivity can be compromised by the inherent limitations of the nanomaterials used and the design of the sensor. Additionally, selectivity issues arise when the sensor responds to non-target substances, leading to false positives or reduced accuracy. Developing nanobiosensors with enhanced sensitivity and selectivity often involves optimizing the surface chemistry and sensor design, which can be a complex and resource-intensive process [102].

Nanobiosensors can suffer from interference caused by other substances present in the sample. For example, in environmental or food samples, the presence of organic compounds, ions, or particulate matter can affect the performance of the sensor. Such interference can eventuate inaccurate readings or reduced reliability of the sensor. Addressing these issues typically requires the incorporation of advanced filtering or separation techniques, which can complicate the sensor design and increase costs. Ensuring the stability and reproducibility of nanobiosensors is another significant challenge [4]. The performance of nanobiosensors can be affected by environmental conditions such as temperature, humidity, and pH, which can lead to degradation of the nanomaterials or the biosensing elements. Additionally, achieving consistent results across multiple sensors and experiments is crucial for reliable data interpretation. Researchers must develop robust fabrication methods and storage conditions to maintain the stability and reproducibility of the sensors over time, which can be challenging and may involve additional development efforts. The cost of producing nanobiosensors and scaling up their manufacture remains a considerable limitation. While laboratory-scale production of nanobiosensors can be relatively straightforward, scaling up to commercial levels often involves high costs and technical complexities [53]. The use of specialized materials, intricate fabrication processes, and quality control measures can drive up production costs, making it challenging to develop affordable and widely accessible sensors. Additionally, achieving scalability while maintaining the performance and quality of the sensors presents a significant hurdle for their broader adoption and commercialization. Addressing these challenges requires ongoing research and innovation to enhance the performance, affordability, and practicality of nanobiosensors for heavy metal detection. Advances in materials science, sensor technology, and manufacturing processes will play a crucial role in overcoming these limitations and expanding the application of nanobiosensors in various fields [103].

8 Conclusion

The detection of heavy metals in food is critical due to the toxic effects these contaminants can have on human health, even at trace levels. Traditional methods like atomic absorption spectroscopy (AAS) and inductively coupled plasma mass spectrometry (ICP-MS) are effective but come with drawbacks, including high costs, time-consuming processes, and the need for complex equipment. This has prompted the exploration of alternative methods, with nanobiosensors emerging as a promising solution. These sensors render high sensitivity, selectivity, and the potential for rapid, on-site detection, leveraging the unique properties of nanomaterials. Electrochemical nanobiosensors, in particular, have shown significant potential for detecting trace levels of heavy metals in food due to their simplicity, affordability, and adaptability.

Various types of nanomaterials have been employed in these sensors, including inorganic nanomaterials, organic polymers, and biomaterials. Inorganic nanomaterials, such as carbon-based nanomaterials and metal nanoparticles, are valued for their conductivity and catalytic activity. Organic polymers offer stability and biocompatibility, while biomaterials, like enzymes and DNA, contribute high selectivity, making them ideal for biosensing applications. Despite these advancements, challenges remain that hinder the wider application of nanobiosensors. Issues like reproducibility, long-term stability, and selectivity in complex food matrices need to be addressed. Additionally, interference from other substances in food and the difficulty of integrating these sensors into portable, user-friendly devices are ongoing hurdles. Addressing these limitations is key to enabling broader commercial and industrial use of nanobiosensors. Looking to the future, innovations in nanomaterials, hybrid systems, and microfabrication techniques are expected to further enhance the performance of nanobiosensors. As these technologies evolve, nanobiosensors have the potential to become indispensable tools in food safety, providing rapid, accurate, and cost-effective detection of heavy metals. This progress will be vital in safeguarding public health and ensuring a safer, more reliable food supply chain.

References

1. Rai PK et al (2019) Heavy metals in food crops: health risks, fate, mechanisms, and management. Environ Int 125:365–385
2. Scutarașu EC, Trincă LC (2023) Heavy metals in foods and beverages: global situation, health risks and reduction methods. Foods 12(18):3340
3. Nkwunonwo UC, Odika PO, Onyia NI (2020) A review of the health implications of heavy metals in food chain in Nigeria. Sci World J 2020(1):6594109
4. Akhavan-Mahdavi S et al (2024) Electrospun nanofiber-based sensors for the detection of chemical and biological contaminants/hazards in the food industries. Adv Coll Interface Sci 325:103111
5. Wang L et al (2020) Recent advances in the development of electrochemical aptasensors for detection of heavy metals in food. Biosens Bioelectron 147:111777
6. Rai PK et al (2018) A critical review of ferrate (VI)-based remediation of soil and groundwater. Environ Res 160:420–448
7. El Hamiani O et al (2015) Metal concentrations in plants from mining areas in South Morocco: health risks assessment of consumption of edible and aromatic plants. CLEAN Soil Air Water 43(3):399–407
8. Kohzadi S et al (2019) Concentration, source, and potential human health risk of heavy metals in the commonly consumed medicinal plants. Biol Trace Elem Res 187:41–50
9. Mahurpawar M (2015) Effects of heavy metals on human health. Int J Res Granthaalayah 530(516):1–7
10. Hu H (2002) Human health and heavy metals exposure. Life Support Environ Hum Health 4:1–12
11. Collin MS et al (2022) Bioaccumulation of lead (Pb) and its effects on human: a review. J Hazard Mater Adv 7:100094
12. Charkiewicz AE et al (2023) Cadmium toxicity and health effects—a brief summary. Molecules 28(18):6620
13. Hu Y et al (2019) Health risks to local residents from the exposure of heavy metals around the largest copper smelter in China. Ecotoxicol Environ Saf 171:329–336

14. Hassan MU et al (2022) The role of zinc to mitigate heavy metals toxicity in crops. Front Environ Sci 10:990223
15. Genchi G et al (2022) Arsenic: a review on a great health issue worldwide. Appl Sci 12(12):6184
16. Jyothi NR, Farook NAM (2020) Mercury toxicity in public health. In: Heavy Metal Toxicity in Public Health, pp 1–12
17. Iyer M et al (2023) A review of chromium (Cr) epigenetic toxicity and health hazards. Sci Total Environ 882:163483
18. Rehman K et al (2018) Prevalence of exposure of heavy metals and their impact on health consequences. J Cell Biochem 119(1):157–184
19. Genchi G et al (2020) Nickel: human health and environmental toxicology. Int J Environ Res Public Health 17(3):679
20. Manzoor J, Sharma M, Wani KA (2018) Heavy metals in vegetables and their impact on the nutrient quality of vegetables: a review. J Plant Nutr 41(13):1744–1763
21. Shaheen N et al (2016) Presence of heavy metals in fruits and vegetables: health risk implications in Bangladesh. Chemosphere 152:431–438
22. Mawari G et al (2022) Heavy metal accumulation in fruits and vegetables and human health risk assessment: findings from Maharashtra, India. Environ Health Insights 16:11786302221119152
23. Yan M et al (2022) Heavy metal levels in milk and dairy products and health risk assessment: a systematic review of studies in China. Sci Total Environ 851:158161
24. Ziarati P et al (2018) An overview of the heavy metal contamination in milk and dairy products. Acta Sci Pharm Sci 2(7):1–14
25. Singh M et al (2020) Assessment of contamination of milk and milk products with heavy metals. Indian J Dairy Sci 72(6)
26. Kuramshina N et al (2019) Heavy metals content in meat and milk of Orenburg region of Russia. Int J Pharm Res 11(1):09752366
27. Emami MH et al (2023) A review of heavy metals accumulation in red meat and meat products in the Middle East. J Food Prot 86(3):100048
28. González-Torres P et al (2023) Comparative study of the presence of heavy metals in edible vegetable oils. Appl Sci 13(5):3020
29. Shah YA, Lachenmeier DW (2022) Metals in alcoholic beverages and public health implications. In: Handbook of substance misuse and addictions: from biology to public health, pp 1131–1150
30. Elafify M et al (2023) Heavy metal residues in milk and some dairy products with insight into their health risk assessment and the role of Lactobacillus rhamnosus in reducing the lead and cadmium load in cheese. Food Chem Adv 2:100261
31. Özbay S, Dikici E, Soylukan C (2023) Evaluation of biological (feed, water), seasonal, and geological factors affecting the heavy metal content of raw milk. J Food Compos Anal 121:105401
32. Dağcilar K, Gezer C (2021) Heavy metal residues in milk and dairy products produced in Northern Cyprus. Prog Nutr 23(1):e2021019
33. Pećanac B, Golić B, Knežević D (2023) Assessment of contamination with heavy metals in food of animal origin. In: Book of proceedings
34. Barone G et al (2021) Trace metals in pork meat products marketed in Italy: occurrence and health risk characterization. Biol Trace Elem Res 199:2826–2836
35. Hoha GV et al (2014) Heavy metals contamination levels in processed meat marketed in Romania. Environ Eng Manag J 13(9):2411–2415
36. Ciont C et al (2022) Beer safety: new challenges and future trends within craft and large-scale production. Foods 11(17):2693
37. Donadini G, Spalla S, Beone G (2008) Arsenic, cadmium and lead in beers from the Italian market. J Inst Brew 114(4):283–288
38. Soares LMV, Moraes AMMD (2003) Lead and cadmium content of Brazilian beers. Food Sci Technol 23:285–289

39. Nascentes CC et al (2005) Direct determination of Cu, Mn, Pb, and Zn in beer by thermospray flame furnace atomic absorption spectrometry. Spectrochim Acta Part B 60(5):749–753
40. Orescanin V et al (2003) Heavy metals in soil, grape, and wine. J Trace Microprobe Tech
41. Jin M et al (2020) Review of the distribution and detection methods of heavy metals in the environment. Anal Methods 12(48):5747–5766
42. Rehan I et al (2021) Nutritional and toxic elemental analysis of dry fruits using laser induced breakdown spectroscopy (LIBS) and inductively coupled plasma atomic emission spectrometry (ICP-AES). Saudi J Biol Sci 28(1):408–416
43. Zhong W-S, Ren T, Zhao L-J (2016) Determination of Pb (Lead), Cd (Cadmium), Cr (Chromium), Cu (Copper), and Ni (Nickel) in Chinese tea with high-resolution continuum source graphite furnace atomic absorption spectrometry. J Food Drug Anal 24(1):46–55
44. Li M et al (2013) Nanostructured sensors for detection of heavy metals: a review. ACS Publications
45. Bua DG et al (2016) Heavy metals in aromatic spices by inductively coupled plasma-mass spectrometry. Food Addit Contam Part B 9(3):210–216
46. Liu Y et al (2017) Progress on sensors based on nanomaterials for rapid detection of heavy metal ions. Sci China Chem 60:329–337
47. Gumpu MB et al (2015) A review on detection of heavy metal ions in water–an electrochemical approach. Sens Actuators B Chem 213:515–533
48. Cui L, Wu J, Ju H (2015) Electrochemical sensing of heavy metal ions with inorganic, organic and bio-materials. Biosens Bioelectron 63:276–286
49. Guruprasath N et al (2024) Review on emerging applications of nanobiosensor in food safety. J Food Sci
50. Jurado-Sánchez B et al (2020) Nanobiosensors for food analysis. Handbook of food nanotechnology. Elsevier, pp 415–457
51. Sabaghi M, Jamali SN (2024) Advancements, challenges, and future prospects of nanobiosensors in food packaging for allergen detection. J Food Meas Charact 18(5):3444–3457
52. Singh K, Arya P (2024) Nanobiosensor potentialities for food toxin detection. Nanobiotechnology for food processing and packaging. Elsevier, pp 173–190
53. Valenzuela-Amaro HM et al (2023) Emerging applications of nanobiosensors in pathogen detection in water and food. Biosensors 13(10):922
54. Thakur PS, Sankar M (2022) Nanobiosensors for biomedical, environmental, and food monitoring applications. Mater Lett 311:131540
55. Meng R et al (2023) The innovative and accurate detection of heavy metals in foods: a critical review on electrochemical sensors. Food Control 150:109743
56. Ding Q et al (2021) Electrochemical detection of heavy metal ions in water. Chem Commun 57(59):7215–7231
57. Qian W, Hong-Mei B, Xiao-Jun H (2021) Research progress of electrochemical detection of heavy metal ions. Chin J Anal Chem 49(3):330–340
58. Patel G et al (2020) Application of nanosensors in the food industry. Nanosensors for smart cities. Elsevier, pp 355–368
59. Taghdisi SM et al (2016) An electrochemical aptasensor based on gold nanoparticles, thionine and hairpin structure of complementary strand of aptamer for ultrasensitive detection of lead. Sens Actuators B Chem 234:462–469
60. Podešva P, Gablech I, Neužil P (2018) Nanostructured gold microelectrode array for ultrasensitive detection of heavy metal contamination. Anal Chem 90(2):1161–1167
61. Li S-S et al (2018) Surface Fe (II)/Fe (III) cycle promoted ultra-highly sensitive electrochemical sensing of arsenic (III) with dumbbell-like Au/Fe_3O_4 nanoparticles. Anal Chem 90(7):4569–4577
62. Kusuma SAF et al (2023) Gold nanoparticle-based colorimetric sensors: properties and application in detection of heavy metals and biological molecules. Sensors 23(19):8172
63. Amiri S et al (2018) Ultrasensitive and highly selective FRET aptasensor for Hg^{2+} measurement in fish samples using carbon dots/AuNPs as donor/acceptor platform. New J Chem 42(19):16027–16035

64. Song M, Khan IM, Wang Z (2021) Research progress of optical aptasensors based on AuNPs in food safety. Food Anal Methods 14(10):2136–2151
65. Du H et al (2022) Nanomaterial-based optical biosensors for the detection of foodborne bacteria. Food Rev Intl 38(4):655–684
66. Niu X et al (2018) A "turn-on" fluorescence sensor for Pb^{2+} detection based on graphene quantum dots and gold nanoparticles. Sens Actuators B Chem 255:1577–1581
67. Liu X et al (2019) Recent advances in nanomaterial-enabled screen-printed electrochemical sensors for heavy metal detection. TrAC Trends Anal Chem 115:187–202
68. Ye W et al (2020) Electrochemical detection of trace heavy metal ions using a Ln-MOF modified glass carbon electrode. J Solid State Chem 281:121032
69. Cheng Z, Li G, Liu M (2015) Metal-enhanced fluorescence effect of Ag and Au nanoparticles modified with rhodamine derivative in detecting Hg^{2+}. Sens Actuators B Chem 212:495–504
70. Zhu H et al (2015) Graphene nanodots-encaged porous gold electrode fabricated via ion beam sputtering deposition for electrochemical analysis of heavy metal ions. Sens Actuators B Chem 206:592–600
71. Dai B et al (2022) Recent advances in carbon nanotube-based biosensors for biomolecular detection. TrAC Trends Anal Chem 154:116658
72. Chopade RL et al (2023) Carbon nanotube-based nano-biosensors for detecting heavy metals in the aquatic environment. Environ Sci Pollut Res 30(5):11199–11209
73. Yi W et al (2019) Sensitive electrochemical sensor based on poly (L-glutamic acid)/ graphene oxide composite material for simultaneous detection of heavy metal ions. RSC Adv 9(30):17325–17334
74. He X et al (2011) Differential pulse anodic stripping voltammetric determination of Cd and Pb at a bismuth glassy carbon electrode modified with nafion, poly (2, 5-dimercapto-1, 3, 4-thiadiazole) and multiwalled carbon nanotubes. Microchim Acta 173:95–102
75. Alizadeh T, Amjadi S (2011) Preparation of nano-sized Pb^{2+} imprinted polymer and its application as the chemical interface of an electrochemical sensor for toxic lead determination in different real samples. J Hazard Mater 190(1–3):451–459
76. Zhang X et al (2020) Single-step electrochemical sensing of ppt-level lead in leaf vegetables based on peroxidase-mimicking metal-organic framework. Biosens Bioelectron 168:112544
77. Rui X-Y et al (2024) Applications of electrospun nanofibers in food safety and quality analysis: a review. Trends Food Sci Technol, 104686
78. Du Y, Yu D-G, Yi T (2023) Electrospun nanofibers as chemosensors for detecting environmental pollutants: a review. Chemosensors 11(4):208
79. Zhang C et al (2019) Poly (aspartic acid) electrospun nanofiber hydrogel membrane-based reusable colorimetric sensor for Cu (II) and Fe (III) detection. ACS Omega 4(11):14633–14639
80. Song L et al (2016) A novel biosensor based on Au@Ag core–shell nanoparticles for SERS detection of arsenic (III). Talanta 146:285–290
81. Bagheri H et al (2013) Simultaneous electrochemical determination of heavy metals using a triphenylphosphine/MWCNTs composite carbon ionic liquid electrode. Sens Actuators B Chem 186:451–460
82. Guo W et al (2021) Advances in aptamer screening and aptasensors' detection of heavy metal ions. J Nanobiotechnology 19:1–19
83. Magar HS et al (2017) Highly sensitive choline oxidase enzyme inhibition biosensor for lead ions based on multiwalled carbon nanotube modified glassy carbon electrodes. Electroanalysis 29(7):1741–1748
84. Wang N et al (2016) Functionalized gold nanoparticles/reduced graphene oxide nanocomposites for ultrasensitive electrochemical sensing of mercury ions based on thymine–mercury–thymine structure. Biosens Bioelectron 79:320–326
85. Dai X, Wu S, Li S (2018) Progress on electrochemical sensors for the determination of heavy metal ions from contaminated water. J Chin Adv Mater Soc 6(2):91–111
86. Zhou X, Pu H, Sun D-W (2021) DNA functionalized metal and metal oxide nanoparticles: principles and recent advances in food safety detection. Crit Rev Food Sci Nutr 61(14):2277–2296

87. Tang S et al (2013) Ultrasensitive electrochemical detection of Pb^{2+} based on rolling circle amplification and quantum dotstagging. Biosens Bioelectron 42:608–611
88. Liu S et al (2016) Ultrasensitive electrochemical DNAzyme sensor for lead ion based on cleavage-induced template-independent polymerization and alkaline phosphatase amplification. Biosens Bioelectron 83:33–38
89. Zhou Q et al (2016) Highly sensitive electrochemical sensing platform for lead ion based on synergetic catalysis of DNAzyme and Au–Pd porous bimetallic nanostructures. Biosens Bioelectron 78:236–243
90. Kannan P, Guo L (2020) Nanosensors for food safety. Nanosensors for smart cities. Elsevier, pp 339–354
91. Jinadasa KK et al (2022) Smart materials for mercury and arsenic determination in food and beverages. Microchem J 179:107472
92. Abu-Salah KM et al (2015) DNA-based nanobiosensors as an emerging platform for detection of disease. Sensors 15(6):14539–14568
93. Mao K et al (2020) Nanomaterial-based aptamer sensors for arsenic detection. Biosens Bioelectron 148:111785
94. Sharma R et al (2015) Recent advances in nanoparticle based aptasensors for food contaminants. Biosens Bioelectron 74:612–627
95. Pelossof G et al (2011) Amplified surface plasmon resonance based DNA biosensors, aptasensors, and Hg^{2+} sensors using hemin/G-quadruplexes and Au nanoparticles. Chem A Eur J 17(32):8904–8912
96. Miao P, Tang Y, Wang L (2017) DNA modified Fe_3O_4@ Au magnetic nanoparticles as selective probes for simultaneous detection of heavy metal ions. ACS Appl Mater Interfaces 9(4):3940–3947
97. Wen S-H et al (2018) Electrochemical sensor for arsenite detection using graphene oxide assisted generation of Prussian blue nanoparticles as enhanced signal label. Anal Chim Acta 1002:82–89
98. Elsebai B et al (2017) Catalase based hydrogen peroxide biosensor for mercury determination by inhibition measurements. J Hazard Mater 340:344–350
99. Shen T et al (2013) A reusable and sensitive biosensor for total mercury in canned fish based on fluorescence polarization. Talanta 117:81–86
100. Sun C et al (2018) Utilization of aptamer-functionalized magnetic beads for highly accurate fluorescent detection of mercury (II) in environment and food. Sens Actuators B Chem 255:775–780
101. Sreekanth S et al (2021) Multi-walled carbon nanotube-based nanobiosensor for the detection of cadmium in water. Environ Res 197:111148
102. Thakur M, Wang B, Verma ML (2022) Development and applications of nanobiosensors for sustainable agricultural and food industries: recent developments, challenges and perspectives. Environ Technol Innov 26:102371
103. Kulkarni MB, Ayachit NH, Aminabhavi TM (2022) Recent advancements in nanobiosensors: current trends, challenges, applications, and future scope. Biosensors 12(10):892

New Insights on Nano Biosensors Applications for Chemical and Adulterant in Foods

Renu Agarwal, P. Harini, and J. Sri Varshni

Abstract Food safety and quality have received increasing attention in recent years, and they greatly impact people's health. Food adulteration is a problem with food quality and safety that affects people worldwide and is challenging to identify. Pesticides are chemicals used for the control of pests in different doses eventually leading to damage of foods produced from plants and crops due to their toxic nature. Their detection is always important in hazard analysis. Different techniques are involved in detecting chemicals and adulterants present in food. The types and concentrations of adulterants in various food types can be precisely determined using modern detection techniques such as high-performance liquid chromatography, and gas chromatography-mass spectrometer. However, the use of these techniques is limited due to their cost, complex sample preparation, and operating characteristics. The use of nanostructures in biosensors is known as nano biosensors which bring the food freshness or contamination monitoring. This chapter thus discusses the nature of nano biosensors and their applications in identifying infections, toxins, and chemical contamination, adulteration which is present in commodities, thus increasing their shelf life and assuring the end user's safety.

Keywords Food safety · Adulteration · Chemicals · Nano biosensors · Food freshness · Shelf life

---

R. Agarwal (✉)
Assistant Professor, Department of Business Administration, Shrimathi Devkunvar Nanalal Bhatt Vaishnav College for Women, Chromepet, Affiliated to the University of Madras, Chennai, Tamil Nadu, India
e-mail: goldie_agarwal2002@yahoo.com

P. Harini
Department of Horticulture and Food Science, VIT School of Agricultural Innovations and Advanced Learning, Vellore Institute of Technology, Vellore, Tamil Nadu, India

J. Sri Varshni
Project Associate, IIT Madras BS Degree Program, CODE IITM, ICSR Building, Indian Institute of Technology - IITM, Vellore, Tamil Nadu, India
e-mail: srivarshnij2907@gmail.com

D. Sillu et al. (eds.), *Nanobiosensors for the Food Industry*, Smart Nanomaterials Technology, https://doi.org/10.1007/978-981-95-0136-6_9

1 Introduction

A device known as a sensor is capable of detecting and quantifying the presence of analytes in samples. The sensor is comprised of a detection system called a receptor, a transducer, and a readout system. The transducers (sensors) are produced with nanomaterials such as AuNP (gold nanoparticles), graphene oxide, and quantum dots [1]. The other variations include semiconductor nanomaterials, metal nanomaterials, carbon nanomaterials, polymeric nanomaterials. These due to their super surface impact and little size impact have got extensive consideration [2, 3]. Nanosensors are any point of biochemical or surgical sensing used to convey information about nanoparticles to the macroscopic world. It is a reimagined chemical sensor or biosensor that utilizes nanomaterials in analytical devices. The utilization of nanoclusters is expanding with their uncomplicated composite routes [4], reduced harmfulness, unchallenging operation, prompt reciprocation, solubility, and compatibility [5].

Nanosensors are classified into various types such as nano biosensors, chemical nanosensors, electrometers, and deployable nanosensors [6]. In accordance with the theory of structure, nanosensors are categorized either by the kind of transduction mechanism employed for detection or by the bioreceptor utilized for molecular recognition [7, 8]. The first category is an optical nanosensor, which employs spectroscopy for absorption, illumination, phosphorescence, dissipation, and refraction to determine characteristics including energy, amplitude, polarization, and decomposing time [9, 10]. Potentiometry and amperometry are the two electrochemical transduction processes used by the other type, known as an electrochemical nanosensor. Microbalance, microcantilever, and surface acoustic waves are all part of the mass transduction process [11]. The degree of electron gains or loss that exists during bio interactions can be determined with electrochemical signals, such as current, voltage, and resistance [1].

Nanosensors are categorized into deployable, chemical, biosensor and electrometer, and biosensor, based on the applications' principles. The chemical sensor analyzes the signal using electronics and capacitive readout cantilevers. A single chemical or biological molecule can be captured by this kind of sensor because of its exceptionally high sensitivity. On documentation, various nanosensors are employed in the measurement of entities such as ion concentrations [12]. Another type of sensor that is small and portable and would be employed in the military or other branches of national security is the deployable nanosensor [13]. The most recent kind, known as biosensors, is one of the most heavily sponsored subfields of nanosensor research. This is mostly because of the potential for this technology to aid in the early diagnosis of cancer as well as other ailments.

Certain kinds of DNA can also be detected by the biosensors. Many traditional PCR techniques, such as multiplex PCR, conventional PCR, or PCR combined with other techniques like PCR-RFLP, sequencing, and analysis of single-strand conformation polymorphisms (PCR-SSCP), which is based on the electrophoretic mobility of the amplified fragments in their single strand and folded conformation, have been used in various food authentication studies [14]. Additionally, single sequence repeats (SSR), sometimes referred to as microsatellites, and random amplified polymorphic DNA (RAPD), which employ single and brief arbitrary primers, were employed in food authentication [15]. The implementation of DNA-based techniques, such as PCR and real-time PCR, allows for rapid and accurate detection of foodborne pathogens and adulterants. This is crucial for ensuring that food products meet safety standards and are free from harmful contaminants, thereby protecting consumer health [16]. The ability to detect multiple targets simultaneously using multiplex PCR enhances the efficiency of food testing. This means that a single test can identify various contaminants or adulterants, streamlining the testing process and reducing the need for multiple assays [17].

According to [18], the first hybrid biosensor based on GNPs that is species-specific was created. As little as 1% pig adulteration in ready-to-eat beef meatball preparations was successfully detected by the sensor. The great specificity and sensitivity of the sensor for the pork DNA were further demonstrated by the cross-testing findings using several commercial meatball formulas. For the first time, the sensor's hybridization kinetics were examined, and a hyperbolic correlation between target concentration and fluorescence intensity was found. Potential targets in processed mixed meat products were quantified with greater than 90% accuracy using a linear curve that was realized across the target's moderate concentration. Target DNA authentication was no longer dependent on costly RT-PCR, time-consuming electrophoresis, or tedious blotting procedures thanks to this technology.

In biosensors, there are multiple methods by which the acceptor, a biological element, is attached to the converter. This is highly capable of detecting both biological interactions and analytes. The connection of the bioreceptor layer (biological component) to the transducer is crucial for the success of biosensors. The primary objective is to establish a connection between the biological element and the sensory surface while maintaining the biological component's stability. The use of nanostructures in the structure of biosensors has led to valuable advances in this field for various reasons. The purpose of incorporating nanomaterials into the structure of biosensors is to enhance the level of stabilization required for biomaterials. Thus, increasing the sensitivity, catalysis, and possibility of reacting at low potentials and facilitating the speedy transfer of electrons from the active reaction center to the electrode surface in electrochemical nanosensors [19, 20].

1.1 Features and Operating Principles of Sensors

Sensors are assessed based on several important features. The selectivity of the system is a significant indicator of their ability to distinguish the analyte from other materials in the sample. The more selective the analyte is, the more effective the identification and measurement can be with the least interference from other materials. Sensitivity is the next crucial aspect. With minimal changes in analyte concentration, the

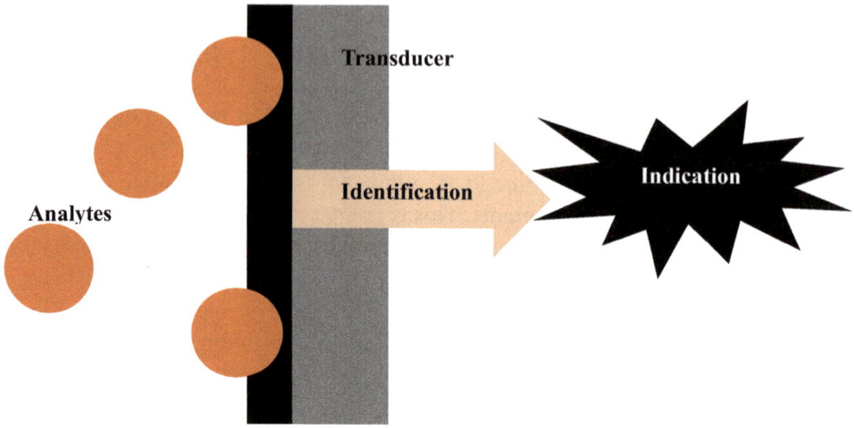

Fig. 1 Operating principle of biosensor. The main units include analytes and a transducer involving identification and indication

sensor's output signal has undergone a significant change and a high sensitivity can be achieved. Another important feature that is defined by accuracy is repeatability. Accuracy implies that the measurement results are comparable to the real value.

High precision implies that the results of repeated measurements are closely aligned [20]. The specific interaction between the biological layer and the analyte is caused by the latter, but the transducer transforms the information resulting from this interaction into a quantifiable effect. The amount of these changes is measured by the reading system. The interaction-induced physical phenomenon is measured by the reading system. The basic principle is shown in Fig. 1.

The elimination of chemical intermediates in electron transfer through the use of nanomaterials in biosensors can result in device simplification. This is crucial for the development of third-generation biosensors. The surface adsorption of biomolecules can be significantly impacted by nanoparticles due to their significant specific surface area and high free surface energy.

Nano biosensors are sophisticated instruments that detect and examine particular chemical, biological, or physical characteristics of target molecules by integrating the concepts of nanotechnology with biotechnology. Utilizing the special qualities of nanomaterials, these sensors operate at the nanoscale to provide exceptionally sensitive and selective monitoring. Nano biosensors have drawn a lot of interest and are being used in an array of industries, including food safety, environmental surveillance, and health. A readout system, transducer, and detector are the essential parts of a nano biosensor. The target probe interacts selectively with the sensing agent, which is usually an artificial compound or biomolecule that can bind to one specific target. The transformation of the signals on result of the binding is transformed by the converter into a valuable output the result, giving crucial information about the target molecule's existence and amount [21].

2 Nanotechnology and Biosensors in Food Industry

Food safety and food quality have received increasing attention in recent years, and they have a big impact on people's health. Food adulteration is a problem with food quality and safety that affects people all over the world and is challenging to identify. Detection of chemicals and adulterants present in food is important. The types and concentrations of adulterants in various food types can be precisely determined using modern detection techniques such as high-performance liquid chromatography and gas chromatography mass spectrometer. However, the use of these techniques is limited due to their cost, complex sample preparation, and operation characteristics. In the context of food safety, nano biosensors can be utilized to monitor pesticide residues, which are a significant concern in agricultural products. For instance, certain biosensors have been developed to detect specific pesticides like chlorpyrifos and carbofuran with very low detection limits, indicating their effectiveness in ensuring food safety [22–24].

Electrochemical DNA biosensors have been employed to study the damage caused by pesticides to DNA, showcasing their application in both food analysis and environmental monitoring. Moreover, aptamer-based sensors, which are a type of nucleic acid sensor, have been developed for the detection of various food contaminants, including pesticides. These sensors leverage the high affinity of aptamers for their targets, allowing for sensitive detection of harmful substances in food products [25]. Biosensors are considered to be potential tools for monitoring environmental and food pollutants. They are high-sensitivity, fast, accessible, and field-convenient devices that can recognize and measure targets by converting the target recognition into a physically identifiable sign, such as magnetic, optical, or electrical signals [26].

Nanotechnology is experiencing rapid growth and is a diverse field with both technical and clinical aspects. Nanoparticles, quantum dots, nanotubes, nanorods, nanowires, nanochips, and many other components are increasingly utilized to manufacture nanosensors and nano biosensors, aiming to enhance the precision and sensitivity of reactions. The use of nanostructures in biosensors is known as nano biosensors which bring the food freshness or contamination monitoring. Analytical efficiency has a significant impact on the LOD (Limit of Detection), selectivity, sensitivity, and signals to noise ratio when nanotechnology and biosensors were combined [27]. They are described as a possible substitute and are a classical or partially quantitative, nanometric scale instrument that identifies chemical compounds by using specific biochemical reactions as sensing elements [28].

Food safety is a crucial concern in the food industry due to its direct impact on human health [29]. In the food industry, nano biosensors are effectively employed in certain crucial aspects such as the detection of adulterants, allergens, chemicals, microorganisms, toxins, vitamins, antioxidants, and packaging [30–32]. The ability to detect contaminants quickly and accurately is facilitated by them, which are more sensitive and selective than traditional sensors [33]. Using nanomaterials is a common approach in food production, processing, packaging, labeling, transporting, tracing, and maintaining excellent food products. Numerous elements, including biological, microbiological, chemical, biochemical, and mechanical components, might affect a food's nutritional profile. In order to prevent neurotoxicity conditions affecting the human brain, kidney, and liver, an enzyme-based biosensor that contains glutamate dehydrogenase and glutamate oxidase has been designed to gauge the amounts of certain toxins, such as monosodium glutamate [34]. The promising technology of nano biosensors ensures high-quality and protected merchandise. Due to their high sensitivity, selectivity, and quick reaction, they are ideal for detection. Numerous advantages are offered by them over traditional strategies, such as rapid response and miniaturization.

3 Nano Biosensors, Chemicals, and Adulterants in Food Industry

Particularly in the food industry, precise recognition and quantification are critical to productive management. Chemical analysis using the conventional technique is usually complex and intensive. This approach could not be appropriate for real-time monitoring, which would make it challenging to react swiftly to modifications in product reliability or protection. Another difficulty is the absence of defined techniques for estimating the duration of food storage. Establishing uniform quality control policies and processes can be challenging because different production stages may call for different processes [32, 35, 36].

Additionally, shelf life may be impacted by interactions between food merchandise and its packaging. Moisture permeability, air permeability, and volatiles can all have an impact on process durability and efficacy. Effective shelf-life preservation requires an understanding of and ability to control these interactions. Development and implementation of revolutionary technology, including nano biosensors, are necessary to meet these problems. The accuracy, efficiency, and dependability of inspection procedures can be improved by incorporating this technology into food packaging design. By addressing the drawbacks of conventional techniques, nano biosensors offer real-time precision monitoring capabilities, prompt intervention, and improved quality control. They offer insightful information about shifts in the food chain, which improves decision-making and lowers food waste [30, 36–40].

Figure 2 describes the following points including the contaminants of food [6];

- Toxic substances generated by specific fungi, such as species of Aspergillus, Penicillium, and Fusarium, found in foodstuffs that are warm and humid, such as grains, nuts, and other foods, can result in illnesses like immunosuppression, cancer, and liver damage.
- Heavy metals from industrial discharge, contaminated water or soil, and environmental pollution, such as lead, mercury, cadmium, and arsenic. Reasons developmental issues, organ failure, and neurological damage can result from accumulation in the body.

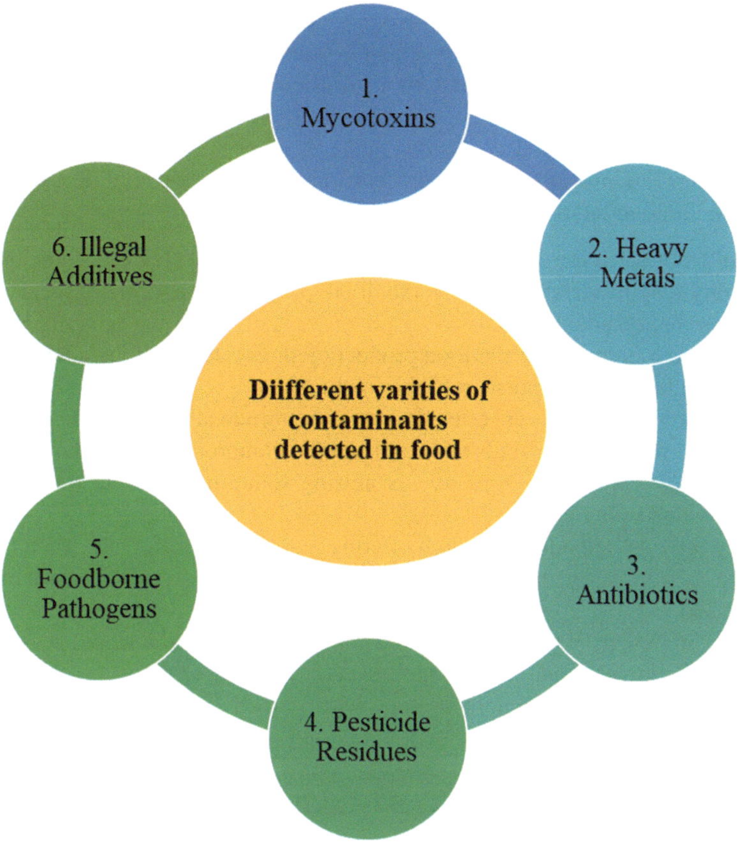

Fig. 2 Different varieties of contaminants detected in food

- Antibiotics: The leftovers of antimicrobial medications used in aquaculture and livestock. excessive use in animal husbandry or improper withdrawal times. Human allergic reactions and antibiotic resistance may result from it.
- Traces of chemicals used to shield crops from pests and diseases are known as pesticide residues. Reasons reproductive problems, cancer, and hormonal disruption can result from prolonged exposure.
- Foodborne pathogens are dangerous microorganisms that can be found in tainted food, such as Salmonella, E. Coli, and Listeria. It results from unsafe food handling practices, poor hygiene, and tainted water or ingredients. It can result in serious infections, gastrointestinal disorders, and food poisoning.
- Illegal Additives: Ingredients that are combined with food without authorization or in excess of what is allowed (e.g., unauthorized dyes, preservatives). Fraudulent practices in food processing is the term for it. It may be toxic, cause allergic reactions, or pose long-term health hazards.

Numerous crucial reasons apply to having a superior tracking system, nano biosensors in particular. Their advanced tracking systems include accurate shelf-life determination. They track the essential parameters such as temperature and humidity, thus giving a precise assessment of the foods' shelf life [41]. The confidence and transparency between food production and consumer is enhanced. Consumers are enabled to make informed purchasing decisions thus building confidence. They are useful in the food industry for the following reasons;

- Adulterant detection such as starch powder preservative chemical dyes and urea.
- Quality control by monitoring the food products' quality and detection of contamination adulterants.
- Shelf-life monitoring of the food product by the detection of any changes in the desired chemical composition of the food.
- Traditional approaches are often time-consuming and labor-intensive. This delays the identification of hazards and results in the compromisation of human health. Thus, nano biosensors come as a promising solution with their improved food quality and safety [33].
- They provide immediate research offering real-time detection and monitoring.
- With their high sensitivity, trace containments of pathogens in food samples are detected.
- The miniature nano biosensors are integrated into portable handheld devices. This results in on-site monitoring of the entire food supply chain.
- Informed making throughout the food supply chain.

3.1 Detection of Chemicals and Adulterants with Nano Biosensors

Detection of chemicals during food processing techniques is essential for protecting human health. It may also be useful for controlling the global trade of contaminated food products [42]. Chemical contamination is regarded as unwanted substances that may intentionally or unintentionally be added to food products during food production. A few of the contaminants include chemicals such as furan, melamine, pesticides, drug residues, heavy metals, dyes, and food additives [43]. The various categories of chemical contaminants include agrochemicals, industrial contaminants, environmental contaminants, pesticides, heavy metals, and melamine. They can be given under sub-classifications as shown below in Fig. 3 [44].

Monitoring chemical contaminants is essential for protecting human health and ensuring safe global trade of food products. Chemical contaminants are unwanted substances that can be intentionally or unintentionally added during food production, potentially affecting safety and quality. Chemicals include furan, melamine, pesticides, drug residues, heavy metals, dyes, and food additives. Chemical contaminants

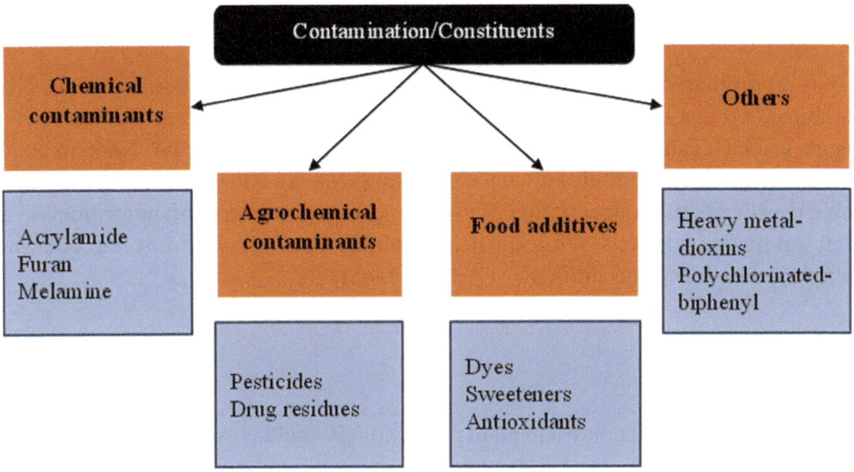

Fig. 3 Classification of chemical constituents. The major divisions include chemical contaminants, agrochemical contaminants, food additives, and other such contaminants

include acrylamide, furan, and melamine. Agrochemical contaminants include pesticides and drug residues from agricultural practices. Food additives are dyes, sweeteners, and antioxidants, which may be added for specific purposes. Others include heavy metals (e.g., lead, mercury), dioxins, and polychlorinated biphenyls (PCBs), which often arise from environmental or industrial sources.

The conventional methods for the detection of chemical contaminants included,

- HPLC—High-Performance Liquid Chromatography
- LC-MS—Liquid Chromatography/Mass Spectrometry
- GC-MS—Gas Chromatography/Mass Spectrometry
- ELISA—Enzyme-Linked Immuno Sorbent Assay
- CE—Capillary Electrophoresis.

These methods are of high sensitivity and with good accuracy. However, they were complicated and time-consuming with an additional need for expensive instruments. Trained personnel were required. Their application in detection was also strongly limited. These led to the development of novel methods that are simple, selective, sensitive, and rapid [45]. Techniques based on nano biosensors can detect, process, and generate precise signals [46]. One benefit of nano biosensors is their ability to determine compounds early on in the processing and storage stages. Customized plasmonic nanoparticles, silver and gold nanoparticles with deoxy ribonucleic acid, that identify changes in optical characteristics, and cantilever detection of changes in mass resonance frequency are the principal analytical approaches.

3.1.1 Fertilizers

They are similarly abused as much as pesticides, which is a problem. *Melamine* is an architectural fertilizer that is also added to protein-containing commodities such as eggs, biscuits, candies, coffee drinks, and so forth as a contaminant. Nanomaterials based on silver and gold fluorescence sensors (DNA Ag NP and DNA Au NP) were able to estimate the concentration of melamine in cow milk and infant formulas even if it was picomolar levels. While the nanomaterials forming cobalt nitroprusside were employed to determine sulfites by cyclic voltammetry.

3.1.2 Pesticides

They are used for pest control in plant protection. Pesticides used include fungicides, insecticides, herbicides, and rodenticides. The mentioned types of products consist of chemicals which generally put in application for the prevention of spoilage of food due to pests, fungi, and rodents. The chemical pesticides employed include organophosphates, carbamates, organochlorines, etc. These pesticides if present in food material, may pose a health risk and should be monitored. Standards for the acceptable amounts of pesticides in food products have been established by the Environmental Protection Agency (EPA) [21]. Nevertheless, the bioaccumulated pesticide residues in human tissue cause cytotoxicity and function as carcinogens.

The residues of pesticides from agricultural products, with a toxic nature, can pose a threat to the food industry [47]. The enzyme acetylcholinesterase (AChE) is used in biosensors for detecting pesticides [41]. This enzyme binds to the pesticides. A graphene-based nanomaterial with polymers imprinted uses an electrochemical sensor for pesticide detection [48]. Whole-cell biosensors based on several potentiometric, amperometric, and conductometric detection techniques have been employed for the screening for pesticides [49]. However, research into revolutionary nanoparticle-based detection techniques has been spurred by biosensors' low precision and constrained detection range. Thus, there is a pressing need for innovative techniques to identify pesticide residues in food products in order to verify food safety.

Since biosensors have high sensitivity and selectivity and can identify dangerous compounds in real time, there has been an increasing amount of synergy between nanotechnology and biosensors [50]. The applications of nano biosensors in food safety for detecting specific types of pesticides, such as organophosphorus and carbamate compounds, as well as various mycotoxins, including aflatoxin, deoxynivalenol, zearalenone, fumonisin, ochratoxin A, and patulin. The integration of nanomaterials like gold nanoparticles, magnetic nanoparticles, carbon nanotubes, and graphene enhances the performance of these sensors by improving their sensitivity, selectivity, and stability [47, 51]. For instance, gold nanoparticles (AuNPs) are widely used due to their unique properties that facilitate the immobilization of biomolecules, which is essential for the detection process. Similarly, magnetic nanoparticles (MNPs) enhance electron transfer between analytes and electrodes, making them effective in biosensor applications. Metal-organic frameworks (MOFs) and graphene derivatives also play a crucial role in developing highly selective sensors for electrochemical detection [52]. MNPs, particularly iron oxide nanoparticles, are used to enhance electron transfer in sensors. Acetylcholinesterase (AChE) was immobilized on MNPs to create a sensor for detecting organophosphorus pesticides in milk and water, demonstrating good sensitivity and reusability [52].

Various mycotoxins such as T-2 toxin, deoxynivalenol (DON), aflatoxins (AFB1), and ochratoxin A (OTA) can be detected using different types of nanomaterials-based fluorescent biosensors. Contaminants like mercury (Hg^{2+}), lead (Pb^{2+}), and cadmium (Cd^{2+}) are also detectable through specific biosensing methods, utilizing materials like quantum dots. The technologies can identify antibiotic residues, including tetracycline (TET), oxytetracycline (OTC), and doxycycline (DOX), through various fluorescent biosensing approaches. The biosensors can be employed to detect pesticide residues in food products, ensuring compliance with safety standards. Detection of pathogens that can cause foodborne illnesses is another critical application of these biosensing technologies. The biosensors can also identify illegal additives that may be present in food products, contributing to food safety [6].

3.1.3 Organophosphates

They are in many of the pesticide formulations. These phosphates are detected using the aid of the gold nanomaterial. The most regularly employed type of nanoparticles in the application of biosensors includes gold nanoparticles [53]. *Carbofuran* is a pesticide used in the control of insects and nematodes in crops. The excessive use of this pesticide may result in vomiting, nausea, abdominal cramps, blood vision, etc., and at high doses may lead to the failure of the respiratory system. *Dichlorodiphenyltrichloroethane* (DDT) is a widely used pesticide around the world due to its effective insecticidal properties. However, they are banned in many countries due to their deleterious effects on wildlife and humans. A nano biosensor based on the immune variety of gold nanoparticle (AuNP) format was developed to detect DDT. Recently, for the identification of pesticides on a minute scale called the nanoscale, a quantum dot nanosensor was discovered. A nanosensor device for the detection of tetradifon, aldrin, atrazine, paraoxon, and glyphosate was created [54] using non-toxic zinc oxide quantum dots [55].

3.1.4 Preservatives

They are very much connected to modern food technology due to the growing demand for processed foods. The group of additives that protect consumers from microbiological risk include benzoates. Their detection and quantification are very important to control food quality. Certain dyes are organic chemicals and few are from petroleum products. Synthetic dyes are mixed with food due to their lower cost, intense color, and increased stability [56]. A few of the dyes include;

- Brilliant blue
- Sunset yellow
- Erythrosine
- Allura red
- Tartrazine.

Some of them may cause abortions of chromosomes in human cells. The synthetic dyes are determined by nanosensoric methods such as graphene oxide-modified multi-walled carbon nanotubes (MWCNT) and polypyrrole-functionalized single-walled carbon nanotubes (SWCNT). The SWCNT is also employed as a toxin-detection sensor in food.

3.2 Adulterants

The act of purposefully combining, replacing, altering, or misrepresenting the original material with non-authentic, exogenous, similar looking, inexpensive, natural or synthetic compounds in order to give the impression of freshness, conceal low quality, and maximize financial gains is known as adulteration [41]. Common adulterants [27] identified in the food sector include the following three categories as shown in Fig. 4.

Nowadays, businesses frequently contaminate food with less costly alternatives to increase revenue as it becomes necessary for establishing and sustaining a balanced diet for people of all ages. An increasing threat that concerns consumers about the safety and shelf life of food products is food adulteration. Adulteration of food has become common in various ways, involving all food groups from cereals, milk, meat, vegetable oils, fruit juices, processed foods, and packed foods. Thus, a well-known concern in the food processing sector is known to be Adulteration. For the safety of the consumers, the food standards and safety authorities of various countries have

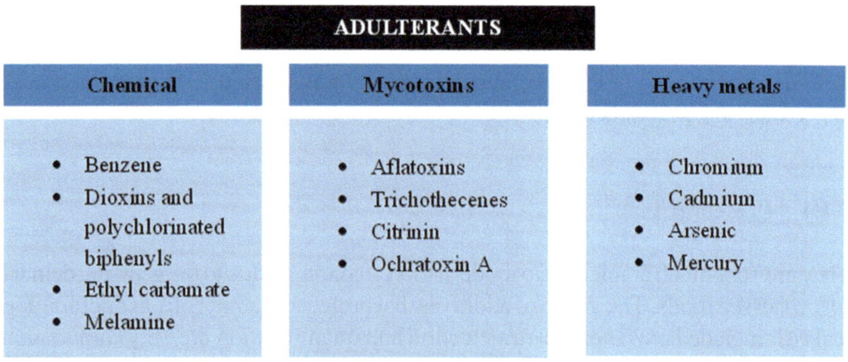

Fig. 4 Common adulterants in the food sector include the classifications of chemicals, mycotoxins, and heavy metals

set guidelines and standard rules under various acts and laws in most of the nation for most of all food products. These regulatory rules and guidelines are intended to guarantee the safety and purity of the food [57]. Dairy products and milk are staple foods that are regularly consumed throughout the world due to their rich source of lipids, vitamins, minerals, and high-quality proteins, casein, and whey proteins essential for human nutrition and the production of amino acids. Dairy products that have traditionally been consumed in large quantities include cheese and butter.

The adulterants in milk can be classified into four categories;

- With the nitrogen contents
- With the milk's fat content and the addition of detergents
- With water and other substances
- To extend the dairy product's shelf life.

Among the most popular types of milk adulteration is, economically driven adulteration, which includes adding vegetable protein, water, whey, and milk from different species. However, these adulterations do not have a major negative impact on health. However, some milk adulterants can seriously harm health and should not be disregarded, like urea, melamine, formalin, detergents, ammonium sulfate, benzoic acid, salicylic acid, H_2O_2, and sugars. A naturally occurring ingredient in milk, urea often originates from the feed or grass that dairy animals are fed. Urea should normally be administered externally if it is present in excess of the normal acceptable range of 10–16 mg/dl [58]. Urea is frequently used to adulterate milk in order to extend its shelf life and SNF. Despite giving milk a creamy white appearance, adding it makes it toxic for the user because it can lead to serious health issues such ulcers, acidity, and indigestion [59].

Additionally, it was discovered that glycerol monocarpylate and a detergent called Sinopol were added to milk in order to improve its volume. It was discovered that these compounds were dangerous for human intake, especially for young children and expectant mothers. In another case, it was discovered that every day, 2000–3000 L of water, syrup liquid, illegal malto-dextrin powder, and other additives were added to pure milk. One of the causes of detergents in milk was also discovered to be the handling and packaging sectors' poor sanitation, hygiene, and cleaning practices. These detergents' high alkaline levels harm bodily tissue and can result in gastrointestinal issues and food poisoning.

Spices are a category of expensive food. They are also included in medicinal commodities. These result in a high level of adulteration with substances such as sudan dye, lead chromate, sawdust, and other coloring agents during processing leading to a degradation of its quality in addition to exposing the consumers to hazardous impacts. Hence, the analysis of the spice products is crucial. The contamination was primarily assessed with conventional methods such as chromatographic and spectroscopic techniques. However, due to few restrictions in this modern world, they are replaced with nanotechnological tools [48]. The tools used with the principle of nanotechnology include nano biosensors, nanobarcodes, nanoprobes, nanocantilevers, e-tongue, nanocarbon, nanotracers, and nanocomposite. The nanosensors used in the detection of spice contamination are effective due to their luminescent, colorimetric, and fluorescent properties for tracing and detection.

The nanomaterials used in nano biosensors include;

- Gold nanoparticles (AuNP), to detect Sudan dyes due to their increased surface area and improved functionalization.
- Carbon nanotubes, to detect heavy metals in spices on-site.
- Graphene-based sensors, to detect illicit dyes used in chili, turmeric, and other spice powders with their remarkable conductivity property.

Melamine is added to protein-containing commodities such as eggs, biscuits, candies, coffee drinks, and so forth as a contaminant. The DNA Cu NP estimated the concentration of melamine in cow milk and infant formulas even in picomolar levels. Aflatoxins comprise some toxic and cancerogen biological agents, which may be found in food, spoiled with molds such as *Aspergillus flavus* and *Aspergillus parasiticus*. The detection of aflatoxin has been done using the immuno-chromatographic strip method of nanomaterials. This aflatoxin present in peanuts and rice can be detected with a carbon nanomaterial using a surface plasmon resonance-based sensor [60]. Ochratoxins are produced by *Aspergillus*. This contaminant is detected with a rapid immunological method.

Antigen modified magnetic nanoparticles are used as signal probes [41]. An immune and chromatographic type chemosynthetic peptide sensor is also employed in the detection of ochratoxin A [61, 62]. Biofilm nano biosensors with DNA aptamers were used as well [63]. Heavy metals such as copper (Cu), mercury (Hg), and arsenic (Ar) are detected using carbon nanotubes [64]. In addition, pork meat and a few processed meat samples were tested for adulteration using a fluorescent nano biosensor [65]. Peanuts, sunflower, soybean, sesame, palm, wheat, corn, potato, rice flour, maize, chicken, etc., were tested with colorimetric biosensors for adulteration [66].

Nano biosensors have emerged as a significant tool in the detection of chemical adulterants in foods, particularly in fruit juices. In fruit juice adulteration, nano biosensors can detect specific adulterants such as formaldehyde in orange juice, with detection ranges as low as 0.01 to 0.3 mg/L. Moreover, advancements in biosensor technology have improved detection capabilities, enabling the identification of various types of adulteration, including the addition of other juices or substances. For instance, biosensors have been developed to monitor microbial growth in grape juice, which can indicate fraudulent activities related to contamination or adulteration. The integration of nanostructures, such as α-Fe2O3, with electrochemical enzymatic biosensors has shown promising results in enhancing the detection of analytes, further underscoring the potential of nano biosensors in food safety applications. These developments highlight the critical role of nano biosensors in the food industry, particularly in the rapid and reliable detection of adulterants, thereby ensuring compliance with safety standards and protecting consumer health [67]. The use of AuNPs in combination with antibodies has been employed to measure chlorpyrifos levels in fruits and vegetables, showcasing the ability to detect pesticides at trace levels [25].

4 New Developments

Food that is fresh and with no health deteriorating substances or contaminants is considered to be safe for human consumption. Hence, analysis and quality control of the produced food is necessary. The *electronic nose* has an olfactory system that is similar to mammals which can detect smells. It is employed with fragrance detection. This gas sensor can identify any change in odor, flavor, or quality. The nanoparticles present help in an improved gas absorption on sensor surfaces [24]. The *hTAS2R38* is a *nano bioelectronic tongue sensor* used for bitter taste detection as well as to detect chemicals such as phenylthiocarbamide and propylthiouracil.

The wireless sensor networks employ nano biosensors to communicate wirelessly with monitors regarding the parameters associated with the food produced. The alert system is a generator for the detection of any deviation in the parameters. The data is transmitted to the central control system for further management of food products. The sensor data collected are stored in the cloud. This eliminates the need for local storage and resources. With the centralized software, other stakeholders can access the data through dashboards or specified applications. Nanotechnology makes it possible to use biosensors in food processing and applications.

The increase in the degree of sensitivity of these nanomaterials was successfully shown by the one-step "turn-on" homogenous experiment for multiplexed pathogen detection conducted utilizing polydimethylsiloxane (PDMS)/paper/glass hybrid microfluidic biochips. In this case, functionalized graphene oxides (rGO) were used [41]. Surface-enhanced Raman spectroscopy (SERE) is a notable advanced sensing technique. This technique is a combination of Raman spectroscopy with nanomaterials which provide reliable detection without labeling. The quick response time is due to the employment of plasmonic sensing. These detections result in improved efficiency and accuracy. The nanosensor devices differ from other sensor technologies on the market today due to a number of important characteristics, such as direct, label-free, and real-time electrical signal transduction, ultrahigh sensitivity, exquisite selectivity, and the possibility of massively integrating addressable arrays.

5 Conclusion

The increased public demand and a complex chain of food supply make the food produce vulnerable to adulteration and chemical contamination. This adversely affects the human health. Thus, standards and policies are established to control these crimes. The conventional methods are useful however in this technical world a superior technology is required. This led to the employment of nano biosensors which are more reliable and faster than traditional sensors when it comes to detecting toxins and adulterants. The food industry's usage of novel nanocomposite materials has greatly improved food quality. The shelf life of perishable foods can be increased, food waste can be decreased, and the need for chemical preservatives can be decreased with the help of nano composite materials' desirable qualities, which include higher mechanical strength, antimicrobial capabilities, barrier qualities and thermal stability to prevent migration in food packaging systems. Continuous innovation and improvement in this industry are fueled by the demands and specifications of customers [37–40].

References

1. Srivastava AK, Dev A, Karmakar S (2018) Nanosensors and nanobiosensors in food and agriculture. Environ Chem Lett 16:161–182. https://doi.org/10.1007/s10311-017-0674-7
2. Afreen S, Muthoosamy K, Manickam S, Hashim U (2015) Functionalized fullerene (C60) as a potential nanomediator in the fabrication of highly sensitive biosensors. Biosens Bioelectron 63:354–364. https://doi.org/10.1016/j.bios.2014.07.044
3. Liang R, Wei M, Evans DG, Duan X (2014) Inorganic nanomaterials for bioimaging, targeted drug delivery and therapeutics. Chem Commun 50(91):14071–14081. https://doi.org/10.1039/C4CC03118K
4. Tao Y, Li M, Ren J, Qu X (2015) Metal nanoclusters: novel probes for diagnostic and therapeutic applications. Chem Soc Rev 44(23):8636–8663. https://doi.org/10.1039/c5cs00607d
5. Campos A, Troc N, Cottancin E, Pellarin M, Weissker HC, Lermé J et al (2019) Plasmonic quantum size effects in silver nanoparticles are dominated by interfaces and local environments. Nat Phys 15(3):275–280. https://doi.org/10.1038/s41567-018-0345-z
6. Zhou J, Gui Y, Lv X, He J, Xie F, Li J, Cai J (2022) Nanomaterial-based fluorescent biosensor for food safety analysis. Biosensors 12(12):1072. https://doi.org/10.3390/bios12121072
7. Bahadır EB, Sezgintürk MK (2017) Biosensor technologies for analyses of food contaminants. In: Nanobiosensors. Academic Press, pp 289–337. https://doi.org/10.1016/B978-0-12-804301-1.00008-4
8. Dumitru LM, Irimia-Vladu M, Sariciftci NS (2016) Biocompatible integration of electronics into food sensors. In: Comprehensive analytical chemistry, vol 74. Elsevier, pp 247–271. https://doi.org/10.1016/bs.coac.2016.04.009
9. Fehr M, Frommer WB, Lalonde S (2002) Visualization of maltose uptake in living yeast cells by fluorescent nanosensors. Proc Natl Acad Sci 99(15):9846–9851. https://doi.org/10.1073/pnas.142089199
10. Kim YP, Daniel WL, Xia Z, Xie H, Mirkin CA, Rao J (2010) Bioluminescent nanosensors for protease detection based upon gold nanoparticle–luciferase conjugates. Chem Commun 46(1):76–78. https://doi.org/10.1039/b915612g
11. Di Francia G, Alfano B, La Ferrara V (2009) Conductometric gas nanosensors. J Sens 2009(1):659275. https://doi.org/10.1155/2009/659275
12. Göpel W (1996) Nanosensors and molecular recognition. Microelectron Eng 32(1–4):75–110. https://doi.org/10.1016/0167-9317(96)00004-4
13. Yonzon CR, Stuart DA, Zhang X, McFarland AD, Haynes CL, Van Duyne RP (2005) Towards advanced chemical and biological nanosensors—an overview. Talanta 67(3):438–448. https://doi.org/10.1016/j.talanta.2005.06.039
14. Gil LA (2007) PCR-based methods for fish and fishery products authentication. Trends Food Sci Technol 18(11):558–566. https://doi.org/10.1016/j.tifs.2007.04.016
15. Nizar NNA, Zainal IH, Bonny SQ, Pulingam T, Vythalingam LM, Ali ME (2018) DNA and nano biosensor technology for the detection of adulteration and microbial contamination in religious food. In: Elsevier eBooks, pp 409–431. https://doi.org/10.1016/b978-0-08-101892-7.00022-5
16. Pérez-López B, Merkoçi A (2011) Nanomaterials-based biosensors for food analysis applications. Trends Food Sci Technol 22(11):625–639. https://doi.org/10.1016/j.tifs.2011.04.001

17. Hossain MM, Ali ME, Hamid SBA, Asing N, Mustafa S, Desa MNM, Zaidul I (2016) Targeting double genes in multiplex PCR for discriminating bovine, buffalo, and porcine materials in the food chain. Food Control 73:175–184. https://doi.org/10.1016/j.foodcont.2016.08.008
18. Ali M, Hashim U, Mustafa S, Man YC, Adam T, Humayun Q (2012) Nanobiosensor for the detection and quantification of pork adulteration in meatball formulation. J Exp Nanosci 9(2):152–160. https://doi.org/10.1080/17458080.2011.640946
19. Ge X, Asiri AM, Du D, Wen W, Wang S, Lin Y (2014) Nanomaterial-enhanced paper-based biosensors. TrAC Trends Anal Chem 58:31–39. https://doi.org/10.1016/j.trac.2014.03.008
20. Huang X, Zhu Y, Kianfar E (2021) Nano biosensors: properties, applications and electrochemical techniques. J Market Res 12:1649–1672. https://doi.org/10.1016/j.jmrt.2021.03.048
21. Chen S, Hwang MT, Wang J, Ganguli A, Park I, Kim Y et al (2022) Advances in electronic nano-biosensors and new frontiers in bioengineering. In: 2022 International electron devices meeting (IEDM). IEEE, pp 1–17. https://doi.org/10.1109/IEDM45625.2022.10019512
22. Bai JA, Rai VR (2018) Effect of nanoparticles on gastrointestinal tract. In: Nanotechnology applications in the food industry. CRC Press, pp 503–516. https://doi.org/10.1201/9780429488870
23. Carvalho FP (2017) Pesticides, environment, and food safety. Food Energy Secur 6(2):48–60. https://doi.org/10.1002/fes3.108
24. Thakur M, Wang B, Verma ML (2022) Development and applications of nano biosensors for sustainable agricultural and food industries: recent developments, challenges, and perspectives. Environ Technol Innov 26:102371. https://doi.org/10.1016/j.eti.2022.102371
25. Christopher FC, Kumar PS, Christopher FJ, Joshiba GJ, Madhesh P (2020) Recent advancements in rapid analysis of pesticides using nano biosensors: a present and future perspective. J Clean Prod 269:122356. https://doi.org/10.1016/j.jclepro.2020.122356
26. Khoshbin Z, Housaindokht MR, Verdian A, Bozorgmehr MR (2018) Simultaneous detection and determination of mercury (II) and lead (II) ions through the achievement of novel functional nucleic acid-based biosensors. Biosens Bioelectron 116:130–147. https://doi.org/10.1016/j.bios.2018.05.051
27. Jha SN (2016) Common adulterants and contaminants. In: Elsevier eBooks, pp 25–61. https://doi.org/10.1016/b978-0-12-420084-5.00002-0
28. Tiwari A, Turner AP (eds) (2014) Biosensors nanotechnology. Wiley. https://doi.org/10.1002/9781118773826
29. Kaur G, Bhari R, Kumar K (2023) Nanobiosensors and their role in detection of adulterants and contaminants in food products. Crit Rev Biotechnol 44(4):547–561. https://doi.org/10.1080/07388551.2023.2175196
30. Biswas R, Alam M, Sarkar A, Haque MI, Hasan MM, Hoque M (2022) Application of nanotechnology in food: processing, preservation, packaging and safety assessment. Heliyon 8(11):e11795. https://doi.org/10.1016/j.heliyon.2022.e11795
31. Rai VR, Bai JA (2018) Nanotechnology applications in the food industry, Rai VR, Bai JA (eds). CRC Press. https://doi.org/10.1201/9780429488870
32. Taleski V, Bosnakovski D, Zdravkovska M, Nasov I, Trajkovska-Petkoska A (2015) Applications of nanotechnology and nanomicrobiology in food safety and health science. https://eprints.ugd.edu.mk/id/eprint/13071
33. Kakimova Z, Orynbekov D, Zharykbasova K, Kakimov A, Zharykbasov Y, Mirasheva G et al (2023) Advancements in nano bio sensors for food quality and safety assurance-a review. Slovak J Food Sci/Potravinarstvo 17(1). https://doi.org/10.5219/1903
34. Arugula M, Simonian A (2016) Nanocarbon-based multi-functional biointerfaces: design and applications. ECS J Solid State Sci Technol 5(8):M3045. https://doi.org/10.1149/2.0091608jss
35. Stavrov V, Stavreva G, Tomerov E, Villani M, Kotsev S, Kostadinov K (2021) Piezoresistive cantilevers' platform for bio-chemical sensing. In: 2021 IEEE international conference on manipulation, manufacturing and measurement on the nanoscale (3M-NANO). IEEE, pp 409–412. https://doi.org/10.1109/3m-nano49087.2021.9599795

36. Thekkethil AJ, Nair R, Madhavan A (2019) The role of nanotechnology in food safety: a review. In: 2019 international conference on computational intelligence and knowledge economy (ICCIKE). IEEE, pp 405–409. https://doi.org/10.1109/iccike47802.2019.9004412
37. Bratovčić A, Odobašić A, Ćatić S, Šestan I (2015) Application of polymer nanocomposite materials in food packaging. Croat J Food Sci Technol 7(2):86–94. https://doi.org/10.17508/cjfst.2015.7.2.06
38. El-Sayed SM, Youssef AM (2023) Eco-friendly biodegradable nanocomposite materials and their recent use in food packaging applications: a review. Sustain Food Technol 1(2):215–227. https://doi.org/10.1039/d2fb00021k
39. Kumar D, Farrukh M, Faisal N (2021) Nanocomposites in the food packaging industry: recent trends and applications. In: Research anthology on food waste reduction and alternative diets for food and nutrition security, pp 122–146. https://doi.org/10.4018/978-1-7998-5354-1.ch006
40. Kumar N, Kaur P, Bhatia S (2017) Advances in bio-nanocomposite materials for food packaging: a review. Nutr Food Sci 47(4):591–606. https://doi.org/10.1108/NFS-11-2016-0176
41. Singh N, Yadav SS (2025) Nanotechnological advancement in spices adulteration detection and authenticity validation. Food Control 167:110806. https://doi.org/10.1016/j.foodcont.2024.110806
42. Xu ML, Gao Y, Han XX, Zhao B (2017) Detection of pesticide residues in food using surface-enhanced Raman spectroscopy: a review. J Agric Food Chem 65(32):6719–6726. https://doi.org/10.1021/acs.jafc.7b02504
43. Kuswandi B, Futra D, Heng LY (2017) Nanosensors for the detection of food contaminants. In: Nanotechnology applications in food. Academic Press, pp 307–333. https://doi.org/10.1016/B978-0-12-811942-6.00015-7
44. Kumar V, Guleria P, Mehta SK (2017) Nanosensors for food quality and safety assessment. Environ Chem Lett 15:165–177. https://doi.org/10.1007/s10311-017-0616-4
45. Kumar V, Agrawal S, Bhat SA, Américo-Pinheiro JHP, Shahi SK, Kumar S (2022) Environmental impact, health hazards, and plant-microbes synergism in remediation of emerging contaminants. Clean Chem Eng 2:100030. https://doi.org/10.1016/j.clce.2022.100030
46. Neethirajan S, Weng X, Tah A, Cordero JO, Ragavan KV (2018) Nano-biosensor platforms for detecting food allergens–new trends. Sens Bio Sens Res 18:13–30. https://doi.org/10.1016/j.sbsr.2018.02.005
47. Wang L, Wang Q, Wang S, Cai R, Yuan Y, Yue T, Wang Z (2022) Bio-control on the contamination of Ochratoxin A in food: current research and future prospects. Curr Res Food Sci 5:1539–1549. https://doi.org/10.1016/j.crfs.2022.09.007
48. Xu L, Granger C, Dong H, Mao Y, Duan S, Li J, Qiang Z (2020) Occurrences of 29 pesticides in the Huangpu River, China: highest ecological risk identified in Shanghai metropolitan area. Chemosphere 251:126411. https://doi.org/10.1016/j.chemosphere.2020.126411
49. Koncki R, Radomska A, Głąb S (2000) Bioanalytical flow-injection system for control of hemodialysis adequacy. Anal Chim Acta 418(2):213–224. https://doi.org/10.1016/S0003-2670(00)00961-2
50. Singh RP (2017) Application of nanomaterials toward development of nanobiosensors and their utility in agriculture. In: Nanotechnology: an agricultural paradigm, pp 293–303. https://doi.org/10.1007/978-981-10-4573-8_14
51. Tai JH, Foster P, Rosales A, Feng B, Hasilo C, Martinez V et al (2006) Imaging islets labeled with magnetic nanoparticles at 1.5 Tesla. Diabetes 55(11):2931–2938. https://doi.org/10.2337/db06-0393
52. Gong Z, Huang Y, Hu X, Zhang J, Chen Q, Chen H (2023) Recent progress in electrochemical nano-biosensors for detection of pesticides and mycotoxins in foods. Biosensors 13(1):140. https://doi.org/10.3390/bios13010140
53. Vohra T, Grover H, Saxena S, Verma DK, Rameshwari R (2020) A review on nanoparticles based biosensors for pesticide detection in water. In: Importance & applications of nanotechnology, pp 1–9
54. Sahoo D, Mandal A, Mitra T, Chakraborty K, Bardhan M, Dasgupta AK (2018) Nanosensing of pesticides by zinc oxide quantum dot: an optical and electrochemical approach for the detection

of pesticides in water. J Agric Food Chem 66(2):414–423. https://doi.org/10.1021/acs.jafc.7b04188
55. Nile SH, Baskar V, Selvaraj D, Nile A, Xiao J, Kai G (2020) Nanotechnologies in food science: applications, recent trends, and future perspectives. Nano Micro Lett 12:1–34. https://doi.org/10.1007/s40820-020-0383-9
56. Viswanathan S, Radecka H, Radecki J (2009) Electrochemical biosensors for food analysis. Monatshefte für Chemie-Chemical Monthly 140:891–899. https://doi.org/10.1007/s00706-009-0143-5
57. Himshweta N, Singh M (2023) Nanosensor platforms for detection of milk adulterants. Sens Actuators Rep 5:100159. https://doi.org/10.1016/j.snr.2023.100159
58. Abdallah MIM, Marouf IIA, Bazalou MS, Dawoud AS (2008) Determination of urea in COW'S milk sold in Damietta GOVERNORATE. J Egypt Vet Med Assoc 68(2):1–7
59. Trivedi UB, Lakshminarayana D, Kothari IL, Patel NG, Kapse HN, Makhija KK et al (2009) Potentiometric biosensor for urea determination in milk. Sens Actuators B Chem 140(1):260–266. https://doi.org/10.1016/j.snb.2009.04.022
60. Ranjan S, Dasgupta N, Chakraborty AR, Melvin Samuel S, Ramalingam C, Shanker R, Kumar A (2014) Nanoscience and nanotechnologies in food industries: opportunities and research trends. J Nanopart Res 16:1–23. https://doi.org/10.1007/s11051-014-2464-5
61. Pola-López LA, Camas-Anzueto JL, Martínez-Antonio A, Luján-Hidalgo MC, Anzueto-Sánchez G, Ruíz-Valdiviezo VM et al (2018) Novel arsenic biosensor "POLA" obtained by a genetically modified E. coli bioreporter cell. Sens Actuators B Chem 254:1061–1068. https://doi.org/10.1016/j.snb.2017.08.006
62. You KH, Luo XE, Hu WJ, Xu Y, Guo JB, He QH (2020) Environmental-friendly gold nanoparticle immunochromatographic assay for ochratoxin A based on biosynthetic mimetic mycotoxin-conjugates. World Mycotoxin J 13(2):267–276. https://doi.org/10.1021/acs.jafc.1c04824
63. Moon YS, Kim HM, Chun HS, Lee SE (2018) Organic acids suppress aflatoxin production via lowering expression of aflatoxin biosynthesis-related genes in Aspergillus flavus. Food Control 88:207–216. https://doi.org/10.1016/j.foodcont.2018.01.017
64. Reza MSA, Hasan MM, Kamruzzaman M, Hossain MI, Zubair MA, Bari L et al (2019) Study of a common azo food dye in mice model: toxicity reports and its relation to carcinogenicity. Food Sci Nutr 7(2):667–677. https://doi.org/10.1002/fsn3.906
65. Ali ME, Razzak MA, Hamid SBA (2014) Multiplex PCR in species authentication: probability and prospects—a review. Food Anal Methods 7(10):1933–1949. https://doi.org/10.1007/s12161-014-9844-4
66. Bai S, Li S, Yao T, Hu Y, Bao F, Zhang J et al (2011) Rapid detection of eight vegetable oils on optical thin-film biosensor chips. Food Control 22(10):1624–1628. https://doi.org/10.1016/j.foodcont.2011.03.019
67. Shokri S, Soltani M, Mazaheri Y, Rezagholizade-Shirvan A, Mohammadi M, Shariatifar N, Zeinali T (2024) New and advanced methods in detection of fruit juice adulteration: focusing on nano-biosensor. J Spectrosc 2024(1):4691816. https://doi.org/10.1155/2024/4691816

Nanobiosensors for Microbial Contaminants and Their Impacts on Food Industry

Abhishek Choudhary, Arshiya Bhati, and Manas Kumar Dhal

Abstract Health and food safety experts continue to have serious concerns about foodborne infections. Worldwide, reports of foodborne-related mortality average 0.42 million respectively. High-concern pathogens include bacteria like Salmonella, Escherichia coli, Campylobacter, and Listeria monocytogenes, which are known to produce Shiga toxins. Among the aquatic pathogens of great concern are Schistosoma japonicum, leptospirosis, Vibrio cholerae, and others. The global incidence of food-borne illnesses remains high, even with significant efforts being made in the areas of food quality control to monitor the presence of these pathogens of concern in various sources. These factors make it necessary to create new, quicker pathogen detection techniques that may be used with real-time surveillance plans. In place of more time-consuming and unsuitable techniques for large-scale surveillance, biosensor-based methods have become unique tools for the quicker identification of food pathogens. In a nutshell, biosensors are devices that detect infections by biochemical reactions involving biorecognition components such as tissues, isolated enzymes, antibodies, genetic materials, or aptamers. The majority of biosensors rely on the correlation between electrical, thermal, or optical signals when pathogen biomarkers are present. The use of molecular and nanotechnology makes it possible to identify pathogens at incredibly low pathogen concentrations quickly and with high sensitivity. As a matter of fact, the incorporation of magnetic, silver, iron, and gold nanoparticles (NP) into biosensors has proven to enhance their detection capabilities. The main use of nanomaterials and biosensor-based instruments for the identification of pathogens in food is the subject of this chapter. Furthermore, it emphasizes how nanoparticles have improved biosensor devices. Particular benefits are provided by nanomaterials for pathogen detection. The sensitivity and selectivity of the biosensors are improved by the more efficient interaction with pathogenic agents made possible by the nanoscale and high specific surface area. Lastly, the capacity of biosensors to bind to certain

Abhishek Choudhary and Arshiya Bhati are contributed equally to the chapter.

A. Choudhary · A. Bhati
School of Biological and Life Sciences, Galgotias University, Greater Noida, Uttar Pradesh, India

M. K. Dhal (✉)
Department of Zoology, Shivaji College, University of Delhi, New Delhi, India
e-mail: manasdhal@shivaji.du.ac.in

D. Sillu et al. (eds.), *Nanobiosensors for the Food Industry*, Smart Nanomaterials Technology, https://doi.org/10.1007/978-981-95-0136-6_10

molecules, like antibodies or nucleic acids, makes it easier to identify the pathogens of interest.

Keywords Nanobiosensors · Nanomaterials · Foodborne diseases · Antibodies · Food safety

Abbreviations

CNs	Carbon nanomaterials
AuNP	Gold nanoparticles
CNTs	Carbon nanotubes
ATP	Adenosine triphosphate
HCG	Human chorionic gonadotropin
H2O2	Hydrogen peroxide
H2S	Hydrogen sulphide
ISE	Ion-selective electrodes
FET	Field-effect transistors
CdSe and ZnS	Cadmium selenide and zinc sulphide
LAMP	Loop-mediated isothermal amplification
FRET	Fluorescence resonance energy transfer
ZEN	Zearalenone
OTA	Ochratoxin
ZnO	Zinc oxide
RFID	Radio frequency identification
MNPs	Magnetic nanoparticles
NaYF4	Sodium yttrium fluoride

1 Introduction

The population of the world is predicted to grow exponentially from its projected 8.5 billion by 2030 to 9.8 billion by 2050, which would pose challenges for the extraction of natural resources and the production of food. Reduced land area, a lack of food and crops, competition for natural resources, and the need to boost crop output in unfavorable environmental conditions are other significant obstacles [1]. Furthermore, food contamination by microbes during processing and storage is a problem that is causing worry on a global scale. In order to increase the effectiveness of natural resources, including plant hormones, a wide range of pathogens, herbicides, fertilizers, and metal ions used in the production of agricultural and food products, it is imperative that new sustainable methods and/or techniques be developed. Advanced functional materials are now available thanks to nanotechnology, which is energizing the agri-food sectors' current procedures. A variety of techniques, including chemical, biological, and physical approaches, have been used to synthesize nanomaterials [2]. Utilizing the special qualities of nanoparticles has allowed for the advancement of traditional technology. The integration of nanomaterials with biosensors, also referred to as nanobiosensors, has enhanced the sensors' capacity for a wide range of environmental applications. Therefore, the creation of reliable, real-time nanobiosensors that integrate highly specialized biological components with distinctive nanomaterial features is a superior alternative to the quick and simple diagnosis and detection of plant diseases [3]. Nanobiosensors are specialized, sensitive, and non-invasive sensors that are made using advanced nanotechnology techniques. Nanobiosensors make it simple to gather and analyze real-time reaction signals [4]. A variety of nanomaterials, including nanowires, nanoparticles, nanocrystals, nanotubes, and nanocomposites, are used in nanobiosensors. The evaluation of ecosystems' natural resource supply, such as soil fertility and groundwater accessibility, is one of the many applications for nanobiosensors [5]. The effects of microbial contamination on nanobiosensors and their implications for the food sector are the main topics of this chapter.

2 Contaminants in Food

Numerous factors, including recently discovered chemical risks, erroneous use of pesticides and veterinary medications, naturally occurring toxicants like phytotoxins, bacterial contamination, and chemical additives added during processing, can lead to food contamination. Food contamination detection is crucial for both safeguarding human health and preventing the international trade of tainted food goods [6]. Toxins and pathogens are examples of biological contaminants found in food, while pesticides, veterinary medicine residues, and food additives are examples of chemical contaminants. These pollutants have the potential to cause major health problems, especially drug-resistant diseases that decrease the effectiveness of medications, especially those involving veterinary drug residues [7]. In actuality, the food business requires appropriate analytical techniques for food product processing and quality assurance. Since chemical and biological pollutants are not visible like physical contaminants are, their identification is crucial to the food's quality. With the exception of a few key analytes such sugars, alcohols, flavors, sweeteners, and amino acids, the majority of food applications concentrate on the identification of chemical and biological contaminants. Consequently, it is imperative to allocate resources toward the advancement of sensors, specifically chemical and biosensors, for the purpose of evaluating food quality. These technologies have demonstrated great potential as substitutes for conventional analytical methods like spectroscopy and chromatography. Still, relatively few nanosensors are widely used in food processing or quality assurance. The development of affordable, dependable, and resilient nanosensors and nanobiosensors that can function in real-world environments requires a significant amount of work [8] (Fig. 1).

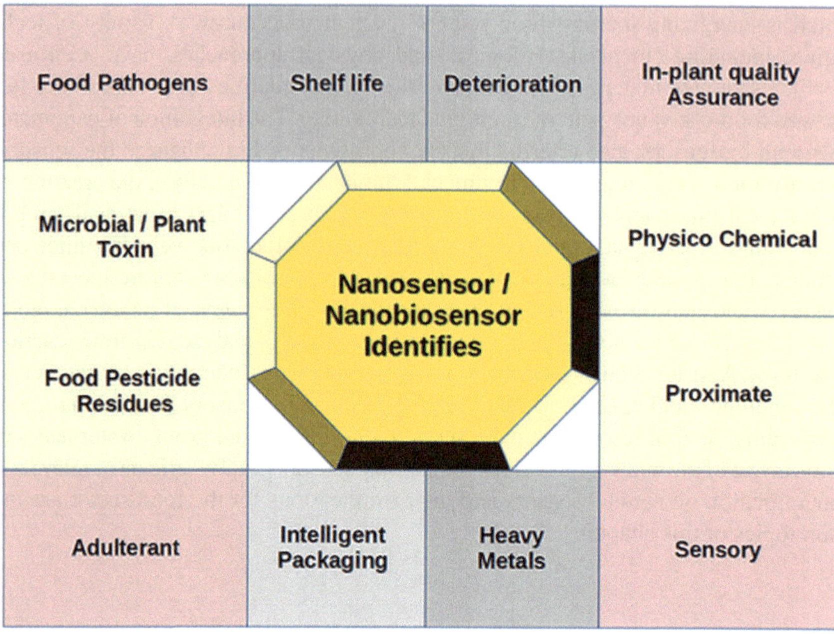

Fig. 1 Uses of nanosensors

2.1 *Biological Contaminants*

Unwanted, hazardous biological substances—such as toxins and bacteria—can originate from naturally occurring toxicants, be foodborne pathogenic bacteria, or develop during the preparation, distribution, and storage of food. These chemicals are referred to as biological contaminants. In order to preserve the health of both humans and animals, it is evident that biological pollutants that cause foodborne illnesses must be identified and prevented. The main pollutants that lead to foodborne illnesses include other foodborne pathogenic bacteria and naturally occurring toxicants like ochratoxins and aflatoxins. Most outbreaks of foodborne illness are caused by pathogenic bacteria, including *Salmonella*, *Campylobacter*, *Listeria monocytogenes*, *Escherichia coli*, *Staphylococcus aureus*, and *Bacillus cereus* [9].

2.2 *Chemical Contaminants*

Unwanted and hazardous chemicals known as chemical contaminants can be added to food either purposefully or accidentally. They might originate from natural sources, environmental contamination, or food processing, transportation, and storage. Chemical contaminants come in many forms, from industrial and environmental pollutants to pesticides, heavy metals, and melamine, which are new concerns that may have negative impacts on the health of humans and animals [10].

3 Toxicants Occurring Naturally

There are lots of toxicants occurring naturally out of which some of these described here.

3.1 Ochratoxin

To detect ochratoxin A (OTA), a quick and sensitive immunological technique was also Created [11]. Aspergillus is a common producer of this mycotoxin. Utilizing antigen-modified magnetic nanoparticles (MNPs) as immunosensing probes and antibody-functionalized rare-earth-doped $NaYF_4$ (sodium yttrium fluoride) upconversion nanoparticles as multicolor signal probes, the approach was based on competitive fluorescence immunoassay for the simultaneous detection of AF B1 and OTA in food items.

The fluorescent signals of the nanocomposites steadily dropped when the quantities of AF B1 or OTA increased, although the fluorescent intensity was maximum in the absence of mycotoxins. Here, the quick separation and purification of the immune complex via the use of MNPs helps to shorten the assay's total duration. Nanostructured zinc oxide (nano-ZnO) film served as the foundation for another immunological technique for the detection of OTA [12]. The film has been applied to a glass plate coated with indium tin oxide (ITO) in order to immobilize BSA and rabbit immunoglobulin antibodies (r-IgGs). In this method, the free amino-terminal sites of r-IgG preferentially bind to the carboxylic group of the OTA molecules, whereas positively charged nano-ZnO interacts electrostatically with the carboxyl groups of r-IgG. Additionally, an aptasensor has been designed as a specific and sensitive fluorescence assay for OTA detection [13]. In the aptasensor, single-walled carbon nanotubes (SWCNTs) serve as a quencher. The aptasensor was used to demonstrate the linear detection range of 25–200 nM and the detection limit of 24.1 nM. N-acetyl-L-phenylalanine, ochratoxin B, and warfarin did not interfere with this nanosensor's ability to react to OTA. A 1% beer-containing buffer solution that had been tampered with at various OTA concentrations was tested in order to confirm the validity of aptasensor. The outcomes demonstrated how well the aptasensor detected OTA in an actual food sample.

3.2 *Aflatoxin*

The most common naturally occurring toxin in food crops is aflaoxin B1 (AF B1). This most common mycotoxin is produced by strains of *Aspergillus parasiticus* and *Aspergillus flavus* that can grow on food crops during production and storage. It exhibits teratogenic, carcinogenic, mutagenic, and immunosuppressive properties. Because of this, it has been classified as a human carcinogen by the International Agency for Research on Cancer (IARC). A novel nanosensor developed utilizing the immunosensor approach has been used to identify this mycotoxin. The immunosensor is a highly quick and efficient way to detect AF B1 because it is based on antibody-attached gold nanoparticles (AuNPs) and has a reaction time of 60 s. It can detect AF B1 in the range of 10e100 ng/dL. Here, the 4-mercaptobenzoic acid (4-MBA)-based self-assembled monolayer (SAM) on gold electrode (MBA/Au) was immobilized with the aflatoxin B1 antibody (aAF B1), which is covalently connected to cysteamine functionalized gold nanoparticles (C-GNPs) [14]. A lateral flow strip was also used to identify AF B1 in food [15]. Here, the immunosensor employs AgAu—gold-shelled nanoparticles with a silver core and a monoclonal antibody bound on them—as the detecting reagent. The membrane-based immunodipsticks are composed of a control line containing goat antimouse IgG and a test line containing AF B1 conjugated to bovine serum albumin (BSA). Red AgAu nanoparticles coated with anti-AF B1 are used as the signaling reagent in the strip to create one or two colored lines on the membrane, making it simple to visually identify AF B1.

4 Foodborne Pathogens

E. coli can cause food poisoning symptoms including bloody diarrhea, stomach pains, nausea, and vomiting, whereas Salmonella can cause symptoms like diarrhea, fever, and abdominal cramps. Carbon nanotubes (CNTs) have been integrated into a nanobiosensor to detect Salmonella [16]. Here, the high surface area of CNTs was immobilized onto a glassy carbon electrode (GCE) after covalently attaching to Salmonella monoclonal antibodies via diimide-activated imidation as the coupling mechanism. The change in charge-transfer resistance and impedance before and after the creation of an antigen-antibody combination served as the pathogen detection sensing mechanism. This is because the behavior of carbon nanotubes (CNTs) as molecular wires allows electrical communication between the conjugated antigen-antibody combination and the underlying electrode. Magnetic nanoparticles (MNPs) coated with antibodies were used to detect Salmonella bacteria in milk. The samples and MNPs were separated using an external magnetic field [17].

This technique involved dispersing the MNPeSalmonella clusters in a buffer solution and exposing them to the UV-absorbent TiO2 nanocrystals (TNs) for antibody binding. After the MNPeSalmonella TN complexes were separated from the solution, the UV visibility spectrum for the solution without the attached TNs was measured. Since the UV visibility spectrum was inversely correlated with the Salmonella concentration, the test demonstrated a high degree of sensitivity toward minute levels of Salmonella germs. Bacterial pathogens were eliminated from urine, food, and water samples using amine-functionalized magnetic nanoparticles (AF-MNPs), with removal rates ranging from 88.5 to 99.1% [8]. Strong electrostatic interactions with the negatively charged regions on the bacterial surfaces resulted from the binding of at least eight distinct gram-positive and gram-negative bacterial species to the positive charges of AF-MNPs. These interactions were made possible by the ionic strength, the pH of the phosphate buffer solution, and the quantity of AF-MNPs. This method is intriguing, but more study is needed to determine how specific it is to harmful and helpful microorganisms. Furthermore, it has been reported that several arrays of SWCNT field-effect transistors (SWCNT-FETs) functionalized with aptamers were created for the purpose of detecting *E. coli* [18]. This nanobiosensor's great sensitivity was demonstrated by a conductance drop of more than 50% in the presence of E. coli. A sensor device created by researchers enables the quick and accurate identification of E. coli. Gold nanoparticles were directly placed to carbon nanotubes (CNTs) that had been produced on a graphite surface. To finish the assembly, *E. coli*-specific thiolated RNA was affixed to the nanoparticles. Because gold nanoparticles were added, this mixed nanomaterial had the advantage of preserving the electrical characteristics of the CNTs. The CNTs' larger surface area allowed for more places to attach analytes, which improved the sensitivity of the method. The detection of E. coli was shown to be 189% higher with RNA-coated gold nanoparticles than with untreated ones. A piezoelectric nanobiosensor was developed for the quick identification of *E. Coli* O157:H7 utilizing oligonucleotide-coated gold nanoparticles [19].

5 Types of Biosensors

For agriculture and food corporations, nanosensors provide a promising new method of ensuring the quality and safety of food [20]. A biosensor is a sophisticated and potent analytical instrument that uses biological, chemical, and physical sensing techniques [21, 22]. Since the goal of biosensors is to uncover the diversity found in biomolecular complexes like oligonucleotides, antigen-antibody interactions, living cells, tissue-level interactions, and interactions involving enzyme substrates and lectin-glycoproteins, it can be concluded that nanobiosensors are used for the detection of urea, glucose, pathogens, and different microorganisms. Nanosensors' small size, stability, repeatability, durability, and mobility make them superior to other kinds of sensor [23]. Nanomaterial or carbon nanotube (CNT) electrodes are used to connect electrochemical sensors. Electrochemical sensors are related to carbon nanotube-based electrode designs. To utilize its enhanced optical property, the metallic nanoparticle can stick to other nanoparticles. Other nanoparticles that can be incorporated into optical nanosensors to increase their flexibility and efficiency are quantum dots and upconversion nanomaterials [24]. Certain mass nanosensors fall into a third group that has restricted uses in the food and agriculture industries because of their mechanical and piezoelectric properties, which are attributed to microcantilevers and crystals, respectively [25] (Table 1).

5.1 Electrochemical Sensors

The biological components and the target analytes interact with the chemical and biomolecular nanofabrication [26]. The quantity of electron gain or loss that happens during biointeractions can be measured using electrochemical signals such as voltage, current, and impedance. These electrochemical signals are quantitatively generated and are related to the presence of an analyte in a sample. Extremely sensitive electrochemical transducers must have minimal energy requirements, dependability, low cost, low maintenance, speed, low detection limits, and simplicity that are compatible with current miniaturization and nanofabrication technologies in order to be used in practical applications. Based on the fundamental principle of measurement, electrochemical nanosensor devices can be categorized into three groups: potentiometry, voltammetry, and amperometry. One kind of electrochemical sensor that continuously measures the current produced by the redox reactions of electroactive molecules in a biological reaction is an amperometric sensor. The faradaic current at a given voltage indicates the concentration of electroactive substances. Current forms a link with electroactive species because it is typically proportional to voltage. The medical and diagnostic fields make extensive use of amperometric nanosensors. The ATP development sensor and sophisticated pregnancy tests that use beta-HCG (human chorionic gonadotropin) sensors are two examples [27]. Because of their great sensitivity, low cost, and reliability—all made possible by nanofabrication technology—amperometric biosensors are used in the food and agriculture sectors, as well as in environmental and biomedical applications. Numerous compounds, including organophosphate [28]. Sulfonamide [29] amperometric bio-nanosensors can detect hydrogen peroxide (H_2O_2) [30, 31], hydrogen sulfide (H_2S), and fructose concentration [32]. *Salmonella typhi* [6], carbosulfan in rice [33], and heavy metal contamination in food samples [34] have all been detected using cyclic voltammetry. Glass electrodes, liquid-membrane electrodes, metal-metal oxide electrodes, gas electrodes, ion-selective electrodes (ISE), and field-effect transistors (FET) are among the many electrode types utilized in potentiometric biosensors [35]. Potentiometric biosensors are used in the fermentation industry to monitor the amount of alcohol in wine and beer [36].

Table 1 Different types of biosensors used for detecting various microbial contaminations

Types of biosensors	Microbial contamination detected	Principle	Target microorganisms	Detection method
Electrochemical biosensors	Waterborne pathogens	Measures electrical signals due to microbial activity	*E. coli*, *Salmonella*, *Vibrio cholerae*	Amperometry, Potentiometry, Conductometry
Optical biosensor	Bacterial contamination in food and water	Fluorescence, absorbance	*Listeria*, *Salmonella*, *Campylobacter*	Electrochemical optical
Piezoelectric biosensor	Pathogenic bacteria in food and beverages	Detects changes in acoustic waves or mass due to microbial binding	*E. coli*, *Salmonella*, *Staphylococcus aureus*	Quartz crystal microbalance (QCM)
Magnetoelastic biosensor	Bacterial biofilms in medical devices	Measures shifts in mechanical resonance induced by magnetic fields	*Pseudomonas aeruginosa*, *S. aureus*	Resonance frequency shift
Thermal biosensor	Microbial growth in food products	Detects heat changes associated with microbial metabolism	*Bacillus cereus*, *E. coli*, *Listeria monocytogenes*	Calorimetry
Impedimetric biosensor	Bacterial contamination in liquids	Measures changes in electrical impedance due to microbial presence	*E. coli*, *S. aureus*, *Vibrio cholerae*	Impedance spectroscopy
DNA-based biosensor	Detection of microbial genetic material	Hybridization between microbial DNA/RNA and complementary probes	*E. coli*, *Salmonella*, *Norovirus*	DNA hybridization, PCR-based techniques
Enzyme-based biosensor	Microbial toxins or metabolic byproducts	Uses enzymes to catalyze reactions related to microbial byproducts	*Botulinum* toxin, *Cholera* toxin, Enterotoxins	Enzyme-linked reactions (e.g., ELISA)

(continued)

Table 1 (continued)

Types of biosensors	Microbial contamination detected	Principle	Target microorganisms	Detection method
Cell-based biosensor	Environmental monitoring of microbial contamination	Uses living cells to detect microbial stress or toxins	*E. coli*, *Bacillus subtilis*, *Pseudomonas*	Whole-cell biosensing (bioluminescence, fluorescence)
Antibody-based biosensor	Bacterial pathogens in food and water	Uses specific antibodies to detect microbial antigens	*Salmonella*, *Listeria*, *E. coli*	Immunosensing (SPR, Electrochemical)

5.2 Optical Sensors

Absorption, phosphorescence, fluorescence, Raman, and surface-enhanced Raman scattering are just a few of the many spectroscopic measurements that optical biosensors may detect changes in the optical signal. Changes in the intensity, polarity, phase, and wavelength of light can be used to calculate the metrics. The concept of sensing-based quantum dots has progressed through selective functionalization, opening the door for their use in food processing and agriculture [37]. Coming up with this approach looks like a smart idea. The silane group, for instance, alters the surfaces of ZnS and CdSe quantum dots. Additionally, these CdSe and ZnS quantum dots are combined with a methylacrylate-functionalized molecularly imprinted polymer to form a quantum-dot-based optosensor for dicyandiamide detection in milk. The concentration of dicyandiamide correlates linearly with the fluorescence quenching detected by quantum [38].

6 Pathogen Detection with Nanosensors

Today's food industry places a high premium on creating innovative technologies that are more sensitive and resilient. Techniques based on nanobiosensors can generate, decipher, and recognize exact signals. One advantage of employing nanobiosensors is the early identification of chemicals, diseases, and hazards during processing and storage. Functionalized plasmonic nanoparticles, gold and silver nanoparticles coupled with DNA to assess changes in optical properties, and cantilevers to monitor changes in mass resonance frequency are the main analytical techniques. Modern methods that have improved detection include loop mediated isothermal amplification (LAMP), inventive packaging, and microfluidics. Most nanobiosensors developed to date have increased the sensitive solution's detection limit [39]. To ensure the safety and quality of food items, the food industry has developed a range of aptamers and microfluidic approach-based nanobiosensors to meet their diverse needs for identifying toxins, heavy metal residues, diseases, and pesticides. Aptamers that weigh less than 25 kDa are typically needed by biosensors to detect infections. One benefit of employing these aptamers in nanobiosensors is their selectivity, which allows them to selectively couple to chemicals, ions, proteins, bacteria, and viruses [40]. By sensing package oxygen, a photoactivated indicator dye based on methylene dye has been developed to aid in the identification of bacteria, demonstrated a methylene blue-based photoactivated indicator ink that may be used to gauge oxygen concentration, a marker of aerobic pathogen growth. Numerous bacteriophage-based nanobiosensors developed to far for the detection of infections like Staphylococcus aureus have received approval from the US Food and Drug Agency [41]. Most bacterial and fungal illnesses emit toxins that can seriously affect human health both immediately and over time. These adverse consequences include decreased immunity, altered protein metabolism, convulsions, neurotoxicity, and liver cancer [42]. Until now, optical, piezoelectric, and/or electrochemical sensing techniques have been used to identify toxins in food products [43].

Among the several forms of toxicity present in food products are heavy metals. The detrimental effects of heavy metals, such as lead, arsenic, cadmium, and mercury, on metabolic pathways have been shown in several clinical trials. Novel biosensors using genetically modified bacterial cells and green fluorescent signals were developed to detect the presence of these heavy metals in food products. Aflatoxin limits for graphene-based nanosensors based on gold nanoparticles ranged from 0.1 to 2.5 mg [44]. Human health is impacted by foods tainted with antibiotic residues, and the amount of particular foods that are allowed has been determined. It is necessary to check for sulfonamide residues in meat and poultry products. A nano enzyme-labeled immunoassay has been created to detect sulfadiazine in food residue nanosensors [45]. Gold, platinum, and silica have been used as biomimetic materials to construct molecularly imprinted polymers and nanoenzyme conjugates. The accuracy of the sensor has been evaluated for milk and honey with different sulfadiazine concentrations (0.5–12.5 mg/L) using the enzyme-linked immunosorbent assay method. The nanosensor has a limit of detection of IC15 0.09 and an IC50 of 6.1 mg, making it incredibly sensitive. Melamine is widely used in the paper, textile, leather, and other industries (2,4,6-triamino-1,3,5 triazine). It has been used to increase the quantity of artificial protein in milk due to its high nitrogen content. DNA-Cu-NPs were produced using ascorbic acid and DNA as a template, and they were then identified using fluorescence-based detection. A new approach is the use of the AS1411 template for fluorescent copper nanoparticles.

With a limit of detection ranging from 50 to 120 μmol/L, melamine easily detects fluorescence by quenching it. This detection method has shown itself to be less complex, incredibly sensitive, and cost-effective [46]. Food safety is aided by these kinds of nanosensor devices. The most common method for detecting ochratoxin A, a mycotoxin with a high risk of cancer, is immunochromatographic tests. However, this strategy is expensive. To cut costs, a chemosynthetic mimotope peptide was developed, which has a limit of detection of 0.187 ng/m and can detect mycotoxin with the unaided eye in 15 min [47]. Mycotoxin is one of the primary contaminants that results in crop losses during storage. DNA aptamers, which are tiny nucleotides that glow, are detected by FRET [48]. A novel graphene oxide-based steganographic aptasensor has been developed for the detection of ochratoxin (OTA) and zearalenone (ZEN) mycotoxins using the Alexa fluor 488 aptamer and the capture probe Cy3.

ZEN and OTA have detection limits of 1.79 and 1.484 ng/mL, respectively, according to Aptasensor [49]. Mycotoxin zearalenone (ZEN) has been detected using an optically based nanobiosensor that uses a silicon dioxide layer and ZEN-specific antibodies. With a detection limit of 0.01 ng/mL, ZEN-specific antibodies immobilize onto the polyelectrolyte layer upon attaching to mycotoxin, altering the refractive index and verifying the toxin's existence [50]. Performance improvements for electrochemical sensors based on nanomaterials have been explored [51]. Toxins and foodborne pathogens produced by *E. coli* and Vibrio cholera can be detected using a variety of electrochemical biosensors.

Foods contaminated with Campylobacter jejuni are the source of widespread food outbreaks. The use of DNA aptameric-based nanobiosensors for disease detection is a crucial method for guaranteeing food safety. High affinity single-stranded DNA aptamers with a detection limit of 10 CFU/mL and 59 nucleotides have been developed to detect Campylobacter jejuni. Screening of milk samples has validated the findings' correctness [19]. Traditional techniques for diagnosing foodborne germs include time-consuming culture operations. E. coli is thought to be one sign of spoiled food. With a detection limit of 101 CFU/mL after 15 min, ZnO nanoparticles produced from sol-gel and functionalized multiwalled carbon nanotubes have been developed to detect *E. coli* using β-galactosidase.

7 Nanosensors in Packaging of Food

Packaging for food has an impact on food safety. Food packaging plays a critical role in the food processing industry in preserving product quality and improving food safety throughout storage and transportation. Food items with appropriate packaging also have extended shelf life for a number of internal and external factors. Food products' nutritional value and organoleptic qualities are significantly influenced by their packaging and monitoring methods. Using traditional packing techniques to keep food from going bad has a number of disadvantages. With the development of packaging technology, food companies started using newly created active and bioactive containers as well as a variety of techniques to monitor and regulate temperature, pH, moisture, humidity, and gas generation during production and storage. However, given the growing concerns about food safety and the threat presented by bioterrorism, a rapid and sensitive monitoring system is needed to detect them [52]. Current conventional methods, however, are insufficient for this. However, a sensitive and quick monitoring system is required to identify them as worries about food safety and the possibility of bioterrorism grow. However, current traditional approaches fall short of this.

7.1 Detection of Internal Conditions of Packaged Foods Using Nanosensors

Based on internal and exterior variables, the smart packaging system senses and provides information about packed meals, such as the quality of food throughout transport and storage. Customers benefit from being able to get fresh, packaged items with better food safety. By detecting different analytes, smart packaging systems provide real-time food quality monitoring, hence lowering the risk of foodborne illness. The following elements make up the smart packaging system: (i) Indicators: freshness indicators (pathogen or microbial spoilage), integrity or gas indicators, and time–temperature indicators, (ii) RFID tags and barcodes, and (iii) Sensors: fluorescence-based oxygen sensor (gas sensor) and biosensor. Because of their tiny detecting surfaces, nanobiosensors in smart packaging can generate a response signal at low analyte concentrations and simplify assay procedures. From the moment of creation to the customer, the impacts of temperature, humidity, moisture content, and solar intensity on food quality may be tracked using nanosensors thanks to developments in computer and nanotechnology. Salespeople may determine the freshness of the products by interpreting the data that the nanosensors have collected. During storage and shipping, fresh fruits are mixed with nanosensor-based measurement devices to keep an eye on temperature, moisture content, and other atmospheric variables. When RFID tags and nanosensors are combined, it becomes easier to keep an eye on the quality and freshness of perishable goods like meat, fish, fruits, and vegetables while they are being transported and stored. RFID system consists of transponder and reader with their associated antennas [53]. RFID tags are categorized into active and passive tags based on the usage of power source. Active tags require an internal power source for energizing microchips and they are used for long-range communication (20–100 m). Moreover, these devices are more expensive as well as bulky owing to the embedded power source. Alternatively, the passive tags are more compact in size (desirable for food packaging) due to the absence of internal power source. Because of their low cost and endless lifespan, they are appropriate for use in food packaging applications. The RFID tags with low operating frequencies (9–135 kHz) are preferable for monitoring the packaged food products containing high water content as compared to high-frequency RFID tags (0.3–300 GHz).While tropical fruits are being stored, sold, and transported, their ripening progress has been monitored using an RFID system with a metal oxide sensor. Biosensor-enabled RFID tags allow food distributors to improve their supply chain and drastically reduce the amount of food degradation before it reaches the retailers by tracking the freshness and quality of the food products [54].

7.2 Detection of External Conditions of Packaged Foods Using Nanosensors

Nanosensors present in the electronic tongue detect released gases from the deteriorated food products, which indicate the freshness of the products [55]. The freshness of the minced caspian (Clupeonella cultriventris caspia) sprat using bromophenol blue-$GONH_2$ nanosheet and a colorimetric nanosensor. This nanosensor detected the liberation of ammonia and formaldehyde, which was indicated by a change in color from brown to blue at the storage temperature of 4 °C. Mills (2005) used nanosensors with oxygen indicators to determine spoilage in uncooked bacon that was packed under CO_2 conditions [56]. This nanosensor uses O_2 as an indicator. In industrial level packaging, nanosensors play a stringent role in maintaining proper packaging. CNT-based nanosensors are used in fruit juice industries for quick detection of spoilage of antioxidant vitamins such as folic acids [57]. AuNPs attached to nanosensors were used in the beverage industry to assess the level of phenolic content. It is also used to estimate the sugar and ascorbic acids in ripened fruits [3, 58]. The advancement of nanosized sensors in smart and intelligent packaging for food packaging is a significant development for food security. A nanosensor-enabled innovative and intelligent packaging system will aid in the identification, analysis, and documentation of the supply chain. The TiO_2 composite biosensor was designed to analyze the freshness of meat products.

8 Control of Nutrients Value and Quality of Food

Numerous factors, including mechanical, chemical, microbiological, biological, and biochemical ones, may influence the nutritional value of a product. In particular, many pollutants such as insects, mites, fungi, bacteria, and so on fall under the category of biological and microbiological elements many biological and chemical reactions, such as lipid oxidation, mechanical damage, and the Maillard reaction, can take place during post-harvest storage. Germination, sprouting, and senescence are additional mechanical factors that contribute to spoiling. To prevent neurotoxicity illnesses that impact the human kidney, liver, and brain, biosensors based on glutamate dehydrogenase and glutamate oxidase have been developed to assess the levels of specific toxins, such as monosodium glutamate in order to prevent neurotoxicity disorders that affects the kidney, liver, and brain [59]. Multiplex sensing is a development in nanobiosensors that allows for the simultaneous detection of numerous analytes. A nanobiosensor that can identify and quantify the levels of several sweeteners, including sucrose, glucose, cyclamate, and saccharin, was created. The detection of genetically modified organisms using nanobiosensors in food supplies validates the various uses of nanobiosensors in the food sector.

9 Conclusion and Future Prospects

Biosensor technology has greatly improved over time, especially for detecting harmful bacteria in food and water that cause illnesses. Despite these accomplishments, the advancement of technology to create biosensors for detecting pathogens is still a very promising field of research. Different nanomaterials have special qualities that help detect specific diseases and enhance the devices' efficiency. Biosensors use nanoparticles to detect pathogens, providing unique advantages. Their small size and wide surface area help them interact better with pathogen biomarkers, making the biosensors more sensitive and selective. The devices have improved selectivity and sensitivity by being able to be modified with molecules like aptamers, nucleic acids, or antibodies. Aptamers are a versatile and easy tool for detecting infections accurately. Due to their chain-like structure, they are highly sensitive and selective. Increased specificity and selectivity in monitoring pathogens could have interesting implications for aptamers, functionalized aptamers, and other genetically-based biomarker detection systems. Compared to antibody techniques, this technology has advantages like improved specificity and the ability to identify genetic markers instead of protein markers. Identification of pathogens using biosensors greatly benefits public health. The change shows how nanobiosensor technology can be used to quickly and accurately detect harmful microorganisms in food and water. When used correctly, this technology can prevent disease outbreaks and initiate the required steps to ensure consumer safety. In conclusion, there are many advantages to using nanomaterials in biosensors for detecting pathogens. Biosensors are able to provide more precise, sensitive, and selective measurements due to their small size, high specific surface area, functionalization capacity, and signal amplification characteristics. Aptamer technologies and nanoparticles with DNA probes are future applications of DNA-based technologies combined with nanoparticle formulation. This will improve detection techniques to be more sensitive and specific. This type of biosensor can distinguish between infections and different strains with high accuracy. These traits, combined with the use of AI models and fluorescence detection technologies, can enhance monitoring systems' ability to detect issues. Improving food safety and public health can be done by developing tools for disease diagnosis and monitoring, such as biosensors using nanomaterials and aptamers.

References

1. Alvarado K, Bolanos M, Camacho C, Quesada E, Baudrit JV (2019) Nanobiotechnology in agricultural sector: overview and novel applications. Biomater Nanobiotechnol 10:120–141
2. Pola-López LA, Camas-Anzueto JL, Martínez-Antonio A, Luján-Hidalgo MC, Anzueto-Sánchez G, Ruíz-Valdiviezo VM, Grajales-Coutiño R, González JHC (2018) Novel arsenic biosensor POLA obtained by a genetically modified E. coli bioreporter cell Sens. Actuators B Chem 254:1061–1068
3. Jawaheer S, White S, Bessant C, Rughooputh SDDV, Cullen D (2002) Determination of fruit status using a disposable multi-analyte biosensor array and principle component analysis. In: Proceedings of the 7th world congress on biosensors, Kyoto, Japan, pp 15–17
4. Nasirizadeh N, Shekari Z, Ghaani M (2015) A novel electrochemical biosensor based on a modified gold electrode for hydrogen peroxide determination in different beverage samples. Food Anal Methods 8(6):1546–1555
5. Kaphle A, Navya PN, Umapathi A (2018) Nanomaterials for agriculture, food and environment: applications, toxicity and regulation. Environ Chem Lett 16:43–58
6. Singh A, Choudhary M, Singh MP, Verma HN, Singh SP, Arora K (2015) DNA functionalized direct electro-deposited gold nanoaggregates for efficient detection of Salmonella typhi. Bioelectrochemistry 105:7–15
7. Li Z, Li X, Jian M, Geleta GS, Wang Z (2019) Two-dimensional layered nanomaterial-based electrochemical biosensors for detecting microbial toxins. Toxins 12(1):20
8. Huang YF, Wang YF, Yan XP (2010) Amine-functionalized magnetic nano particles for rapid capture and removal of bacterial pathogens. Environ Sci Technol 44:7908–7913
9. Gandhi M, Chikindas ML (2007) *Listeria*: a foodborne pathogen that knows how to survive. Int J Food Microbiol 113:1e15

10. Kuswandi B, Futra D, Heng LY (2017) Nanosensors for the detection of food contaminants. In: Nanotechnology applications in food. Academic Press, pp 307–333
11. Wu S, Duan N, Zhu C, Ma X, Wang M, Wang Z (2011) Magnetic nanobead-based immunoassay for the simultaneous detection of aflatoxin B1and ochratoxin A using up conversion nanoparticles as multicolor labels. Biosens Bioelectron 30:35–42
12. Ansari AA, Kaushik A, Solanki PR, Malhotra BD (2010) Nano structured zinc oxide platform for mycotoxin detection. Bioelectrochemistry 77:75–81
13. Guo Z, Ren J, Wang J, Wang E (2011) Single-walled carbon nanotubes based quenching of free FAM-aptamer for selective determination oochratoxin A. Talanta 85:2517–2521
14. Sharma A, Matharu Z, Sumana G, Solanki PR, Kim CG, Malhotra BD (2010) Antibody immobilized cysteamine functionalized-gold nanoparticles for aflatoxin detection. Thin Sol Films 519:1213–1218
15. Liao JY, Li H (2010) Lateral flow immune dipstick for visual detection of aflatoxin B1 in food using immuno-nanoparticles composed of a silver core and a gold shell. Microchim Acta 171:289–295
16. Jain S, Singh SR, Horn DW, Davis VA, Ram MK, Pillai S (2012) Development of an antibody functionalized carbon nano tube biosensor for food borne bacterial pathogens. J Biosens Bioelectron S11(002):1–7
17. Joo J, Yim C, Kwon D, Lee J, Shin HH, Cha HJ, Jeon S (2012) A facile and sensitive detection of pathogenic bacteria using magnetic nanoparticles and optical nano crystal probes. Analyst 137:3609–3612
18. So HM, Park DW, Jeon EK, Kim YH, Kim BS, Lee CK, Choi SY, Kim SC, Chang H, Lee JO (2008) Detection and titer estimation of Escherichia coli using aptamer functionalized single-walled carbon-nano tube field-effect transistors. Small 4:197–201
19. Chen SH, Wu VCH, Chuang YC, Lin CS (2008) Using oligonucleotide-functionalized Au nano particles to rapidly detect food borne pathogens on a piezoelectric biosensor. J Microbiol Methods 73:7–17
20. Jafarizadeh-Malmiri H, Sayyar Z, Anarjan N, Berenjian A (2019) Nano-sensors in food nanobiotechnology. In: Nanobiotechnology in food: concepts, applications and perspectives. Springer, Cham, pp 81–94
21. Lin VSY, Motesharei K, Dancil KPS, Sailor MJ, Ghadiri MR (1997) A porous silicon-based optical interferometric biosensor. Science 278(5339):840–843
22. Sumitha MS, Xavier TS (2023) Recent advances in electrochemical biosensors—a brief review. Hybrid Adv 2:100023. https://doi.org/10.1016/j.hybadv.2023.100023
23. Agrawal S, Rathore P (2014) Nanotechnology pros and cons to agriculture: a review. Int J Curr Microbiol App Sci 3(3):43–55
24. Monavari SM, Marsusi F, Memarian N, Qasemnazhand M (2023) Carbon nanotubes and nanobelts as potential materials for biosensor. Sci Rep 13:3118. https://doi.org/10.1038/s41598-023-29862-9
25. Srivastava AK, Dev A, Karmakar S (2017) Nanosensors and nanobiosensors in food and agriculture. Environ Chem Lett 16(1):161–182
26. Chaubey A, Malhotra B (2002) Mediated biosensors. Biosens Bioelectron 17(6–7):441–456
27. Santandreu M, Alegret S, Fabregas E (1999) Determination of β-HCG using amperometric immunosensors based on a conducting immunocomposite. Anal Chim Acta 396(2–3):181–188
28. Yan C, Wang Q, Yang Q, Wu W (2020) Recent advances in aflatoxins detection based on nanomaterials. Nanomaterials 10(9):1626
29. Xu Y, Ding J, Chen H, Zhao Q, Hou J, Yan J, Ren N (2013) Fast determination of sulfonamides from egg samples using magnetic multiwalled carbon nanotubes as adsorbents followed by liquid chromatography–tandem mass spectrometry. Food Chem 140(1–2):83–90
30. Neethirajan S, Ragavan V, Weng X, Chand R (2018) Biosensors for sustainable food engineering: challenges and perspectives. Biosensors 8:23–31
31. Zhang Z, Lin M, Zhang S, Vardhanabhuti B (2013) Detection of aflatoxin M1 in milk by dynamic light scattering coupled with superparamagnetic beads and gold nanoprobes. J Agric Food Chem 61(19):4520–4525

32. Antiochia R, Vinci G, Gorton L (2013) Rapid and direct determination of fructose in food: a new osmium-polymer mediated biosensor. Food Chem 140(4):742–747
33. Nesakumar N, Sethuraman S, Krishnan UM, Rayappan JBB (2016) Electrochemical acetylcholinesterase biosensor based on ZnO nanocuboids modified platinum electrode for the detection of carbosulfan in rice. Biosens Bioelectron 77:1070–1077
34. Yavuz S, Erkal A, Kariper İA, Solak AO, Jeon S, Mülazımoğlu İE, Üstündağ Z (2016) Carbonaceous materials-12: a novel highly sensitive graphene oxide-based carbon electrode: preparation, characterization, and heavy metal analysis in food samples. Food Anal Methods 9(2):322–331
35. Walker NL, Roshkolaeva AB, Chapoval AI, Dick JE (2021) Recent advances in potentiometric biosensing. Curr Opin Electrochem 28:100735. https://doi.org/10.1016/j.coelec.2021.100735
36. Aggarwal S, Sehgal S, Tandon AW (2021) Biosensors-types and application in food processing industry. J Postharvest Technol 9(1):1–19
37. Ameta SK, Rai AK, Hiran D, Ameta R, Ameta SC (2020) Use of nanomaterials in food science. In: Ghorbanpour M, Bhargava P, Varma A, Choudhary DK (eds) Biogenic nano-particles and their use in agro-ecosystems, pp 457–488. https://doi.org/10.1007/978-981-15-2985-6_24
38. Liu C, Liu W, Yang J, Chen Y, Lei Z (2016) Non-destructive detection of dicyandiamide in infant formula powder using multispectral imaging coupled with chemometrics: non-destructive detection of dicyandiamide in infant formula. J Sci Food Agric 97(7):2094–2099
39. Caon T, Martelli SM, Fakhouri FM (2017) New trends in the food industry: application of nanosensors in food packaging. In: Grumezescu AM (ed) Nanobiosensors. Academic Press, pp 773–804
40. Sharma R, Ragavan KV, Thakur MS, Raghavaro KSMS (2015) Recent advances in nanoparticle based aptasensors for food contaminants Biosens. Bioelectron 74:612–627
41. Silva NFD, Magalhães JMC, Freire C, Delerue-Matos C (2018) Electrochemical biosensors for Salmonella: state of the art and challenges in food safety assessment. Biosens Bioelectron 99:667–682
42. Alghuthaymi MA, Kalia A, Bhardwaj K, Bhardwaj P, Abd-Elsalam KA, Valis M, Kuca K (2021) Nanohybrid antifungals for control of plant diseases: current status and future perspectives. J Fungi 7(48). Ahadır EB, Sezgintürk MK (2017) Biosensor technologies for analyses of food contaminants nanobiosensors. Elsevier, Amsterdam, the Netherlands, pp 289–337
43. Bahadır EB, Sezgintürk MK (2017) Biosensor technologies for analyses of food contaminants nanobiosensors. Elsevier, Amsterdam, the Netherlands, pp 289–337
44. Bhardwaj H, Pandey M, Sumana G, Rajesh (2019) Electrochemical aflatoxin B1 immunosensor based on the use of graphene quantum dots and gold nanoparticles. Microchim Acta 186(8):592
45. He J, Zhang L, Xu L, Kong F, Xu Z (2020) Development of nanozyme labelled biomimetic immunoassay for determination of sulfadiazine residue in foods. Adv Polym Technol 7647580:1–8
46. Ga L, Ai J, Wang Y (2020) AS1411 templated fluorescent copper nanomaterials synthesis and its application to detect melamine. J Chem 2020(1):4067578
47. You KH, Luo XE, Hu WJ, Xu Y, Guo JB, Heo H (2020) Environment-friendly gold nanoparticle immunochromatographic assay for ochratoxin a based on biosynthetic mimetic mycotoxin conjugates. World Mycotoxin J 13(2):267–276
48. Zhao Y, Zheng F, Ke W, Zhang W, Shi L, Liu H (2019) Gap-tethered Au@AgAu raman tags for the ratiometric detection of MC-LR Anal. Chem 91:7162–7172
49. Wang L, Huo X, Qi W, Xia Z, Li Y, Lin J (2020) Rapid and sensitive detection of Salmonella typhimurium using nickel nanowire bridge for electrochemical impedance amplification. Talanta 211:120715
50. Nabok A, Al-Jawdah AM, Gémes B, Takács E, Székács A (2021) An optical planar waveguide-based immunosensors for determination of Fusarium mycotoxin zearalenone. Toxins 13(2):89
51. Gupta R, Raza N, Bhardwaj SK, Kumar V, Kim KH, Bhardwaj N (2021) Advances in nanomaterial-based electrochemical biosensors for the detection of microbial toxins, pathogenic bacteria in food matrices. J Hazard Mater 401:12337

52. Kim SJ, Gobi KV, Iwasaka H, Tanaka H, Miura N (2007) Novel miniature SPR immunosensor equipped with all-in-one multi-microchannel sensor chip for detecting low-molecular-weight analytes. Biosens Bioelectron 23:701–707
53. Gerst M, Bunduchi R, Graham I (2005) Current issues in RFID standardization. University of Edinburgh
54. Vanderroost M, Ragaert P, Devlieghere F, De Meulenaer B (2014) Intelligent food packaging: the next generation. Trends Food Sci Technol 39:47–62
55. Lamri M, Bhattacharya T, Boukid F, Chentir I, Dib AL, Das D, Djenane D, Gagaoua M (2021) Nanotechnology as a processing and packaging tool to improve meat quality and safety. Foods 10(11):2633
56. Naghdi S, Rezaei M, Bahramifar N, Kuswandi B (2022) Preparation and characterization of intelligent color-changing nanosensor based on bromophenol blue and $GONH_2$ nanosheet for freshness evaluation of minced Caspian sprat (Clupeonella cultriventris caspia) stored at 4°C. Chem Pap 76(5):3133–3146
57. Wei S, Zhao F, Xu Z, Zeng B (2006) Voltammetric determination of folic acid with a multi-walled carbon nanotube-modified gold electrode. Microchim Acta 152:285–291
58. Chelliah R, Wei S, Daliri EBM, Rubab M, Elahi F, Yeon SJ, Jo KH, Yan P, Liu S, Oh DH (2021) Development of nanosensors based intelligent packaging systems: food quality and medicine. Nanomaterials 11(6):1515
59. Arugula MA, Simonian AL (2016) Biosensors for detection of genetically modified organisms in food and feed genetically modified organisms in food. Elsevier, Amsterdam, the Netherlands, pp 97–110

Food Packaging and Nanobiosensors: Friends or Foe

Neelam Vishwakarma, Shubham Kumar Patial, and Suman Singh

Abstract The need for high-quality, safe, and long-lasting food products is crucial in addressing food scarcity issues in modern systems. Indicators of food quality and safety include the detection of pathogens, food particle deterioration, physical damage, and contamination. Advancements in nanotechnology have revolutionized fields such as food packaging, quality control, and monitoring, leading to the development of smart and active food packaging solutions that surpass traditional methods. This chapter focuses on the general introduction of nanobiosensors and their applications in detecting food pathogens and quality, offering real-time monitoring solutions. It examines the advantages of nanobiosensors while also discussing potential risks, challenges, consumer concerns, environmental impacts, and regulatory considerations. Thus, this chapter provides a balanced perspective on integrating nanobiosensors into food packaging, exploring their scientific principles, current applications, and implications, and assessing their role as either friends or foes.

Keywords Food safety · Nanobiosensors · Food packaging · Shelf-life · Food quality

1 Introduction

The food and agriculture sector plays a pivotal role in driving the economic growth of developing nations, through activities such as crop cultivation, livestock rearing, aquaculture, food processing, and packaging. A critical aspect of this industry is the regulation of production and the maintenance of quality and safety standards throughout the supply chain, where process control and quality monitoring are essential [1]. Recently, the integration of nanotechnology has provided a new edge

N. Vishwakarma · S. K. Patial · S. Singh (✉)
Academy of Scientific and Innovative Research (AcSIR), Ghaziabad, India
e-mail: ssingh@csio.res.in

Applied Materials and Instrumentation (AMI), CSIR-Central Scientific Instruments Organisation, Chandigarh, India

D. Sillu et al. (eds.), *Nanobiosensors for the Food Industry*, Smart Nanomaterials Technology, https://doi.org/10.1007/978-981-95-0136-6_11

to agrotechnology, improving irrigation efficiency, optimizing fertilizer use, and enhancing food production, processing, packaging, and storage. Nanotechnology-based sensing has gained significant momentum, offering broad-spectrum applications that further strengthen the efficiency, quality, and sustainability of the food and agriculture sector [2].

In today's world, food packaging goes beyond mere storage and containment; it plays a crucial role in preserving freshness and quality, extending shelf-life, ensuring safety, and preventing contamination [3]. The worldwide food security crisis has not only mandated the need for effective food packaging but also promoted innovative approach explorations in packaging materials and techniques to avoid microorganisms and chemical contaminations, hostile physical conditions such as temperature, moisture, and air disturbances. The primary requisite of good-quality food packaging is attributed to least interaction between food and packaging and effective barrier between the outer environment and food material along with lowest chemical and preservative additions. Advanced packing technological inventions have aimed to encounter the increasing customer requirements with minimum processing requisite with low chemical preservation additions and strict adherence to regulatory guidelines, globalized food demands, food security terrors etc. [4].

Nowadays, innovative packaging techniques not only provide increased shelf-life and flavor retention but also aid in monitoring the quality levels and provide indications in drop in consumability [5]. These indicators could point the consumer toward any possible allergens, detrimental chemical and pesticidal accumulations, pathogen contamination and quality drops [2, 6, 7].

The advent of nanotechnology has massively transformed and revolutionized the outlook for variety of sectors including disease diagnosis and treatments, environmental challenges and remediation capabilities, agricultural and food production through translation of theoretical innovations into practical, real-world applications [8–10]. Nanotechnology has played a pivotal role in advancing biosensors and nanosensors, thereby offering enhanced sensitivity and functionalities with robust sensing mechanisms [11].

Biosensors play a critical role in ensuring food safety by detecting contaminants, pathogens, toxins, and allergens. Biosensors are specialized analytical devices used for the quantitative detection of specific analytes [12]. These devices rely on biologically active elements, such as antibodies, enzymes, oligonucleotides, or receptors, which are immobilized on the surface of a transducer. When the biological recognition molecule interacts with its target compound, the transducer converts this interaction into a detectable signal [13–15]. This signal is often quantified through electrochemical, optical, acoustic, mechanical, or electronic methods, and it correlates with the concentration of the analyte [16, 17]. These biosensors, when incorporating nanostructured materials for the sensing or signal enhancing mechanisms, can be defined as the nanobiosensors. Nanobiosensors, which is integration of nanomaterials into the transducer design, represent a significant leap forward in biosensor technology [18]. Their nanoscale dimensions allow for enhanced sensitivity and specificity, enabling the rapid detection of even trace amounts of harmful substances

[19, 20]. These nanostructured sensors provide an efficient alternative to conventional laboratory-based methods, offering advantages such as faster response times, robustness, and portability. By integrating nanobiosensors into food safety protocols, the complexity of standard detection methods is reduced, paving the way for scalable, high-precision solutions that enable real-time monitoring and quality assurance in the food and agriculture sector.

The introduction of nanostructures and nanotechnology has significantly advanced food safety by driving the development of nanosensors and nanobiosensors, revolutionizing every stage of the food production chain, from enhancing agricultural practices to optimizing food processing, packaging, and storage [21, 22]. The integration of nanotechnology into biosensor platforms is redefining point-of-use applications by minimizing the burden of traditional laboratory protocols while providing superior performance in real-time diagnostics [23].

In this chapter, we will explore the relationship between food packaging and nanobiosensors, examining how these technologies can interact to enhance food safety and quality. It aims to assess whether they function as complementary technologies, working together to improve detection and monitoring capabilities in food systems, or if they present conflicting interests in terms of practical application, scalability, and regulatory challenges (Fig. 1).

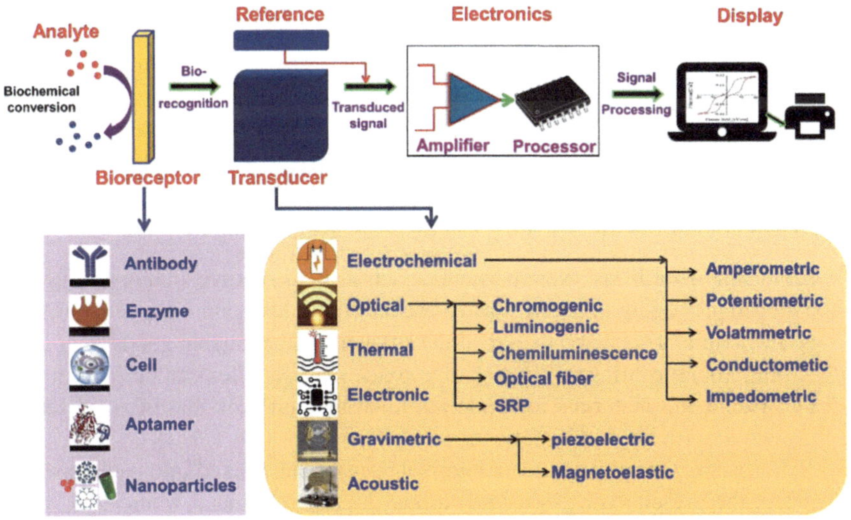

Fig. 1 Overview of biosensors and nanobiosensors (reprint with permission) [20]

2 The Role of Nanotechnology in Food Packaging

Growing global demand to ensure food supply and quality has paved way for several creative innovations in the field of agriculture, food production, packaging and transport, etc. Smart or intelligent packaging systems has attracted special attention among these innovations and has several research and industrial projects dedicated to its development. Intelligent or smart food packaging refers to packaging systems that not only protect and preserve food but also actively monitor, communicate, or respond to changes in the product's condition or environment [24, 25]. These systems often incorporate sensors, indicators, or nanotechnology to detect spoilage, contamination, temperature changes, or freshness. Smart packaging can provide real-time information about the food's quality, safety, or shelf-life, helping consumers and supply chain stakeholders ensure food safety and reduce waste [26–30]. These include examples like time–temperature indicators, freshness sensors, and gas or moisture detection systems.

Nanotechnology was first applied in food packaging in the early 2000s [31]. The initial developments focused on improving the mechanical, thermal, and barrier properties of packaging materials through the incorporation of nanomaterials like nanoclays and nanocomposites [32, 33]. These innovations enhanced packaging durability and extended food shelf-life by providing better protection against moisture, oxygen, and UV light.

The use of nanotechnology in packaging has since evolved, with more advanced applications, such as antimicrobial coatings and nanosensors for spoilage detection, emerging over the following decades [34]. Due to their antimicrobial activity, metal and metal oxide nanoparticles have been used in food packaging. Another advantage of using nanomaterials is reduction in packing materials costs by replacing traditional packing materials with cheaper passive packing [35, 36]. Few commonly used of nanotechnology in smart packaging can be noted as following [37, 38]:

a. Nanocomposite films: Nanoparticles, such as clay or silver nanoparticles, are added to packaging films to improve barrier properties against oxygen, moisture, and UV light. This helps to extend the shelf-life of perishable goods.
b. Antimicrobial coatings: Silver or zinc oxide nanoparticles can be incorporated into packaging materials to create antimicrobial surfaces that inhibit bacterial growth, improving food safety.
c. Nanosensors for spoilage detection: Nanosensors integrated into packaging can detect the presence of gases like carbon dioxide, produced when food spoils. These sensors can trigger color changes or digital indicators that alert consumers to spoilage.
d. Time–temperature indicators (TTIs): Nanosensors can monitor temperature fluctuations during transportation and storage. These TTIs change color to show whether the product has been exposed to temperatures that could compromise its safety or quality [39].
e. Edible nanocoatings: Thin layers of edible nanocoatings, such as those made from starch or chitosan, are applied directly to food products to prevent moisture

loss, oxidation, or microbial contamination, enhancing both safety and longevity without the need for traditional packaging [40].

3 Nanobiosensors and Its Classification in Food Storage

3.1 Introduction to Nanobiosensors

Nanobiosensors are integrative technologies developed by combining biosensing elements with nanostructures or nano-dimensional materials in order to detect specific chemical or biological analytes. The nanobiosensors, alike to biosensors, has two core components: a biorecognition element and a transducer. The biological recognition element—such as an enzyme, antibody, nucleic acid, or receptor—is designed to specifically interact with the target analyte (e.g., pathogens, toxins, or biomolecules) [41]. Once the target binds to the recognition element, this interaction triggers a measurable signal that the transducer converts into a quantifiable output.

In nanobiosensors, the use of nanomaterials enhances the surface area and sensitivity of the sensor. Due to their nanoscale dimensions, these sensors can detect even minute concentrations of analytes, making them highly effective for real-time monitoring. Depending on the signal type, the output can be measured through various detection techniques, including electrochemical, optical, or piezoelectric methods [42–45]. Nanobiosensors can be miniaturized and embedded in various devices for portable, rapid, and highly sensitive detection in applications such as food safety, health care, and environmental monitoring.

Nanomaterials play a crucial role in the functioning and performance of nanobiosensors. Some of the most commonly used nanomaterials in food packaging include [37, 46] (Fig. 2):

a. Gold nanoparticles (AuNPs): Known for their excellent conductivity and biocompatibility, gold nanoparticles are often used in optical and electrochemical biosensors to enhance sensitivity.
b. Silver nanoparticles: Due to their antimicrobial properties and strong optical characteristics, silver nanoparticles are commonly used in sensors designed for pathogen detection and spoilage monitoring.
c. Carbon nanotubes (CNTs): These cylindrical nanostructures have exceptional electrical and mechanical properties, making them ideal for electrochemical biosensors. Their high surface area also allows for immobilizing large quantities of biological recognition elements.
d. Graphene and graphene oxide: With their excellent electrical conductivity, high surface area, and flexibility, graphene-based nanomaterials are increasingly used in biosensors to detect various foodborne pathogens and contaminants.
e. Quantum dots (QDs): These semiconductor nanocrystals are employed in fluorescence-based biosensors. They offer high sensitivity and can be tuned to emit specific wavelengths of light, enabling precise optical detection of analytes.

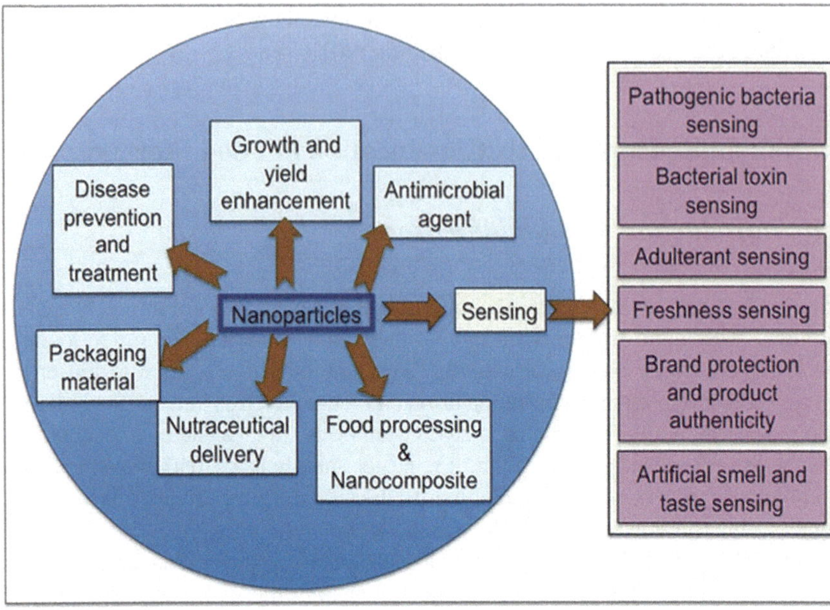

Fig. 2 Nanoparticles in food packaging (reprint with permission) [47]

3.2 Detection of Pathogens, Contaminants, and Spoilage Indicators

Nanobiosensors have significant applications in food safety, where they are used for the detection of pathogens (e.g., Salmonella, *E. coli*, Listeria), contaminants (e.g., pesticides, heavy metals, toxins), and spoilage indicators (e.g., gases like ammonia and carbon dioxide) [34, 48, 49]. The enhanced sensitivity and rapid response time of nanobiosensors make them ideal for detecting even trace amounts of harmful substances that could compromise food safety. Traditional food safety methods like culturing pathogens or performing chemical analysis can take hours or days, but nanobiosensors can deliver results in minutes or even in real time [50].

Nanobiosensors can be deployed directly on food products, in processing environments, or at different stages of the supply chain to monitor food quality continuously. For instance, they can detect changes in the microbial content of a food product, alerting producers or consumers to potential contamination before it becomes harmful. Some biosensors are also capable of multiplexed detection, where multiple pathogens or contaminants can be detected simultaneously, offering a more comprehensive safety assessment [51].

3.3 Integration into Packaging to Ensure Real-Time Monitoring

One of the most exciting applications of nanobiosensors is their integration into smart packaging systems. Smart packaging, enhanced by nanotechnology, goes beyond traditional food packaging by incorporating sensors that can monitor the internal conditions of the packaging and the quality of the food in real time [52]. Nanobiosensors embedded in packaging materials can continuously monitor for spoilage indicators, such as gases produced by microbial activity or chemical reactions within the food. For instance, nanosensors can detect the buildup of carbon dioxide or volatile organic compounds (VOCs) that signal food spoilage, allowing for immediate feedback on the food's condition [8, 53].

This real-time monitoring capability not only ensures food safety but also helps reduce food waste by providing accurate information about the product's freshness. Consumers can be informed through visual indicators (e.g., color changes) on the packaging, or more advanced packaging systems may include digital readouts or smartphone integration to alert consumers or supply chain managers.

Nanobiosensor-enabled packaging can provide real-time temperature tracking to ensure that perishable foods have been stored and transported under appropriate conditions. This is particularly useful in cold-chain logistics, where temperature fluctuations can significantly affect food safety and quality. By integrating nanobiosensors into food packaging, producers and consumers can have greater confidence in the safety, quality, and freshness of the food they handle [54].

4 State-of-the-Art Nanotechnological Integration in Food Packaging and Storage

Nanotechnology has been integrated into food packaging in both the Indian and international markets, offering enhanced food preservation, safety, and quality monitoring. Here are some examples:

a. Freshwrapp by Hindalco Industries: Hindalco, part of the Aditya Birla Group, has developed Freshwrapp, an aluminum foil that incorporates nanotechnology for better barrier properties and enhanced food preservation. It's been widely used in homes, restaurants, and the food industry to package and store various food items, keeping them fresh for longer durations [55].
b. MataNano® by NanoPack (EU Funded Project): NanoPack utilizes natural nanomaterials to create antimicrobial packaging films. These films contain halloysite nanotubes with essential oils that have antibacterial properties. This packaging has been intensively tested in packaging baked goods, dairy products, and meat to extend shelf-life and prevent the growth of harmful bacteria [56].

c. Nanobarrier technologies by European commission: BASF, a global leader in chemicals, developed nanobarrier technology for food packaging that integrates nanoclay particles into plastic films to improve their barrier properties [57].
d. Durethan® by LANXESS: This polyamide nanocomposite material offers improved mechanical and barrier properties, making it ideal for flexible packaging solutions. This nanocomposite has been utilized for vacuum packaging of meats, cheese, and processed foods in the international market [58].
e. Nanocellulose films by American Process Inc. (USA): Nanocellulose-based films are created by incorporating nanofibers derived from cellulose, which improve the mechanical strength and barrier properties of packaging. These films have been used for eco-friendly packaging of dry goods, cereals, and snack foods to prevent moisture infiltration [59].
f. Smart packaging with time–temperature indicators by Insignia Technologies: Insignia technologies has developed smart packaging with embedded time–temperature indicators that change color when food has been exposed to unsafe temperatures. These indicators had been integrated in perishable food items such as meat, fish, and dairy, giving consumers visual cues about the product's freshness [60]
g. OxyDot® is another nanosensor which is commercially available and is used for quantification of dissolved O_2 in sealed drinks and packed food materials. The sensor uses fluorescence intensity and lifetime quenching of the metallic–organic fluorescent dye for successfully estimating O_2 molecules, by immobilizing the materials on a hydrophobic polymer dot. This system has been widely accepted with sensitivity upto 5% of readings, durable structure, fast (<0.1 s response time), and a useful tool for detection of O_2 molecules in real time [61].

5 Synergies Between Food Packaging and Nanobiosensors

Nanoscale technology contributes to solving some problems in food packaging, specifically encapsulating nutritious and uncontaminated food commodities. Incorporating nanomaterials in food packaging systems and integration with nanobiosensors offers the revolutionary potential to improve the safety, shelf-life, and quality control in food supply environments.

5.1 Enriching Food Packaging Functionality with Nanobiosensors

Developing methods to detect food contamination with genetic substances such as microbial growth, spoilage, and chemical contaminants will require a significant technological thrust in the form of nanobiosensors. These responses can later be used in developing food packaging materials to monitor the quality of foodstuffs.

Using biological material as nanosensors with materials like silver nanoparticles, zinc oxide, or titanium nitride makes these low-invasive and susceptible nanomodified devices able to detect changes in the food matrix at earlier time points. For instance, nanosensors could identify variations in temperature, pH, or the discharge of gases created by spoiled food. They are ensuring a dynamic and responsive food safety management approach [62, 63].

5.2 Synergistic Influence in Food Safety, Quality Management System, and Shelf-Life

The combined use of nanobiosensors with the packaging of nanomaterials enhances dual effects: preservation and monitoring of food. Nanomaterials such as silver nanoparticles, nano-zinc oxide, and nano-titanium dioxide have a wide range of applications in food packaging systems for preventing contamination and spoilage by inhibiting the growth of bacteria, yeasts, fungi, and mold. For instance, Luo et al. [64] demonstrated that a nano-$CaCO_3$-based low-density polyethylene (LDPE) packaging could inhibit microbial proliferation in fresh-cut Chinese yams, preserving this vegetable's freshness and appearance [65]. One of the most frequent applications regards nanocomposite films being able to function as a barrier, thus highly influencing the decaying delay and browning processes occurrence in fresh produce. Further, red grapes can be preserved for as long as 15 days at ambient temperature in silver nanoparticles embedded in chitosan or gelatin nanocomposites compared to conventional packaging techniques. Indeed, it counteracts one of the biggest challenges in food packaging, which includes preventing oxidation from exposure to light or temperature, causing food to spoil. Edible films infused with nanomaterials, including protein-polysaccharide-based films, provide enhanced mechanical properties and effectively protect against oxidation and contamination to maintain the freshness of the food for a more extended period [66].

5.3 Successful Utilization of Nanobiosensors and Nanomaterials in Packaging

Several case studies elucidate the successful integration of food packaging with nanobiosensors, evince food safety and consumer confidence improvements. For example, silver nanoparticle-based chitosan films have been tried for their antimicrobial efficacy against common foodborne pathogens such as *S. aureus* and *E. coli*. Such nanocomposites possess a dose-dependent antibacterial performance that ensures food safety during storage and distribution [66].

Biodegradable bilayer films, made from gelatin and agar infused with titanium dioxide (TiO_2), were also developed for use in light nutrition-photo delicate materials (e.g., light-sensitive food items) to minimize deleterious effects of exposure to harmful light while maintaining mechanical strength and water vapor barriers. Results in successful in decreasing the rates of spoilage and also maintaining the food quality for more extended amounts [67]. Manikantan and Varadharaju [68] also revealed that polypropylene-based nanocomposite packaging films improved the storage stability of banana chips. The nanomaterial-based films significantly extended the shelf-life of the banana chips by exhibiting better barrier properties than commercialized packaging systems, producing good-quality banana chips in terms of pH measurement, moisture content, and texture [69] (Figs. 3 and 4).

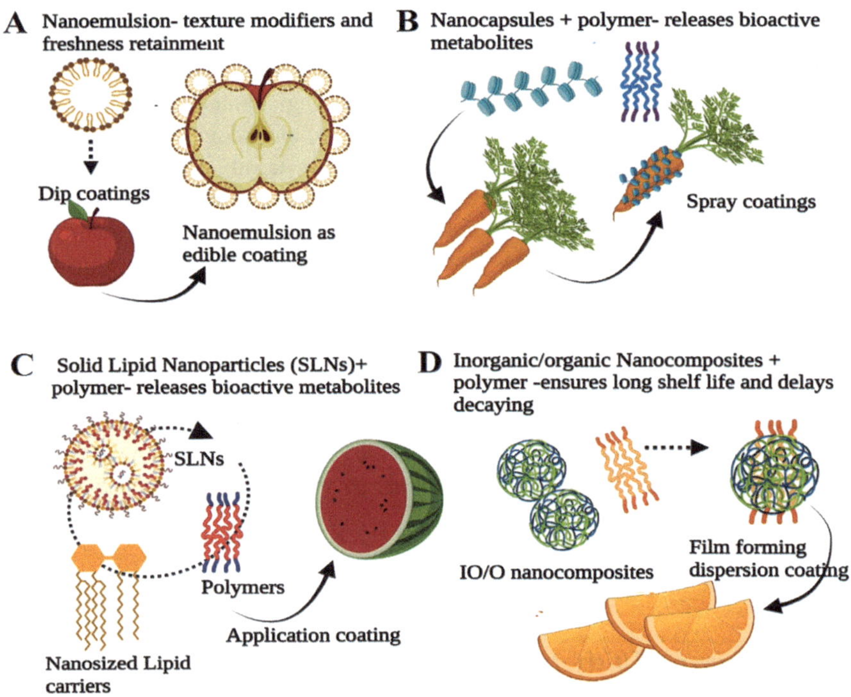

Fig. 3 Applications of nanocomposites in the preservation and packaging of fruits and vegetables: **a** Interaction of foods with nanoemulsions incorporated into edible coatings with a polymer matrix, enhancing bioavailability. **b** Nanoparticles, such as nanospheres or nanocapsules carrying active ingredients, are dispersed within a polymer matrix on the food surface, releasing the active compounds through diffusion. **c** Solid lipid nanoparticles (SLN) and nanostructured lipid carriers embedded in a polymer matrix within edible coatings. These SLNs are key systems for delivering nutraceuticals, offering improved stability, bioavailability, and dispersibility. **d** Inorganic nanocomposites (comprising polysaccharides, proteins, and natural lipids) are applied in small amounts as edible coatings to improve food products' overall quality and shelf-life [70]

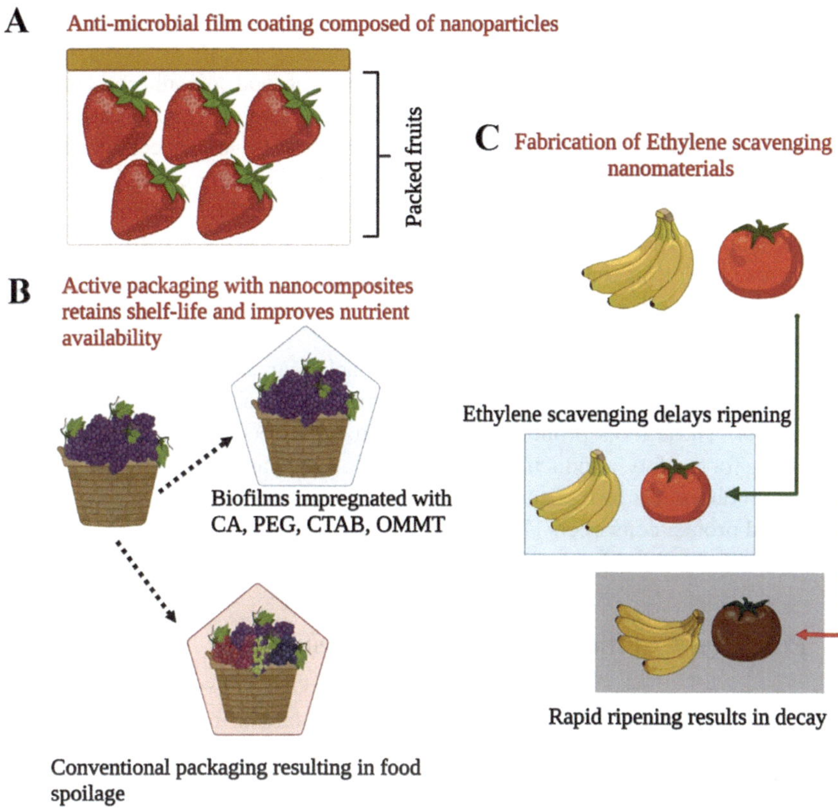

Fig. 4 Diverse applications of antimicrobial nanofilms in active food packaging. **a** Integration of antimicrobial component-loaded nanocarriers into the packaging structure; **b** development of non-toxic nanocomposite films using cellulose acetate (CA), polyethene glycol (PEG), and cetyltrimethylammonium bromide (CTAB)-modified montmorillonite (OMMT), exhibiting significant antimicrobial properties; **c** creation of ethylene scavengers for active packaging to prolong the freshness and delay the ripening of food products [70]

5.4 *Impact on Food Safety and Consumer Trust*

Incorporating nanomaterials and nanobiosensors in food packaging enhances safety and promotes consumer trust by offering concrete advantages, such as prolonged shelf-life and decreased spoilage. In conclusion, active packaging incorporating natural antimicrobial agents has been a potential alternative to improving preserved food products' quality and storage life. By encapsulating these natural antimicrobial agents within nanocarriers via nanoencapsulation processes, the agents can be controlled during storage to maintain the quality of food products throughout their entire shelf-life. For instance, Kaur et al. developed chitosan nanocomposite films impregnated with antimicrobial nanomaterials, which exhibited bacterial growth inhibition [71].

This development dramatically minimizes the risk of food poisoning and increases the safety of packaged foods. Nanomedicine and diet have many advantages, but on the other hand, there is much concern about health safety issues, especially due to the migration of engineered nanoparticles (ENP) into food [72]. For example, studies by Cushen et al. have shown that ENPs such as silver nanoparticles may migrate to the food, especially when storage at changing temperatures is involved. Regulations and greater research efforts into developing safe nanomaterials will be required to ensure that we can take advantage of the benefits of nanotechnology in food packaging without compromising human health [73].

The application of nanotechnology in food packaging and nanobiosensors based on nanomaterials shows promise for overcoming many obstacles observed in the food sector. These tools promise to revolutionize the food packaging industry by increasing food safety, conducting better quality control, and extending the shelf-life of food products. Yet further study and monitoring are needed to minimize possible harms and protect consumers [74].

6 Potential Conflicts, Challenges, and Safety

While many reports emphasize the benefits and promise of nanotechnology-based sensors, several issues still need further exploration:

a. The cost of a single unit of nanosensors or nanobiosensors is already significantly high compared to traditional sensing methods, especially during development. This cost is likely to persist during scaling, depending on the fabrication technique [75].
b. A significant issue is the long-term durability of nanomaterial (NM)-based sensors, especially during extended storage. The performance of such systems over long periods and repeated uses requires more advanced studies [76].
c. Nanosensors are often exposed to medium environments, such as biological fluids, where they may experience interference from other species, potentially causing false-positive signals or cross-interference [77].
d. Due to their biocompatibility and selective catalytic activity, nanomaterials could be incorporated into portable or miniaturized devices for specific applications [78].
e. Environmental and ecological problems persist, mainly due to the high energy required for NM synthesis, which increases energy demand. Additionally, a shortage of skilled professionals for NM synthesis presents another challenge [79].
f. Some studies also express concern regarding a need for more validation with actual samples and the absence of standardized protocols [77].

g. Potential nanotoxicity associated with using nanomaterials (NMs) is a significant concern hindering their broader implementation [80].

The application of nanotechnology in the food and beverage industries faces challenges due to the complexities of food samples. While advancements are possible, creating disruptive products and processes for the sector requires overcoming significant obstacles. One primary concern is ensuring food safety, particularly nanoparticle leaching and migration from packaging materials into food products. Nanomaterials behave differently at the nanoscale, and our understanding of these behaviors remains limited. For nanotechnology to be safe and effective, it is crucial to understand nanoscale functionalities and nanotoxicity [81]. Potential risks, such as toxicity and environmental impact, must be carefully considered. Nanoparticles can penetrate biological barriers and accumulate in tissues and organs. The chemical processes involved in nanoparticle synthesis may produce toxic byproducts, contributing to environmental contamination. Before manufacturing and distributing nano-based food products, comprehensive risk assessments, regulatory frameworks, and public engagement are necessary. Additionally, in vitro and in vivo studies are essential to assess the interactions between nanoparticles and living organisms, particularly in developing eco-friendly antibacterial nanoparticles.

Both organic and inorganic nanoparticles are being explored to enhance food quality, safety, and nutrition. Organic nanoparticles, composed of lipids, proteins, or carbohydrates, are typically spherical or nanofiber and are easier to digest in the gastrointestinal tract. In contrast, inorganic nanoparticles exhibit more complex behaviors, such as variable dimensions, aggregation states, and gastrointestinal effects, impacting their solubility, biosorption, and bioaccumulation. This inorganic nanoparticle degradation would trigger environmental, ethical, and safety concerns, proving the necessity to impose stricter legal frameworks to define toxicity limits for ecological organisms involved.

To reduce eco-toxicity and bioaccumulation risks, edible biopolymers can serve as the backbone for nanostructures. By integrating bionanomaterials into active and intelligent packaging systems, multifunctional solutions can be created without harmful interactions between active components. Although nanomaterials offer many benefits, concerns about their toxicity in humans and animals persist [82]. Toxicity depends on factors like nanoparticle type, size, and the ability to penetrate biological systems. Monitoring nanoparticle migration in packaging materials and biosensors is critical. While in vitro and in vivo studies show that metals and metal oxides generally have biocompatible properties, they can still trigger pro-inflammatory responses. Nanoparticles' effects are often system-specific, based on their inherent properties, highlighting the need for further research. Toxicity assessments use a variety of analytical testing protocols, including in vitro, in vivo, and physicochemical tests. However, these testing methods vary between laboratories, making direct comparisons difficult. Therefore, standardized operating procedures must be established to assess nanoparticle toxicity accurately under consistent conditions. To mitigate health risks, food safety regulations from organizations like the World Health Organization (WHO), the Food and Drug Administration (FDA), and the European Food Safety

Authority (EFSA) should be made more stringent. Future studies should investigate nanomaterials' interaction, association, and solubility in food packaging and sensing systems, focusing on the gastrointestinal tract (GIT). Detection of chronic exposure to food-contact nanomaterials should be enhanced to ensure a safe and healthy life [70].

7 The Future Scope and Conclusion

Future scope nanotechnology has made tremendous progress in food science, particularly in tracking, tracing, and monitoring food quality. It offers innovative ways to detect contaminants, pathogens, and pesticides. However, several nanosystems are still in the developmental phase and require more extensive research for large-scale applications. Both the opportunities and challenges related to food safety need to be evaluated simultaneously. For example, silver nanoparticles have been used to extend the shelf-life of food products like asparagus and poultry by inhibiting the growth of bacteria and molds. Zinc oxide nanoparticles have been effective in preserving orange juice and liquid egg albumen without compromising their quality, while titanium oxide has shown promise in protecting strawberries from browning and decay. These nanomaterials, along with smart packaging solutions, are evolving rapidly. Concepts like nanocomposite polymeric films and antigen-specific biomarkers are turning into viable technologies. Carbon–neutral, biodegradable nanocomposites are emerging as potential sustainable packaging materials. Additionally, nanosensors are being explored for their ability to quickly detect foodborne pathogens such as bacteria, viruses, and mycotoxins. These sensors, often paired with AI-based wireless technologies, offer real-time data transmission between smart devices and biosensors. Artificial intelligence (AI) is already playing a critical role in food quality control, sorting, and prediction of market trends, with methods like machine learning (ML) and neural networks improving decision-making processes. Machine learning algorithms, such as random forest regression and support vector regression, are being utilized to forecast sales, food waste, and even food insecurity. AI-enabled biosensors and nanosensors are revolutionizing food packaging by incorporating smart elements that detect spoilage gases. Carbon nanotubes are now embedded in packaging materials to identify harmful chemicals and bacteria. This fusion of nanotechnology and AI-driven systems is leading to smarter, more efficient, and sustainable food packaging options. Nanotechnology's potential in the food industry will likely expand significantly in the future. The continuous development of nanomaterials like nanofibers and nanoencapsulation techniques holds promise for enhancing the formulation of functional foods. However, addressing safety concerns through proper regulations is crucial for future growth. By 2050, nanotechnology is expected to dominate food processing, with projections indicating its ability to offer solutions on both macro- and microscales.

Nanotechnology is poised to revolutionize various sectors, including food production, by preventing spoilage and enhancing the quality of food products. Nanomaterials are already being used in preservatives, antimicrobial sensors, and packaging materials, and they have the potential to improve the nutritional content and shelf-life of food. Despite these advantages, challenges remain, especially in developing safe and efficient delivery systems for food products. One of the primary concerns is the migration of nanoparticles from packaging materials into food, which requires a deeper understanding of how materials behave at the nanoscale. More research is also needed to fully comprehend the toxicity and environmental impact of nanomaterials, especially those with antibacterial properties. Future advances in nanotechnology could significantly expand the possibilities for creating functional foods. However, this will require the implementation of stringent safety standards and regulatory measures. If properly regulated, nanotechnology has the potential to transform food production entirely by 2050, becoming a highly sophisticated and versatile technology capable of addressing both large-scale and individual-level food challenges.

References

1. Han JH (2014) A review of food packaging technologies and innovations. Innov Food Packag, pp 3–12
2. Srivastava AK, Dev A, Karmakar S (2017) Nanosensors for food and agriculture. Nanosci Food Agricult 5:41–79
3. Nura A (2018) Advances in food packaging technology: a review. J Postharvest Technol 6(4):55–64
4. Dyckman L (2004) Effective food security strategies. Bioterror Food Saf, p 133
5. Ghoshal G (2018) Recent trends in active, smart, and intelligent packaging for food products. In: Food packaging and preservation. Elsevier, pp 343–374
6. Gallo M et al (2020) Relationships between food and diseases: what to know to ensure food safety. Food Res Int 137:109414
7. Sundramoorthy AK, Kumar THV, Gunasekaran S (2018) Graphene-based nanosensors and smart food packaging systems for food safety and quality monitoring. In: Graphene bioelectronics, Elsevier, pp 267–306
8. Muthukrishnan L (2022) An overview on the nanotechnological expansion, toxicity assessment and remediating approaches in agriculture and food industry. Environ Technol Innov 25:102136
9. Sekhon BS (2014) Nanotechnology in agri-food production: an overview. Nanotechnol Sci Appl, pp 31–53
10. Usman M et al (2020) Nanotechnology in agriculture: current status, challenges and future opportunities. Sci Total Environ 721:137778
11. Momin JK, Jayakumar C, Prajapati JB (2013) Potential of nanotechnology in functional foods. Emirat J Food Agricult 25(1):10–19
12. Rodriguez-Mozaz S, de Alda MJL, Barceló D (2006) Biosensors as useful tools for environmental analysis and monitoring. Analyt Bioanalyt Chem 386:1025–1041
13. Naresh V, Lee N (2021) A review on biosensors and recent development of nanostructured materials-enabled biosensors. Sensors 21(4):1109
14. Mehrotra P (2016) Biosensors and their applications: a review. J Oral Biol Craniof Res 6(2):153–159

15. Metkar SK, Girigoswami K (2019) Diagnostic biosensors in medicine: a review. Biocatal Agricult Biotechnol 17:271–283
16. Gizeli E, Lowe CR (1996) Immunosensors. Curr Opin Biotechnol 7(1):66–71
17. Srivastava AK, Dev A, Karmakar S (2018) Nanosensors and nanobiosensors in food and agriculture. Environ Chem Lett 16:161–182
18. Kulkarni MB, Ayachit NH, Aminabhavi TM (2022) Recent advancements in nanobiosensors: current trends, challenges, applications, and future scope. Biosensors 12(10):892
19. Sharma P et al (2021) A review on biosensors and nanosensors application in agroecosystems. Nanoscale Res Lett 16:1–24
20. Ramesh M et al (2022) Nanotechnology-enabled biosensors: a review of fundamentals, design principles, materials, and applications. Biosensors 13(1):40
21. Nurfatihah Z, Siddiquee S (2019) Nanotechnology: recent trends in food safety, quality and market analysis. Nanotechnol Appl Energy Drug Food 15:283–293
22. Warriner K et al (2014) Developments in nanoparticles for use in biosensors to assess food safety and quality. Trends Food Sci Technol 40(2):183–199
23. Yang T et al (2016) Recent progresses in nanobiosensing for food safety analysis. Sensors 16(7):1118
24. Yousefi H et al (2019) Intelligent food packaging: a review of smart sensing technologies for monitoring food quality. ACS Sens 4(4):808–821
25. Kalpana S et al (2019) Intelligent packaging: trends and applications in food systems. Trends Food Sci Technol 93:145–157
26. Chelliah R et al (2021) Development of nanosensors based intelligent packaging systems: food quality and medicine. Nanomaterials 11(6):1515
27. Rodrigues C et al (2021) Bio-based sensors for smart food packaging: current applications and future trends. Sensors 21(6):2148
28. Osmólska E, Stoma M, Starek-Wójcicka A (2022) Application of biosensors, sensors, and tags in intelligent packaging used for food products: a review. Sensors 22(24):9956
29. Shao P et al (2021) An overview of intelligent freshness indicator packaging for food quality and safety monitoring. Trends Food Sci Technol 118:285–296
30. Mustafa F, Andreescu S (2018) Chemical and biological sensors for food-quality monitoring and smart packaging. Foods 7(10):168
31. Imran M et al (2010) Active food packaging evolution: transformation from micro-to nanotechnology. Crit Rev Food Sci Nutr 50(9):799–821
32. Mihindukulasuriya S, Lim L-T (2014) Nanotechnology development in food packaging: a review. Trends Food Sci Technol 40(2):149–167
33. Wyrwa J, Barska A (2017) Innovations in the food packaging market: active packaging. Eur Food Res Technol 243:1681–1692
34. Mohammadi Z, Jafari SM (2020) Detection of food spoilage and adulteration by novel nanomaterial-based sensors. Adv Colloid Interf Sci 286:102297
35. Zhang S et al (2019) Recent advances in nano-materials for packaging of electronic devices. J Mater Sci Mater Electr 30:13855–13868
36. Mlalila N et al (2016) Transformation of food packaging from passive to innovative via nanotechnology: concepts and critiques. J Food Sci Technol 53:3395–3407
37. Bumbudsanpharoke N, Choi J, Ko S (2015) Applications of nanomaterials in food packaging. J Nanosci Nanotechnol 15(9):6357–6372
38. Emamhadi MA et al (2020) Nanomaterials for food packaging applications: a systematic review. Food Chem Toxicol 146:111825
39. Yang L et al (2024) Exploration and application prospect of the advanced technology of time–temperature indicators. Text Res J, p 00405175241266999
40. Poonia A, Mishra A (2022) Edible nanocoatings: potential food applications, challenges and safety regulations. Nutr Food Sci 52(3):497–514
41. Vestergaard MC, Tamiya E (2015) Nanobiosensors and nanobioanalyses: a review. Nanobiosens Nanobioanal, pp 3–20

42. Hammond JL et al (2016) Electrochemical biosensors and nanobiosensors. Essays Biochem 60(1):69–80
43. Aylott JW (2003) Optical nanosensors—an enabling technology for intracellular measurements. Analyst 128(4):309–312
44. Fahmy HM et al (2022) Recent progress in graphene-and related carbon-nanomaterial-based electrochemical biosensors for early disease detection. ACS Biomater Sci Eng 8(3):964–1000
45. Hwang GT et al (2015) Flexible piezoelectric thin-film energy harvesters and nanosensors for biomedical applications. Adv Healthcare Mater 4(5):646–658
46. Barage S et al (2022) Nanomaterial in food packaging: a comprehensive review. J Nanomater 2022(1):6053922
47. Kumar V, Guleria P, Mehta SK (2017) Nanosensors for food quality and safety assessment. Environ Chem Lett 15:165–177
48. Nanda PK et al (2022) Emerging role of biosensors and chemical indicators to monitor the quality and safety of meat and meat products. Chemosensors 10(8):322
49. Mohammad ZH et al (2022) Application of nanotechnology in different aspects of the food industry. Discov Food 2(1):12
50. Chen J, Park B (2016) Recent advancements in nanobioassays and nanobiosensors for foodborne pathogenic bacteria detection. J Food Protect 79(6):1055–1069
51. McGrath T, Elliott C, Fodey T (2012) Biosensors for the analysis of microbiological and chemical contaminants in food. Analyt Bioanalyt Chem 403:75–92
52. Siddiqui J et al (2022) Nanomaterials in smart packaging applications: a review. Small 18(1):2101171
53. Tiwari S et al (2020) Volatile organic compounds (VOCs): Biomarkers for quality management of horticultural commodities during storage through e-sensing. Trends Food Sci Technol 106:417–433
54. Adeyeye SAO (2019) Food packaging and nanotechnology: safeguarding consumer health and safety. Nutr Food Sci 49(6):1164–1179
55. Hindalco Fb
56. Nanopack
57. Technology-Germany N
58. Lanxess Db
59. (USA), NFbAPI
60. Technologies SPwT-TIbI
61. Sheets O
62. Tager J (2014) Nanomaterials in food packaging: FSANZ fails consumers again. Chain React 2014(122):16–17
63. Duncan TV (2011) Applications of nanotechnology in food packaging and food safety: barrier materials, antimicrobials and sensors. J Colloid Interf Sci 363(1):1–24
64. Luo Z, Wang Y, Wang H, Feng S (2014) Impact of nano-CaCO 3-LDPE packaging on quality of fresh-cut sugarcane. J Sci Food Agric 94(15):3273–3280
65. Kumar S et al (2018) Biodegradable hybrid nanocomposites of chitosan/gelatin and silver nanoparticles for active food packaging applications. Food Packag Shelf Life 16:178–184
66. Abdollahi M et al (2013) Comparing physico-mechanical and thermal properties of alginate nanocomposite films reinforced with organic and/or inorganic nanofillers. Food Hydrocoll 32(2):416–424
67. Vejdan A et al (2016) Effect of TiO_2 nanoparticles on the physico-mechanical and ultraviolet light barrier properties of fish gelatin/agar bilayer film. LWT Food Sci Technol 71:88–95
68. Manikantan M, Varadharaju N (2011) Preparation and properties of polypropylene-based nanocomposite films for food packaging. Packag Technol Sci 24(4):191–209
69. Bahrami A et al (2020) Antimicrobial-loaded nanocarriers for food packaging applications. Adv Colloid Interf Sci 278:102140
70. Pushparaj K et al (2022) Nano- from nature to nurture: a comprehensive review on facets, trends, perspectives and sustainability of nanotechnology in the food sector. Energy 240:122732

71. Kaur J et al (2020) Nanomaterial loaded chitosan nanocomposite films for antimicrobial food packaging. Mater Today Proceed 28:1904–1909
72. Park S et al (2020) Nanoscale manufacturing as an enabling strategy for the design of smart food packaging systems. Food Packag Shelf Life 26:100570
73. Cushen M et al (2012) Nanotechnologies in the food industry: recent developments, risks and regulation. Trends Food Sci Technol 24(1):30–46
74. Zambrano-Zaragoza ML et al (2018) Nanosystems in edible coatings: a novel strategy for food preservation. Int J Molecul Sci 19(3):705
75. Yang J et al (2020) Nanosensor networks for health-care applications. In: Nanosensors for smart cities. Elsevier, pp 405–417
76. Maduraiveeran G, Sasidharan M, Ganesan V (2018) Electrochemical sensor and biosensor platforms based on advanced nanomaterials for biological and biomedical applications. Biosens Bioelectr 103:113–129
77. Kaur R, Sharma SK, Tripathy S (2019) Advantages and limitations of environmental nanosensors. In: Advances in nanosensors for biological and environmental analysis. Elsevier, pp 119–132
78. Jayan H, Devi LB, Anandharamakrishnan C (2019) Nanosensors for food contaminant detection. Food Nanotechnol 54:309–340
79. Lamabam SD, Thangjam R (2018) Progress and challenges of nanotechnology in food engineering. Impact Nanosci Food Ind 65:87–112
80. Sundari SS, Moses J, Anandharamakrishnan C (2019) Understanding the risk. In: Food nanotechnology. CRC Press, pp 45–66
81. Singh R et al (2023) Future of nanotechnology in food industry: challenges in processing, packaging, and food safety. Global Chall 7(4):2200209
82. Triques CC et al (2020) Influence evaluation of the functionalization of magnetic nanoparticles with a natural extract coagulant in the primary treatment of a dairy cleaning-in-place wastewater. J Clean Product 243:118634

Nanobiosensors for Food Storage and Supply Chain Management

G. V. S. Subbaroy Sarma, N. Hari Kumar, J. Bhemarajam, and Murthy Chavali

Abstract Since damaged food lacks essential nutrients for growth and may be dangerous to human health if directly consumed, quality is a crucial concern. Because of their distinct qualities, which set them apart from traditional evaluation methods, such as their huge surface area, high selectivity, compact dimensions, excellent biocompatibility and conductivity, real-time onsite analysis, nanomaterials are fascinating. New biosensors with potential applications in the food business have been the focus of much recent progress in the field of nano biotechnology. Nanobiosensors hold great promise for use in a range of food tests, including pathogen, bio toxin, and agrochemical detection. The fundamental prerequisite for food preparation, circulation, storage, and marketing is the management of food quality and safety. Specific kinds of nanobiosensors are used for monitoring the production of food products up until packaging and to determine the product's freshness until spoiling. Enzyme catalysts, which are extremely sensitive to harsh environmental conditions, are mostly used by them. Food inspection sensing technologies, such as optical, chromatographic, and colorimetric, are used. NPs coupled with aptamers to form luminescent labels have been widely used for the quick identification of foodborne and infectious microorganisms. When contemporary information and communication technologies are integrated with nanosensors, innovative and digital techniques for high-accuracy component detection are made possible. The transducer and bio-recognition component collaborate to convert biological signals into digital signals in the biosensor. In the food processing sector, food safety is ensured by this cutting-edge technology. By utilizing nanobiosensors for dynamic food safety monitoring, harmful foodborne pathogens and spoiling problems can be quickly identified, and immediate information on consumable quality can be obtained.

G. V. S. Subbaroy Sarma · J. Bhemarajam
Department of Applied Sciences and Humanities, M.V.S.R. Engineering College, Hyderabad, Telangana, India

N. Hari Kumar
St. Mary's Group of Institutions Hyderabad, Deshmukhi, Yadadri, Telangana, India

M. Chavali (✉)
Materials, MCERTC, Tenali, Andhra Pradesh, India
e-mail: chavalim@gmail.com

D. Sillu et al. (eds.), *Nanobiosensors for the Food Industry*, Smart Nanomaterials Technology, https://doi.org/10.1007/978-981-95-0136-6_12

Keywords Biosensors · Nano-biosensors · Nanomaterials · Nanotechnology · Food safety

1 Introduction

Practically every scientific field, nanotechnology is a relatively new and quickly expanding concept. Research on nanoparticles is common throughout a wide range of fields, and food sciences and technology are adopting an increasing number of novel nanotechnologies. Even at the molecular level, nanoparticles operate on a very small scale, which opens up numerous potential uses in the fields of agriculture and food. The objective of using nanotechnology is to improve production excellence as well as the efficiency of goods along with derivative packaging, transportation, and storage [1, 2].

The Greek term "SENTIRE," which essentially means "to identify" anything, is where the word "sensor" originates. The sensor comprises a transducer, and an output system, with an identification system called a receptor. The sensor's selectivity is a crucial characteristic that shows how well the device can separate the analyte from other substances in the sample. To put it another way, a sensor's ability to detect and measure the analyte with the least amount of interference from other materials in the sample increases with its selectivity. Sensitivity is the next crucial characteristic. High sensitivity indicates that a notable shift in the sensor's output signal is seen with only slight variations in the analyte concentration. Such a sensor has an extremely high bit resolution. Replicability another crucial aspect that is described by the term accuracy is the measurement's conclusions. A high degree of precision means that the outcomes of multiple measurements are fairly close to one another. When the measurement findings are accurate, they are near to the true value.

In simple terms, nanobiosensors are sensors made of nanomaterials; more intriguingly, they are particular sensors that can identify events at the nanoscale. Pathogens, metabolites, antibodies, and nucleic acids are among the biological agents that can be detected by a nanobiosensor. The fundamental concept involves attaching target bioanalytes with bio-receptors followed by altering physicochemical indication of their interaction [3–5]. In a typical nanobiosensor analytes, bio-receptors, transducers, and displays make up its constituent parts. A substance that is interesting and needs observation is called an "analyte." For example, cholesterol is an "analyte" in a biosensor intended to measure cholesterol levels. Enzymes, cells, aptamers, DNA, RNA, and antibodies are examples of molecules that are known as bio-receptors since they are the ones that primarily recognize the analyte. The information produced through a relationship between the bio-receptor with the analyte—which might manifest as illumination, temperatures, pH, charge, density modifications, etc., helps during the bio-recognition mechanism. One part of a biosensor that primarily transforms one type of energy into another is called a transducer. The transducer's technique of converting its bio-recognition incident toward readily quantifiable signals has been referred to as signalization. This generated analyte-bioreceptor contact signal is processed by this section of a biosensor before being shown on the apparatus. The processed signal is then assessed by the biosensor's display unit. An interface system that produces quantifiable figures or curves that the user can easily understand is called a display [6].

2 Nanobiosensor Fundamentals

Nanobiosensors are cutting-edge instruments that identify and examine particular biological, chemical, or physical substances by combining the concepts of nanotechnology with biotechnology. Several basic operating principles underpin the operation of these sensors. The detector that operates by selectively engaging the target probe is the first component of nanobiosensors. This sensing agent may be a synthetic substance with a particular binding capacity or a biological molecule. The sensor's specificity and selectivity, or its capacity to differentiate among the target analyte and extraneous variables. Second, a transducer is used in nanobiosensors to transform the bonding reaction between the target analyte and the detection element into a signal that can be detected. Electrical, optical, light energy or piezoelectric effects were among the physicochemical concepts that might serve as the foundation for the transducer. Transforms a chemical or biologic discovery into a measurable form which may be precisely identified and examined. Amplification methods are frequently used in nanobiosensors to boost the target molecule's sensitivity and detection capacity. Enzymatic processes, nanomaterial-based signal amplification, and electrical or optical signal amplification are a few examples of these techniques. Amplification techniques enable researchers to detect small quantities of noise by boosting the proportion of signal to noise.

Specialized tools or devices are frequently used to interpret and detect the signals produced by nanobiosensors. These could be remote sensing techniques, imaging systems, optical detectors, or electrical gadgets. The sensor output is transformed into useful data by the readout system so that it can be further examined and evaluated. Nanobiosensors are also distinguished by their capacity for integration and downsizing. These sensors can be made incredibly small by the use of nanoscale components and production processes, which enables their integration into platforms. They can be combined to create wearable sensors, lab-on-a-chip devices, or packaged goods. Mobility, adaptability, and integration into intricate systems are made possible by this downsizing.

The development of biosensors can be divided into three main groups: the initial generation, which depends on electricity response; the second variety, which serves as intermediaries among the transducer and its response; and the third generation, which directly uses mediator diffusion. In many of the more demanding industries, such as the food industry, nanotechnology holds enormous promise. It is creating new methods for plant cropping, food production, fertilizer distribution, packaging along other fields. In the food industry, nanobiosensors have great promise for enhancing food quality and identifying toxicity. In practice, nanotechnology is currently mostly

Table 1 Recent uses of nanotechnology in the food industry

Nanosize	Product	Reason
50–150 nm starch nano spheres	Food packaging	The exterior space associated with these nanoparticles is 400 times bigger than the area of organic starch particles
300 nm of iron	Nutritional drink	Nanosized iron particles are more reactive and bioavailable
Nanoparticles of silver	Connection with food resources	Nanosized silver particles have increased antibacterial properties

used in food processing and packaging, as well as in nanobiosensors that identify pollutants and toxins in food. Applications for nanomaterials and nanobiosensors are also being discovered in the food and agriculture industries. They are greatly improving the agricultural sector by incorporating quick-detection networks, intelligent packing, and nutritious ingredients within food for animals [7]. Table 1 provides recent nanotechnology uses in the food sector.

3 General Characteristics of Nanobiosensors

Rai et al. [8] listed the ideal properties of nanobiosensors, which are shown below:

i. Analytes and nonanalytes need to be distinguishable by nanosensors.
ii. They ought to be durable under typical storage conditions.
iii. The analyte interaction should be specific and unaffected by physical factors such as agitation and pH.
iv. Reactivity must be at the lowest possible level.
v. The nanosensors must provide exact, reliable, and repeatable results.

3.1 Chemical Property-Based Nanosensors

The emergence of biosensors featuring different sizes, surface areas, and electrocatalytic qualities has drawn more and more attention to nanomaterials in the past few years. To enhance the electrocatalytic characteristics of the electrode surface, electro active nanoparticles are typically employed as direct recognition elements for the detection of a variety of analytes. Carbon and graphene-metal nanoparticles (NPs) were the basis of several of these electrochemical sensors for the identification of mycotoxins. Siderophores are generally divided into three major types depending on the chemical nature of their locations and connections with iron: hydroxamates, catecholates, and carboxylates. Single-molecule observation is made easier by the synergistic and adaptable solutions that nanomaterials and biosensing platforms offer for easier, quicker, as well as more accurate identification.

When it comes to food manufacturing, the objective was creating a different climate inside the item's packages by minimizing the entry of oxygen and preserving the inside temperature using inert gases like carbon dioxide or nitrogen. Several methods, including gas chromatography of thermal conductivity and Clark electrodes, may be employed to determine oxygen. However, due to its costly and time-consuming, there is a need to find inexpensive and simple implementation indications. This led to the invention of smart inks using TiO_2 particles, which modify color when exposed to oxygen via redox processes [9].

3.2 Physical Property-Based Nanosensors

Fluorescence technologies are going to be among the greatest options for intracellular detection as well as remote observation when compared with conventional analytical techniques. The analytical substance data might be converted into fluorescent signals using these methods [10, 11]. The two different types of fluorescence-based probes are ratio metric-based and single-wavelength intensity-based. The former have simplified instruments and operations and are frequently used for sensing. However, several interfering potentials typically impair the precision of measurements. These possible causes are mostly linked to signal variances brought on by differences in the reliability of the illumination, the dimension of the optical pathway, photo bleaching, and the amount of the probe. However, ratio metric probes, which measure the relative fluctuation in two separated emission peak levels, were produced with either single-excitation modes or double-excitation modes.

Various well-known nanomaterials, including metal–organic frameworks, graphene oxides, carbon dots (CDs), and NPs, can be used to create fluorescent labels for bacterial detecting in order to increase photostability, facilitate production, and enhance biocompatibility. However, because of their excellent fluorescence qualities, CDs are being studied as fluorophores for the highly sensitive and particular identification of heavy metal cations and anions. On the basis of the catalytic oxidation that the dual-emission silica NPs cause, CDs could be utilized for identifying NO_2 within food samples [12].

3.3 Biological Property-Based Nanosensors

To guarantee optimal performance in microorganism-based fabrication as well as to address food safety concerns, microbiological analysis is required in the food business. To quickly and efficiently recognize microorganisms, technologies must be used; therefore, biological nanosensors become essential for this procedure. Because of its ease along with accuracy, traditional techniques such as microbial culture in solid or liquid medium are widely used to identify viable Saccharomyces cerevisiae throughout industrial goods. Yet, that process takes a lot of time as well and costly equipment and expert workers are needed because of its highly sensitive nature. Alternatively, gated MSN-based small devices activated with enzyme-mediated mechanisms are created with mesoporous silica nanoparticles (MSN) along can be applied to additional applications including particle-to-particle interactions, medicinal transport, along biological sensing [13].

4 Categorization of Nanosensors

4.1 Electrochemical Nanosensors

The theory of electrochemistry serves as the foundation for commonly used electrochemical sensors. Electrochemical signals are produced throughout bio-interstation by the electron's moment. Targeting the analytes to produce or accept electrons and measuring them through voltage, current, or impedance, electrochemical nanosensors depend on chemical processes between nanomaterials, biomolecules, and bio-elements [14].

The quantity of analyte in the specimen is directly correlated with the electrochemical signals produced. Electrochemical nanosensors can be divided into three categories according to their modes of operation: potentiometry, voltammetry, and amperometry. In order to determine the concentration, amperometric sensors continuously detect the current produced during the redox process of an electroactive species, where the voltage is certain at a fixed range. By altering a potential, voltammetry nanosensors measure the current under observation. Monitoring the current flowing between the working and counter electrodes reveals a difference in the voltage characteristics between the reference and functioning electrodes. Consequently, a voltammogram—a plot of current against voltage—is created from the information. In the food and agriculture sector, it is often used to identify the presence of heavy metals, Salmonella typhi, and carbosulfan in rice [15–18]. Ions build up in the sensitivity field affect transistors and electrolytic electrodes as equilibrium happens, producing the output signal. Ion-specific electrodes use variations in the potential of the electrode for identifying ions including Na^+, K^+, Ca^{2+}, H^+, and NH^{4+} in a variety of different biological mediums. Numerous substances, such as heavy metals, chemicals, antibiotics, and additives, have been detected with them [19].

4.2 Optical Nanosensors

An instrument that transforms light beams into electrical impulses is called an optical sensor. Similar to optical resistors, it analyzes the actual quantity of light and transforms it into a form which the device is able to interpret. The capacity to determine variations in a variety of light beams is one of an optical sensor's characteristics. The primary basis for this variation is changes in the brightness of light [20]. Adding nanomaterials to optical sensors offers many benefits, including enhanced contrast, photostability, and spectrum specificity. Significant advancements in quantum mechanics, fluorescence, and luminescence have been shown in nanomaterial-integrated optical sensors [21]. In the food business, these sensors were employed for pH sensing, glucose detection, and foodborne pathogen detection [22–24].

An optical sensor is a device that converts light rays into electronic signals. Similar to additional spectroscopic methods, optical sensors depend on identifying alterations in the signal that light emits. Natural fluorescence characteristics of gold, silver, and quantum dots are dependent on their dimension and form. To identify nitrite, reactive oxygen species, harmful microorganisms, and organophosphates, several fluorescent nanosensors have been created [25–27]. Optical sensors with high sensitivity and quick detection are called surface plasmon resonance (SPR) sensors. Concanavalin A, antibiotics, mycotoxins, and pathogens were all detected by an enhanced SPR detection system using optimized metallic, magnetic, and carbon-based nanoparticles, latex, and liposome nanoparticles.

4.3 Chemical Nanosensors

Chemical sensors convert numerous chemical signals, such as concentration, composition, presence of a specific element or ion, particle pressure, etc., into a usable analytical signal and are used to detect a variety of chemical processes. The chemically sensitive area serves as a conduit among the source and drain regions of a field-effect transistor-based chemical sensor [28]. Inaccurate analyte sensing is aided by the environment's chemical reactions, which alter the charges in the channel region and, consequently, conductance. The incorporation of nanomaterials into chemical sensors enhanced their sensitivity, specificity, and speed. Chemical dangers in food, including heavy metals, dioxins, and agrochemicals along with nutritional supplements including colorants and sweeteners, have been detected using chemical nanosensors [29].

4.4 Nanobarcodes

Nanobarcodes are structures at the nanoscale that are encoded using various coding elements that have various characteristics, such as sizes, forms, and colors. Because of their enhanced properties-such as a greater surface area to volume ratio, unique physical and optical assets, and widely accessible preparation techniques resulting from their nanoscale length, nanobarcodes have a wide range of uses. When creating nanobarcodes, several elements should be taken into account, including decoding performance, synthesis strategies, and encoding techniques. Encoding elements can change shape and dimension and possess optical, graphic, magnetic, and phase modification properties [30]. As explained in the next subsections, nanobarcodes are created using a variety of coding processes and decoding strategies.

Organic dyes, quantum dots, carbon dots, and lanthanide-doped nanocrystals are all commonly employed for this, and fluorescence microscopy, reflectance spectroscopy, and Raman spectrophotometers are utilized for the decoding process. The optical characteristics of the encoding parts, such as fluorescence spectra, lifetime, and light scattering, are utilized. Graphical encoding, on the other hand, employs visual patterns, and the design may be in the form of various synthesis-created blueprints. Decoding is done using optical and fluorescence microscopy. Additionally, peptides and DNA are used as coding components in sequence encoding. Mass spectrophotometry and DNA sequencers are employed for decoding.

It is determined by the particles' dimensions, form, and other physical characteristics. Decoding is done using atomic force microscopy and electron microscopy, respectively. In contrast, magnetic encoding relies on the nanoparticle's magnetic characteristics. For this decoding, a vibration sample magnetometer is employed.

It takes advantage of the solid-to-liquid phase change characteristics of nanoparticles. For decoding, infrared imaging devices and the technique of differential scanning calorimetry (DSC) are employed. In the past 10 years, numerous biocide intelligence platforms have been created, including monitoring and evaluating nematodes in agricultural soils, detecting food pests [31], and ensuring the security and nutritional value of food and vegetables. For the quick and reliable identification of Salmonella enterica serovar enteritidis in samples of food, gold and iron oxide nanoparticles independently attached to two distinct DNA barcodes [32]. To confirm the validity of various meals, unique, undetectable barcodes have just been invented. Regarding recognizing products and detection of diseases in food and biosamples, there are numerous publications on the development of nanobarcodes that employ fluorescent barcode nanorods, opaque nanobarcodes, and fluorescent DNA dendrimers.

4.5 Electronic-Nose (E-Nose)

Throughout the previous 10 years, e-nose devices with sensors have advanced quickly toward chemical detection and illness diagnosis. E-Nose technology for identifying diseases and chemical detection of different gaseous mixes, especially those containing volatile organic compounds, has grown significantly with the introduction of novel electronic approaches and associated operational methodology. Technologies for E-Nose inspection have been improved by developments in artificial neural networks, sensor techniques, diseases-specific databases, data interpretation software, and disease biomarker identification. Utilizing the E-Nose for quality assurance and control is its most important function across a range of industries. Figure 1 shows an example of an E-Nose flowchart.

Following the discovery of synthetic equipment which imitated the human smelling sense, the name "electronic nose," or e-nose, was developed. E-noses are utilized in manufacturing facilities to monitor standards of quality and in the

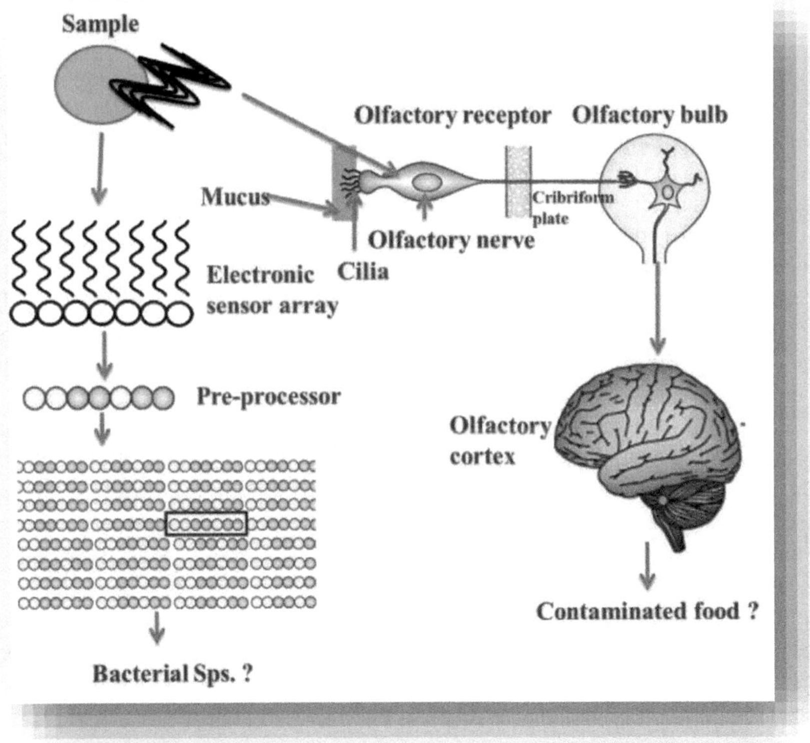

Fig. 1 Comparison and parallels of an electronic nose/tongue through olfactory system versus human brain performance

food manufacturing industries to develop new foods and beverages [33]. Generally, e-noses are made up of four segments.

(1) An electrical circuit,
(2) A sampling point,
(3) A collection of small or large films of semiconductor detectors which overlap slightly and alter the electrical conductance of gases, and
(4) Data processing software.

In order to create a distinctive fingerprint with the array of sensors as an odor trigger, every sensor in the array can uniquely react to particular chemicals. Such signals are often different. To identify the established materials, the structure-recognizer compares the comparing the design of present observation against those that have been saved. In order to provide a comprehensive response to a combination without identifying specific elements, e-noses have been developed recently as an alternate approach to doing fragrance examination [34].

4.6 Electronic Tongue (E-Tongue)

Especially compared to their predecessors, electronic tongues, or e-tongues, have shown to be extremely robust and consistent, leading to increased use. Such sense instruments are employed to assess food's structure and distinguish between distinct dietary tastes [35]. To combine signals from nonspecific and interacting sensors, e-tongues use recognition of patterns protocols. To overcome the ambiguity with human understanding, researchers employ an array of sensors to replicate the taste detection modalities of the human tongue. Additionally, it provides a quick and nondestructive way to evaluate the quality of food both in lab settings as well as electronically throughout the production procedure. Five chemical samples have been studied using a polymer composite sensor based on carbon nanotubes, which is employed as an impedimetric e-tongue [36].

E-tongue instruments are nanostructured components that, upon binding with a biomolecule, may change electrical characteristics like resistance or current. After that, many sensors were created to evaluate and classify the flavor of multi-component assortments resulting in a cutting-edge instrument frequently employed in pharmaceutical, environmental, and food analysis [37, 38]. To produce logical conclusions, the computed electrical signals are run through signal distribution algorithms. This technology uses a variety of low-specificity sensors to detect analytes in spite of specific target receptors that work cross-specifically for the current molecules by detecting electrical signals. Sweet, sour, salty, bitter, and umami are the five basic sensations that were identified by the first produced e-tongue [39].

4.7 Biosensors-On-Chip

With combining sensor technology and microfluidics, the microfluidic biosensors (lab-on-a-chip) system offers reliable and affordable diagnostic instruments. In addition to offering the newest technology for wearable and disposable sensor devices, it enables the simultaneous detection, enhanced precision, and real-time analysis of many analytes in a single tool, allowing the combination of chemical and biological components in just one procedure [40, 41]. This method was used to identify nucleic acids (cDNAs) that had been located in sufficient quantities in cancer patients' blood [42–44].

4.8 Gas Sensors

The objective of the gas nanosensor is to detect certain gases that result from food spoilings, such as NH_3, NO_2, CO_2, CH_4, H_2O_2, $C_6H_{10}O$, and H_2S. Such gas sensors are made of materials such as gold, platinum, and palladium [45]. In order to successfully integrate gas sensing devices into food packaging products, sensors have to sustain packaging efficiency through storage and inspection while also not affecting the functioning or barrier properties of the employed product packing. The implementation of sensors is one of the factors that must be considered within or among covering materials, the endmost characteristics of the established materials, the necessary process surroundings, their impact on packing litter management, etc. [46].

In gas sensors, conductive polymers were often employed. In electroactive conjugated polymer-based sensors, electrically conducting nanoparticles are integrated into insulating polymer matrices. The dominant characteristics of conductive polymers related to their associated π electron pillars include electrical, optical, and magnetic [47]. Using a conducting polymer-based detector, the presence of the gas to be examined causes resonance to increase, resulting in a series of reactions.

4.9 Array Biosensors

Regarding point-of-care utilization in the food sector, a nano porous silicon detector array that utilizes a microfluidic system has recently developed. It is sensitive, fast, and can recognize many antigens. A collection of nonspecific detection elements in a set of sensors that may respond differentially to generated targeting samples is a finger mark enabling processing data in an array sensing technique [48]. For identifying the gasses and aromas released by pathogens throughout food contamination, an array sensor that regulates polymer nanocomposite has been developed [49]. Such gas monitoring, which is made up of polyaniline and carbon black nanostructures, identifies and categorize bacteria by producing distinct behavioral patterns for all of them. Successful responses were confirmed whenever detectors were exposed to bacterial vapor, which altered the sensors' sensitivity. The array exhibits the capability of immediately identifying foodborne pollutants, and it may be utilized in conjunction with portable devices to quantify gas emissions via contaminated packaged foods [50].

4.10 Electro Chemical Immunosensors

An antibody fragment is used by immunosensors to identify its biological component. The foundation of immune-sensor recognition is the particular antibody-antigen combinations. An electrochemical immunity sensor's fundamental elements are a glassy carbon electrode, which is a "working electrode," a "counter electrode," and a "reference electrode." The mechanism underlying these devices has similarities to that of the commonly used biochemical enzyme-linked immunosorbent assay or enzyme-linked immunosorbent assay technique [51]. Several immune sensors follow the same patterns as ELISA; however instead of using substrates resulting in a color conversion as in ELISA, enzyme substrates typically generate an electrochemically active molecule when they combine toward the corresponding enzyme.

The method known as ELISA is difficult to convert to an electrochemical format, even though the development of an electrochemical immune sensor has similarities to that of optical analysis. The result is explained by the fact that each measuring step in an immunological sensor occurs at the electrode/solution interface, whereas the measurement step in an ELISA occurs in a homogeneous solution. Any biological entities that become immobile on an electrode will consequently unavoidably change the measuring signal. A particular type of immunological sensor that do not need labeling was created as a consequence. Because label-free methods reduce the need for reagents and assays, they are seen as beneficial because they save time and money [52].

4.11 Lab-On-A-Chip Device (LOC)

Foodborne infections can be detected using lab-on-a-chip technology, which is a system that combines numerous laboratory processes into an individual compact panel which is often only a few millimeters or centimeters long. The primary benefits they offer include their ease of use, speedy assessment, and high reproducibility as a result of standardized and automated methods. To summarize, the lab-on-a-chip biosensor provides a perfect platform for small and real-time foodborne pathogen applications involving biosensors. These were created using microfluidics, which enables solvent exchange in microchannels. Every stage of the electrochemical immunoassay, including antibody immobilization via electrochemical detection, is capable of being carried out on the system in enclosed microfluidic channels that are waterproof.

Using this instrument with the electrochemical sensor provides the way to more sophisticated systems for analyses. The idea behind the devices is that pre-immobilized chemicals on a nitrocellulose membrane are activated when they come into contact with a fluid sample inflow. Although this device is often qualitative, quantifiable results can be obtained once used in conjunction with an electrochemical sensor. The pre-immobilized Ab on the electrode surface, the selction of the electrochemical technique, and the electrochemical reaction produced according to the reaction between the enzyme and the substrate are the primary determinants of the electrochemical immune sensor's specificity and sensor process.

4.12 Carbon Dots (CD)

A fluorescent sensor array called luminescence carbon dots can be utilized to detect and differentiate between a number of hazardous metal ions, including Hg^{2+}, Cu^{2+}, and Fe^{3+}. The variable fluorescence intensity capabilities, chemical inactivity, nonpoisonous nature, and sufficient biological compatibility of CD, a form of carbon nanomaterial, make it an attractive option for array detection by fluorescence metal ions [53]. Their exteriors are covered by several functional groups that may collaborate with various metal ions to provide nonspecific sensing [54]. Furthermore, the formation of a few functional groups might be changed by changing the manufacturing techniques, interaction circumstances, and raw materials [55].

Biomolecules may be detected and imaged using the fluorescent CD. This is comprised of just nanoscale quantum dots (QD), such as carbon, graphene, and polymer QD. The quenching and restoration of fluorescence is the fundamental concept of the CD. CD can be synthesized in two ways: top-down nano cutting, which reduces macro carbon materials into nanoparticles, and bottom-up organic pathway, which forms precursors by aggregation and dehydration processes [56]. Because CDs may act as both electron donors and acceptors, they might be employed as the primary ingredient in a range of biosensors. Numerous uses for CD-based sensing systems exist in the food industries, such as the identification of diseases, banned additives, nutrients, pesticides, and toxins.

4.13 Nanotest Strips

When the specific Nanotest strip combines with the food item to be evaluated, the results can be seen either qualitatively or quantitatively as changes in color, fluorescence, or magnetic transition. Advantages of this in situ diagnostic tool include flexibility of usage and manufacture, inexpensive price, excellent accuracy and sensitivity, exactness strike, no maintenance or equipment needed, instant findings display, portability, and more. The entire system is expected to incorporate an extremely economical sensing module and demonstrate significant potential in food safety research. In experimentation approaches; the markings of nanoparticles have included organic nanostructures such as nanotubes of carbon, lanthanides, magnetic nanoparticles, quantum dots, and colloidal gold. The current and thoroughly researched approach is the extensively reported Nanotest strip labeled using colloidal gold [57]. The equipment's high density of electrons allows it to trap macromolecules from biology without interfering with their metabolism, changing color from orange to red to purple in this manner.

4.14 Carbon Nanotubes (CNT)

Bucky tubes, another name for carbon nanotubes, are quantum products with a topological geometry which resembles the curved hexagonal shape of graphite. Unlike ordinary nanoparticles having a specific surface area, high conductivity, and significant rigidity, it exhibit quantum effects. It is simpler to recognize qualitatively or semi-quantitatively by the unaided eye due to its conspicuous black shade [58]. The potential uses of CNTs in food processing will be greatly improved or accelerated by their nano-reinforcing effects on polymers, especially biopolymers. The three most common methods for producing CNT were chemical vapor deposition, laser ablation, and arc discharge. Carbon nanotubes are categorized as single-walled (SWCNT) or multi-walled (MWCNT), depending on whether they are composed of a single layer of graphene cylinder or several layers. In addition, SWCNT and MWCNT are both promising active materials for packaging due to their strong antibacterial qualities [59]. Utilizing CNT with great biological compatibility and effective qualities in package design will be made significantly simpler by studying surface modification techniques and CNT toxicity [60].

4.15 Nanocellulose Film (NCF) and Nano Cantilevers (NCC)

Whenever microbial growth impacts the quality of food, nanocellulose films show it. NCF comes from a variety of bio-renewable sources and can be recycled, renewable, and sustainable [61]. Because of its poor moisture resistance, NCC degrades readily if used alone; nevertheless, when combined to form a film, it effectively protects the food product from obstacles like water and gas. Organic NCC generates bioactive substances that prevent the pathogen from growing. These biofilms are produced chemically, which improves their ability to transmit light and makes them more efficient toward microbial activity [62].

Another development in nanosensing technologies is the application of nano cantilevers, which are employed for identifying a variety of chemical and molecular reactions, such as the Ag–Ab reaction. The physical factor/electrochemical transmission brought through biomolecule contact forms the basis of its identification principle. The nanocantilever's interacting interface is composed of up of tiny silicon-based components. The fundamental idea underlying the detection of pathogens is that the organic matter of the pathogen makes the cantilever surface to fluctuate at distinct frequencies. An item called Bio-finger, which is intended for detecting poisons, chemicals, and pathogens in food, is produced by combining cantilevers designed of piezoresistive materials.

4.16 Nanocomposites

Polymers and nanoparticles interact to create nanocomposites with improved polymer characteristics. Maintaining the food's freshness and cleanliness is the primary goal. Since carbonated drinks possess excellent gas barrier qualities, they are widely utilized to prevent CO_2 leaks. The economical and environmentally beneficial nanoclay, sometimes referred to as phyllosilicates, represents one of the most effective examples of nanocomposites. Two categories of nanocomposites that may be utilized to further classify them [63].

1. The polymeric composition is intercalated, or organized, as a result of penetrating in the interfacial area.
2. Exfoliated-dispersed randomly as a result of prolonged penetration.

Amongst the readily accessible nanoclays include Aegis, Durethan, Imper, and Nanocor. Red meat, processed cheese, and candies are covered with nanolaminates for nanoencapsulation. Composites of metal and polymers have antibacterial properties.

4.17 Release-on-Command Concept

One interesting development in packaged foods is the potential of controlled antibacterial discharge via packaging supplies. The use of antimicrobial compounds in packaging supplies may increase the lifespan by preventing microbial activity and contamination. For example, a device known as a "bio switch" discharges antimicrobial chemicals upon demand whenever bacterial growth is detected [64]. The basic concept is that the antimicrobial responds adequately to variations in the environment. Internal or external stimuli enable the product's antibacterial component to be discharged. But specific conditions, including pH, temperature, or UV light, can activate the antibacterial ingredient, which discharges upon demand. This tool may be able to increase food storage's lifespan and sustainability which would lower the amount of pesticides used. A common instance of release-on-command antimicrobial substances in food packaging is polysaccharide molecules having enhanced antibacterial qualities.

4.18 *Humidity Indicator and Nano Bioluminescence Detection Spray*

The humidity level within a sealed food product determines the true state of the food's quality. Nanocrystalline cellulose film was applied to provide an iridescent film for the humidity indicator. One important feature of the humidity indicator is the customizable hue of these films, showing the force of electromagnetic interaction. Whenever the indicator transitions from a dry to a humid state, its color changes via bluish-green to reddish-orange. The distinctive feature of this indicator is that photonic crystals are resistant to photo bleaching-induced degradation, which makes them better suited for determining grade [65].

5 Nanosensors in SCM

In modern highly complicated and constantly changing operational surroundings, supply chain management (SCM) is widely acknowledged as essential to corporate growth and competitiveness [66]. The overall achievement of any supply chain participant is determined by the effective management, integration, and coordination of critical business operations throughout the chain. As nanosensor technology has advanced recently, nanosensors have demonstrated their enormous promise in a variety of fields. Consequently, it is imperative to assess the potential applications of this innovative technology in the management of supply chains. Applications of nanosensors in the supply chain are examined through some important perspectives in the sections that follow.

5.1 *Nanosensors in Supply Chain Management Systems*

Handling specific responsibilities is no longer sufficient for successful supply chain management (SCM); instead, activities must be integrated into essential supply chain processes. Even while current wireless sensing network hardware technologies are suitable for usage in transportation tracking gadgets, the most recent advancement in WSN system programming misses an adequate set of features. Evers et al. introduced a Sensor System, an interface that can meet the demands of logistical situations involving continuous monitoring. Sensor Scheme acts as interpreters that use its

Scheme language of programming to run flexibly imported application code for WSN systems. Further to having high-level programming features like garbage gathering, interaction through automatic information gathering, and co-routines that set up multiple levels for control and allow blocking of I/O calls, it offers a secure execution of the surroundings where failing programs are unable to malfunction the system. Sensor Schemes might be used well across many others, including "traditional" WSN uses in contrast to monitoring logistical procedures.

To enhance the efficiency of the supply chain and provide many advantages to all associated parties, supply chain management (SCM) seeks to connect all supply chain agents to collaborate throughout the company [67, 68]. A new framework for the creation and application of nanosensors within integrated devices was introduced by Envoy et al. The "striped" large-aspect ratio cylindrical forms that are the basis of modern nanosensor technology are manufactured utilizing porous membranes as precursors. Dielectrophoretic forces are used to modify such nanostructures, enabling their unique construction and characterization. Additionally, this assembly provides the design required to carry out nanostructures' post-IC assembly, which makes it possible to create "mixed-mode" integrated circuits that include communication, along processors for signals, and sensing circuits. Therefore, first design for these mixed-mode devices allows architectures that combine basic resistance read-out circuitry via dielectrophoretic assembling sites [69]. In simple terms, nanosensors might be encoded within the packaging or even the product itself, in addition to being affixed to the products and functioning as sophisticated RFIDs.

5.2 Nanosensors in Supply Chain Environmental Detection

It could be possible to utilize nanosensors to detect changes in the atmosphere and item quality that occur throughout storing as well as transporting processes. It is necessary to monitor and record any modifications in the environment that products encounter during transportation, like variations in humidity, temperatures, or smell, as these factors may cause the items' quality to deteriorate.

Oxygen inside food packaging was known as primary source for food deteriorating because it causes lipids and oils become oxidized enabling microorganisms to flourish. Additionally, oxygen speeds up the processes that occur within food packing, resulting in issues with taste, rancidity, texture changes, and discoloration. Using nanotechnology, it is possible to create ethylene-scavenging bags for vegetables and fruit packaging; moisture absorbent strips for fresh meat, poultry and oxygen-capture devices for slicing processed meats, beer, beverages, cooked pastas, and packaged foods. Depending on the nanotechnology strategy, active sealing films are being manufactured to enable specific regulation of enzymes that alter fragrance and oxygen transfer [70]. When dispersion chemicals alter the surface of nanomaterials, they can serve as substrates used by oxidoreductase enzymes. To preserve the beer fresh and extend its lifespan beyond six months, the clay nanoparticles implanted inside the plastic containers harden the packaging, decreasing permeability for gases and minimizing carbon dioxide losses through the beverage and oxygen entry into the bottle.

Numerous innovative advancements are taking place, all of which are pertinent to the capacity of nanosensors to identify malfunctions in storage or packaging settings. The use of nanostructure-engineered chemical sensors for dangerous gases and vapor detecting as well as applications [71] may also be extended to nanosensors for gas and vapor detection in environments or containers for storage. Single-walled carbon nanotubes (SWNTs) are exceptional nanomaterials to identify tiny quantities of damaging and dangerous compounds because of their distinctive electrical characteristics and huge surface area [72].

At low temperatures, an incredibly responding gas nanosensor has been demonstrated by Iliadis and Ali [73]. In CMOS compatibility processing, ZnO nanoparticles self-assemble in a polyacrylic matrix spin-cast on p-type (100) Si at low temperatures to generate the nanosensor. Particularly for low gaseous levels, sensitivity remained always greater for the tiny nanoparticles [73]. A new sensing device made of chains of amino acid-mediated produced gold nanoparticles was presented by Lee et al. It has a sub-ppm lower threshold of detection for H_2S and quick reaction and recovery rates at room temperature. The detection of temperatures in storage is possible with this sensing system. The lateral development of a carbon nanotube (CNT) among two electrodes and its application as a nano temperature monitor were demonstrated by Kuo et al. [74]. MEMS technologies are utilized to fabricate electrodes. By employing microwave plasma chemical vapor deposition across each electrode, the CNT is produced selectively [74].

Regarding uses like humidity control and monitoring, the sensors' exceptional sensing qualities such as their rapid response times and recuperation durations, superior repeatability, the concept of linearity and environmental stability are crucial. Studies on nanosensors that observe the humidity or variations in temperature brought on by humidity consist of:

- Electrically spun LiCl-doped TiO_2 nanofibers are used in extremely delicate and stable moisture nanosensors [75];
- Using electrospinning and calcination, Na(+)-doped ZnO nanofiber membranes have been utilized to create a novel, precise moisture nanosensor [76];
- Low-drift chemical sensor technology for humidity detection that can be used at both micro and nanoscale sizes, based on a novel, all-differential method [77].

5.3 Nano Research in the Food Supply Chain

The production, packing, and preservation of food will be significantly impacted by the new, outstanding materials that can be created due to the special physicochemical characteristics of nanostructured materials. The fabrication of nanoscale food ingredients and additives, bioactive substance methods for delivery, and creative food packaging are the main goals of nanotechnology in food manufacturing at the moment. The establishment of intelligent nutrient or food ingredient delivery systems, quick measurement of chemical and biological pollutants, and nanoencapsulation of nutraceuticals for regulated discharge, filtration of water, and cell-wall breaking for harmful bacteria in agro-food tests or fluids have all been observed recently.

The food company is always searching toward more cost-effective methods of manufacturing and maintaining food. Numerous possibilities are available in the field of nanotechnology for food manufacturing, encapsulation, nanofiltration, etc. Preparation aids in preserving the nutrient content of foods or adjusting the food matrix to meet customer needs. Multifunctional elements utilized in nutritional enhancement are occasionally included in the delivery mechanism rather than being administered simply in their natural state [78]. Nevertheless, such substances have little bioavailability and are frequently broken down during manufacturing. Functional food additives can integrated, absorbed, or distributed in nanostructures by using nanoencapsulation and nanoemulsion formulations. Similarly, active probiotic microorganisms might be included in bread to support gastrointestinal processes, and dietary additives may accumulate nanostructures to cover up the smell, flavor of fish oil, which could incorporated into bread providing additional health advantages. Because nanoemulsions prevent food from breaking down and separating, they eliminate the requirement for stabilizers and may significantly reduce the amount of fat required [79].

One technique for removing low-molecular-mass substances from a fluid media is nanofiltration process. The permeation rate of membranes used for nanofiltration is comparatively low for organic materials but extremely high for salts. Advanced electrical and steric interactions or electroneutrality concepts, like the Donnan effect, dictate the division procedures [80]. The features of the membrane's substrate and the solution being fed have an impact on the functioning of the nanofiltration membrane; therefore, the efficiency of the membranes is primarily determined by three factors: its effective density of charges, the diameter of the effective pores, and the efficient proportional of layer thickness to porosity. Particularly when used for a filtration procedure or for obtaining specific solutes, this approach is extremely desirable for commercial uses in the food and pharmaceutical sectors.

5.4 Food Packaging

Nano reinforcement, nanocomposite active packaging, and nanocomposite sophisticated packaging were new developments in the packaging of food Fig. 2. For food packaging, polymers are currently taking over the role of traditional materials including ceramics, glass, metals, paper, and board because of their practicality, low cost, light weight, and ease of processing. However, when compared to metals and ceramics, they are weaker and are unable to withstand distortion. Additionally, one restricting characteristic of food packaging is its intrinsic permeability to gases and vapors [81]. By bridging the gaps in packaging materials, nano-reinforcing technologies are employed to improve survivability and tensile strength. Food packets with

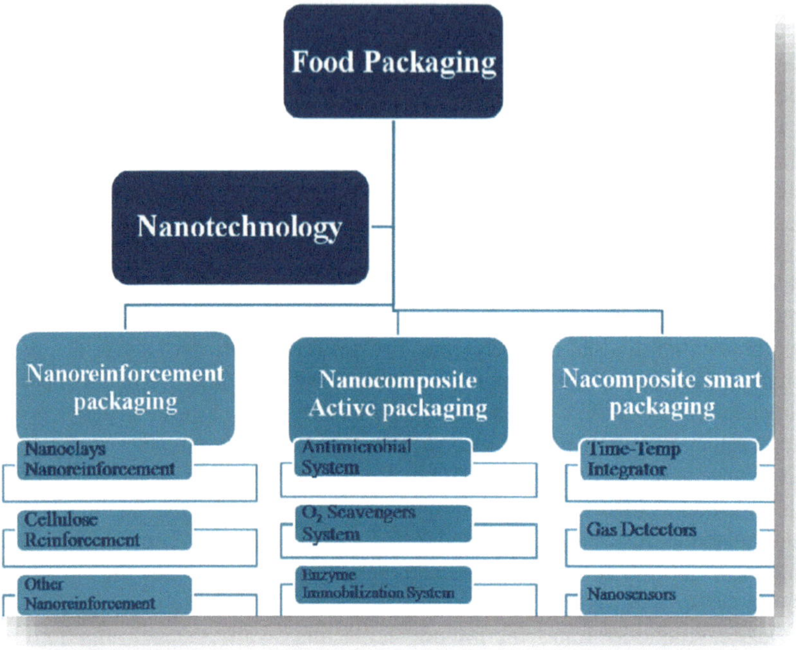

Fig. 2 Recent advances in food packaging studies using nanotechnology

nano reinforcement such as cellulose and nanoclays have increased tensile strength. Nanoclay particles are currently attracting the interest of the packaging sector because of their wide availability, inexpensive price, simplicity of processing, and excellent functionality.

One fascinating substance for creating cheap, portable, and highly effective nanocomposites is cellulose nanoreinforcements (CNRs). Living things create cellulose chains as tiny fibers, that are collections of massive molecules held together by hydrogen bonds. There are crystals and unstructured areas in every microfibril, which is made up of fundamental fibrils. The crystallized components constitute nanocrystals or nano whiskers that can be separated using techniques like acid hydrolysis. The aspect proportions of these components are correlated with the cellulose's region of origin and the circumstances under how it is processed [82].

In contrast to traditional food packaging, an active food package might not just serve as a passively barriers but can also react to the food in a manner that is beneficial, for instance by eliminating a harmful element or emitting substances that are beneficial, like antioxidants or antimicrobials. Advances regarding food stabilization are typically the result of those interactions.

Antimicrobial technologies have become increasingly practical alternatives that can halt the developing of dangerous and decomposing microorganisms on food items. These devices are capable of holding greater numbers of microorganisms and cells because nanoscale substances have a larger surface-to-volume ratio than small-scale products. The antibacterial action of nanoscale substances such as inhibiting growth, killers, or transporters of antibiotics is being studied [83]. Silver has several benefits, like greater stability at high temperatures and reduced volatility, despite its well-known toxicity to a variety of microbes [84]. According to reports, silver nanoparticles might additionally absorb and break down ethylene, which could help them prolong the lifespan of vegetables and fruits. This made it possible to use the resultant NCS-Ag complex as an antibacterial agent in food packaging since it demonstrated strong antimicrobial activity at desirable low silver concentrations of as little as 10 mg/kg.

Oxygen (O_2) contributes to several food degradation processes, including rotten tastes and reactions such as browning. Reduced concentrations of oxygen can be maintained by adding O_2 scavengers to food packing systems. TiO_2 NPs have been effectively added to various polymers to create oxygen scavenger films; however, TiO_2 works via a photocatalytic mechanism; nevertheless, its main drawback lies in the fact it needs ultraviolet rays for operation [85].

There are numerous uses for enzymes in the food sector. However, the straight incorporation of enzymes to meals may not always be applicable due to their reactivity to handling circumstances and/or inhibiting enzymes. Generally speaking, immobilization works well to increase an enzyme's susceptibility to proteases, which along with other reducing agents, have stability to pH and temperature, and the capacity to withstand repeated application or controlled dissolution [86]. Regarding packaging applications, the immobilization of enzymes is being explored. Enzymes such as lactase might be added to packing materials to improve food goods' quality and satisfy the demands of customers who lack certain enzymes. Glutamate dehydrogenase as well as lactate dehydrogenase showed exceptional activity for enzymes following immobilization after being altered to be immobilized by nanoparticles made of SiO_2 [87].

5.5 *Intelligent Labeling and Packaging Using Nanosensors*

Though active packaging will let out chemicals like antimicrobials and colors, tastes, or dietary supplements within the product once it starts to decay, intelligent food packaging may detect whenever its components are going rotten and notify the user. This might be addressed by nanosensors. It involves using a variety of nanosensors which can detect the harmful gases that the food releases when it deteriorates, determining whether the food is "new" or possibly causing the discharge of antioxidants to prolong the duration of its shelf life. Printable inks based on nanotechnology were specially created for this kind of use. Listed below are some business applications for smart packaging for food components.

By altering its color, thermo-sensitive ink put on food packaging serves as a signal of fluctuations in temperature. Just at a particular temperature will the permanent inks be recognizable, and as long as the ambient temperature stays the same, the color will not change. When the temperature rises or falls, the ink's color alters to reflect the alteration. When the temperature drops, reversible inks regain their original hue [88].

The main purposes of RFID labels were to recognize food items, reduce theft, and safeguard food packages from fraud. RFID is an automatic identifying system that includes barcodes, QR codes, and magnetic ink. The RFID system's reader/interrogator and tag/transponder are its two primary parts. Figure 3 shows the schematic representation. Using mobile data transfer, it communicates with the reader's device using radiofrequency that ranges from low to ultrahigh frequencies [89].

Since bacteria thrive in oxygen-rich conditions, research on harmless and permanent nanosensors is essential. By using vacuum and nitrogen releasing, it can ensure that there is no oxygen present in the packing systems. Titania nanoparticles are being used to generate colorimetric oxygen nanosensing employing UV-A illumination. Sensors begin to bleach and turn colorless when under UV light before oxygen is introduced. Using encapsulating polymer, glycerol as a donor of electrons, and methylene blue as a redox dye, tin oxide is being used as a photosensitizer in O_2 colorimetric nanosensors [90]. The colors of the nanosensor sheets will vary depending on the amount of oxygen exposure.

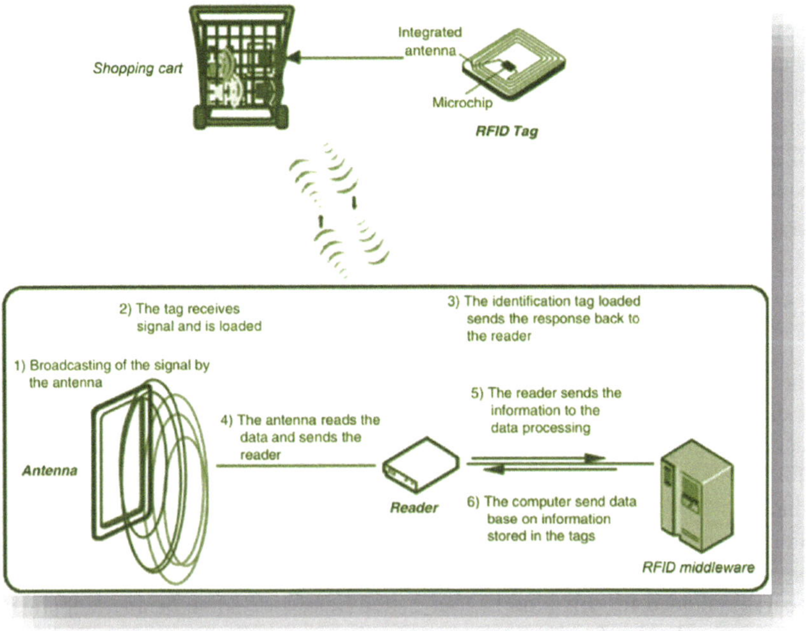

Fig. 3 Nano device with RFID tag system

Time and temperature indicators (TTI) are frequently utilized to translate as well as track the buyer's level of food items because of their accessibility, ease of use, and minimal costs [91]. Many TTI trades were created via enzymatic and polymeric processes. It is critical to monitor temperature as well as time variations to protect goods that require a specific temperature. This is particularly crucial for frozen and refrigerated meals when cold storage is essential.

Within a package, the indicators need to be close to the atmospheric condition that surrounds the food. The presence of oxygen could be a sign that the package might have been altered, wasn't sealed correctly, or is beginning to leak. The decomposition of food produces harmful gasses that can injure customers. When a chemical response occurs in an indicator, a color shift is observed. Removing the seal and observing color variations gradually activates the tag. The ripeness of the fruit is indicated by color variations on its initial intelligent sensing identify, Ripe Sense. The sensor Q created by DSM NV, is a quality indication for seafood deterioration [92]. Regarding monitoring, dynamic labels, and tracing purposes in the smart packaging arrangement, nano devices are developed. Using radio frequency, RFID, an electronic technological device, functions as an active tag attached to food packaging to track and then identify its contents. Micro, inexpensive RFID tags with greater coverage and the ability to operate in harsh environments are made possible by nanosensors.

6 Toxin Detection

Toxicity in the milliliter range can be detected by water-based samples using the help of a microfluidic sensor. Additionally, there are electrochemical and biosensors which might identify toxicity in food using a variety of nanomaterials, such as metallic nanoparticles and supermagnetic nanoparticles. Aflatoxin, a carcinogenic substance, is frequently found in food infected by Aspergillus flavus and Aspergillus parasiticus. Likewise, botulinum neurotoxin type B and brevetoxins found in cooked food were detected using gold nanoparticles functioning as nanoprobes in enzyme-linked immunosorbent and immune-chromatographic assays. Palytoxin, a marine toxin typically found in infected aquatic life, was also identified through carbon nanotube-based electrochemiluminescent sensors [93, 94].

6.1 Food Pathogen Detection

To identify foodborne infections, genetic information is observed. These traditional microbiological techniques for disease control are reliable but challenging. As a result, nanotechnology uses affordable nanosensors in food packaging to identify a variety of harmful microbes. The most efficient of them is the nano bioluminescent spraying, generating a visual growth upon detecting the pathogen strains present in food. The magnetic nanoparticles in this spray interact toward bacteria when food is present. The genetic material of the milking pathogenic bacteria *L. monocytogenes* was successfully isolated using magnetic iron oxide nanoparticles. Surface-enhanced Raman spectroscopy combined with silver nanosensors is an effective method for identifying pathogenic microorganisms [95].

6.2 *Pesticide and Chemical Detection*

Determining the quantity of the use of pesticides as well as the harmful compounds present in food items is equally crucial. To detect carbonate and organ phosphorus insecticides, certain colorimetric and fluorometric nanosensors are being used in conjunction with gold nanoparticles [96]. While core–shell quantum dots composed of cadmium selenide and zinc sulfide were previously studied for the identification of paraoxon, potentiometric detectors using silica composites and MWCNT were currently being examined toward the identification of cadmium ions [97]. Comparably, melamine and parathion insecticides have benefited from the application of voltametric and fluorescence sensors, respectively [98].

6.3 *Identifying the Quality of Food*

Vitamins and other essential food components, like antioxidant chemicals, were especially susceptible to deterioration in packaged foods that must be consumed for the proper operation of numerous metabolic pathways in our bodies. Many nanomaterials are being observed to avoid this; for example, ionic liquid nanocomposites based on carbon nanotubes and nickel oxide nanoparticles have been reported to detect folic acid and ascorbic acid in a variety of milk, flour, and juice samples [99, 100]. The nutritional content of the previously mentioned food constituents has been monitored in a variety of food components using several nanosensors conjugated with silver, iron, gold, tin dioxide, and cuprous oxide nanoparticles.

7 Food Quality Control Using Sensors

The nanosensors might react to microbial contamination, products of decomposition, or variations in the storage atmosphere [101]. By detecting infections, chemical pollutants, and changes linked to spoiling, nanosensors used in food packaging systems can provide immediate food quality status and do away with the requirement for erroneous dates of expiry [102].

7.1 Time–Temperature Integrators (TTI)

These are made to track, file, and interpret food safety. Whenever food is kept under less-than-ideal circumstances, this is especially crucial. A TTI, for example, can reveal whether a product was not sufficiently subjected to a more extreme temperature and how long it endured if it was intended to be frozen. Exploitation measurements, partly temperature record indications, and complete temperature record indicators are the three primary categories into which TTIs are divided [103]. Usually, a color advancement or shift serves as the manifestation of the communications. Timestrip has invented a technique with cold foods utilizing gold nanoparticles, and this becomes red when temperatures are higher than frozen. When gold nanoparticles are accidentally frozen, they accumulate irrevocably and lose their red color [104].

7.2 Detection of Gases

In order to detect product deteriorating caused by bacteria, a variety of sensors for gases have been developed which translate chemical interactions between nanoparticles on the outermost layer through responsive impulses. Metal oxides are typically used by nanosensors to detect gasses. In recent years, conducting polymer-based sensors were employed to identify as well as quantify microorganisms via their gaseous releases. The detectors' variation in resistivity creates an arrangement which matches the gas under study [105].

7.3 O_2 Sensors

The development of harmless and permanent O_2 sensors with vacuum or nitrogen packing is being pursued in response to the growing agreement that oxygen-free food packaging solutions must guarantee the total absence of O_2. The decreasing amount of Methylene Blue (MB) by triethanolamine as a polymer encapsulating medium utilizing UV-A light is photosensitized by TiO_2 NPs in a UV-activated colorimetric O_2 indicator [106]. The detector bleaches whenever subjected to UV light and stays colorless unless it comes into contact with oxygen, which restores its original blue hue. The amount of O_2 absorption directly correlates with the rate of changing the color [107]. Yet another O_2 sensor is a photosensitizer made on nanocrystalline SnO_2 [108].

7.4 Pathogen Detection

Researchers at the University of Rochester Medical Center have shown a new nanotechnology-based technique which quickly and precisely identifies E. coli bacteria in food items to identify the possible presence of harmful organisms. The technological detecting system consists of a digital camera, a tiny silicon chip, and a bacterial protein. Whatever E. coli presents in the test's sample is bound by the chip's protein. The chip's surface alters after the binding process takes place. For evaluation and verification, the digital camera takes a picture of this alteration [109]. Once the particular nanowires are placed on the food, pathogen-specific antibodies are bonded to them, exposing the germs to fluorescent antibodies.

8 Future Prospects, Opportunities, and Conclusions

Currently, every scientific discipline is interconnected with nanotechnology in some manner. Nowadays, society prioritizes manufacturing processes, effectiveness, wellness, and environmental protection; nanosensors and nanobiosensors, with their exceptional efficiency for identifying toxicity phases, may be beneficial to safety for health. By the near future, concepts of "smart" packaging may grow more prevalent, utilizing antigen-specific indicators to create thin layers of nanocomposites made of polymers by integrating nanoparticles into food packaging. Combining biological and chemical elements into an identical structure may open up novel possibilities with nanosensing and nano-based biosensing usage in future, such as enabling multiple measurements of several analytes in one instrument. In order to accomplish previously unobtainable standards of effectiveness, nano-based biosensors and nanosensors have shown remarkable advancements yet overcome numerous obstacles posed by competitors and modern techniques of detecting various characteristics or analytes of concern. With numerous potential applications along with promising prospects, nano-biosensors in food packaging and shelf-life monitoring possess a promising future. Combining artificial intelligence (AI) technologies with nanobiosensors constitutes a promising prospect. Models based on machine learning demonstrate modeling along with linking data to enhance nutritional leadership and choice-making. These simulations use artificial intelligence (AI) techniques for evaluating complicated sensor information, estimate lifespan, improve warehousing circumstances, and offer immediate insight.

Nanotechnology possesses among that has the greatest commercial potential of any recently discovered science or technology. Numerous aspects of the food sector, including manufacturing and packing through ultimate customer usage, have been improved with the ongoing development of nanotechnologies, guaranteeing hygiene and nutritional standards. Still, nanosensors remain at their development and may be widely used upon the properties associated with every exterior chemical which requires to have been recognized. The safeguarding of nutrition hygiene as well as the purity of foods remains an extremely important concern throughout a food supply chain.

References

1. Rameshaiah GN, Pallavi J, Shabnam S (2015) Nano fertilizers and nano sensors: an attempt for developing smart agriculture. Int J Eng Res Gen Sci 3:314–320
2. Sharon M, Choudhary AK, Kumar R (2010) Nanotechnology in agricultural diseases and food safety. J Phytol 2:83–92
3. Long F, Zhu, A, Gu C, Shi H (2013) Recent progress in optical biosensors for environmental applications. In: State of the art in biosensors—environmental and medical applications. InTech, pp 3–28
4. Luka G, Ahmadi A, Najjaran H, Alocilja E, Derosa M, Wolthers K et al (2015) Microfluidics integrated biosensors: a leading technology towards lab-on-a-chip and sensing applications. Sensors 15(12):30011–30031. https://doi.org/10.3390/s151229783
5. Bhalla N, Chiang HJ, Shen AQ (2018) Cell biology at the interface of nanobiosensors and microfluidics. Methods Cell Biol 148:203–227. https://doi.org/10.1016/bs.mcb.2018.09.009
6. Silue TA (2019) Electrochemical analysis and detection of dopamine using chitosan-catechol modified carbon nanotube and Graphene electrodes
7. Omanovic-Miklicanin E, Maksimovic M (2016) Nanosensors applications in agriculture and food industry. Bull Chem Technol Bosnia Herzeg 47:59–70
8. Rai V, Acharya S, Dey N (2012) Implications of nanobiosensors in agriculture. J Biomater Nanobiotechnol 3:315–324. https://doi.org/10.4236/jbnb.2012.322039
9. Lee SK, Mills A, Lepre A (2004) An intelligence ink for oxygen. Chem Commun 17:1912–1913. https://doi.org/10.1039/B406685E
10. El-Safty SA, Shenashen MA, El-Safty SA (2012) Mercury-ion optical sensors. Trends Analyt Chem 38:98–115. https://doi.org/10.1016/j.trac.2012.05.002
11. Fangyuan G, Li Y, Mengmeng F, Ling L, Johnny G, Yuhui A et al (2018) Novel Schiff base (DBDDP) selective detection of Fe(III): dispersed in aqueous solution and encapsulated in silica cross-linked micellar nanoparticles in living cell. J Colloid Interf Sci 514:357–363. https://doi.org/10.1016/j.jcis.2017.12.050
12. Xiang G, Wang Y, Zhang H, Fan H et al (2018) Carbon dots based dual-emission silica nanoparticles as ratiometric fluorescent probe for nitrite determination in food samples. Food Chem 260:13–18. https://doi.org/10.1016/j.foodchem.2018.03.150
13. Jimenez-Falcao S, Villalonga A, Arevalo-Villena M et al (2020) Enzyme-controlled mesoporous nanosensor for the detection of living *Saccharomyces cerevisiae*. Sens Actuat B Chem 303:127197. https://doi.org/10.1016/j.snb.2019.127197
14. Chaubey A, Malhotra BD (2002) Mediated biosensors. Biosens Bioelectron 17:441–456. https://doi.org/10.1016/S0956-5663(01)00313-X
15. Yavuz S, Erkal A, Kariper IA, Solak AO, Jeon S, Mulazımoglu IE, Ustundag Z (2016) Carbonaceous materials-12: a novel highly sensitive graphene oxide-based carbon electrode: preparation, characterization, and heavy metal analysis in food samples. Food Anal Methods 9(2):322–331. https://doi.org/10.1007/s12161-015-0198-3
16. Nesakumar N, Sethuraman S et al (2016) Electrochemical acetylcholinesterase biosensor based on ZnO nanocuboids modified platinum electrode for the detection of carbosulfan in rice. Biosens Bioelectron 77:1070–1077. https://doi.org/10.1016/j.bios.2015.11.010
17. Singh A, Choudhary M, Singh MP, Verma HN, Singh SP et al (2015) DNA functionalized direct electro-deposited gold nanoaggregates for efficient detection of *Salmonella typhi*. Bioelectrochemistry 105:7–15. https://doi.org/10.1016/j.bioelechem.2015.03.005
18. Eftekhari S, Foroughifar N et al (2020) Synthesis and characterization of polymeric nanocomposites based on poly-melamine para formaldehyde and super paramagnetic silicon dioxide

loaded Iron(III) oxide core-shell composite magnetic nanoparticles. Lett Appl Nano Biosci 9(1):914–918
19. Duran N, Marcato PD (2013) Nano biotechnology perspectives: role of nanotechnology in the food industry—a review. Int J Food Sci Tech 48:1127–1134. https://doi.org/10.1111/ijfs.12027
20. Ahuja D, Parande D (2012) Optical sensors and their applications. J Sci Res Rev 1(5):60–68
21. Rong G, Kim EH, Poskanzer KE, Clark HA (2017) A method for estimating intracellular ion concentration using optical nanosensors and ratiometric imaging. Sci Rep 7(1):1–10. https://doi.org/10.1038/s41598-017-11162-8
22. Jafarizadeh-Malmiri H, Sayyar Z, Anarjan N, Berenjian A (2019) Nano-sensors in food nano biotechnology. In: Nano biotechnology in food: concepts, applications and perspectives. Springer, New York, pp 81–91. https://doi.org/10.1007/978-3-030-05846-3
23. Krishna VD, Wu K, Su D, Cheeran MCJ, Wang JP, Perez A (2018) Nanotechnology: review of concepts and potential application of sensing platforms in food safety. Food Microbiol 75:47–54. https://doi.org/10.1016/j.fm.2018.01.025
24. Caon T, Martelli SM, Fakhouri FM (2017) New trends in the food industry: application of nanosensors in food packaging. Nanobiosensors 14:773–804
25. Chen C, Yuan Z, Chang HT, Lu F, Li Z, Lu C (2016) Silver nanoclusters as fluorescent nanosensors for selective and sensitive nitrite detection. Anal Methods 8:2628–2633. https://doi.org/10.1039/C6AY00214E
26. Hu L, Deng L, Alsaiari S, Zhang D et al (2014) "Light-on" sensing of antioxidants using gold nanoclusters. Anal Chem 86(10):4989–4994. https://doi.org/10.1021/ac500528m
27. Dasary SSR, Rai US, Yu H, Anjaneyulu Y, Dubey M, Ray PC (2008) Gold nanoparticle based surface enhanced fluorescence for detection of organ phosphorus agents. Chem Phys Lett 460(1–3):187–190. https://doi.org/10.1016/j.cplett.2008.05.082
28. Bustillo J and S Li (2017) Methods for manufacturing well structures for low-noise chemical sensors. US patent–US9841398B2
29. Kamalii A, Ahilan B, Felix N et al (2019) Applications of nanotechnology in fisheries and aquaculture. J Aquacult Trop 33(3):111–117
30. Shikha S, Salafi T, Cheng J, Zhang Y (2017) Versatile design and synthesis of nano-barcodes. Chem Soc Rev 46(22):7054–7093. https://doi.org/10.1039/C7CS00271H
31. Cho SY, Suh KI, Bae YJ (2013) DNA barcode library and its efficacy for identifying food-associated insect pests in Korea. Entomol Res 43(5):253–261. https://doi.org/10.1111/1748-5967.12034
32. Zhang D, Carr DJ, Alocilja EC (2009) Fluorescent bio-barcode DNA assay for the detection of *Salmonella enterica* serovar Enteritidis. Biosens Bioelectron 24(5):1377–1381. https://doi.org/10.1016/j.bios.2008.07.081
33. Loutfi A, Coradeschi S, Mani GK, Shankar P, Rayappan JBB (2015) Electronic noses for food quality: a review. J Food Eng 144:103–111. https://doi.org/10.1016/j.jfoodeng.2014.07.019
34. Santos JP, Arroya T, Aleixandre M, Lozano J, Sayago I, García M, Fernandez MJ, Ares L, Gutierrez J, Cabellos JM, Gil M, Horrillo MC (2004) A comparative study of sensor array and GC–MS: application to Madrid wines characterization. Sens Actuat B Chem 102(2):299–307. https://doi.org/10.1016/j.snb.2004.04.082
35. Tudor KM, Markovic K, Kalit S, Vahcic N, Havranek J (2014) Application of electronic nose and electronic tongue in the dairy industry. Mljekarstvo 64(4):228–244
36. Pioggia G, Francesco F, Ferro M, Sorrentino F et al (2008) Characterization of a carbon nanotube polymer composite sensor for an impedimetric electronic tongue. Microchim Acta 163(1):57–62. https://doi.org/10.1007/s00604-008-0952-y
37. Ha D, Sun Q, Su K, Wan H, Li H, Xu N, Sun F, Zhuang L, Hu N, Wang P (2015) Recent achievements in electronic tongue and bioelectronic tongue as taste sensors. Sens Actuat B Chem 207:1136–1146. https://doi.org/10.1016/j.snb.2014.00.077
38. Fitzgerald J, Fenniri H (2017) Cutting edge methods for non-invasive disease diagnosis using E-tongue and E-nose devices. Biosensors 7(4):59. https://doi.org/10.3390/bios7040059

39. Hayashi K, Yamanaka M et al (1990) Multichannel taste sensor using lipid membranes. Sens Actuat B Chem 2(3):205–213. https://doi.org/10.1016/0925-4005(90)85006-K
40. Mittal AK, Banerjee UC (2016) Current status and future prospects of nanobiomaterials in drug delivery. In: Nano biomaterials in drug delivery. William Andrew, pp 147–170
41. Luka G, Ahmadi A, Najjaran H, Alocilja E, DeRosa M, Wolthers K, Malki A, Aziz H, Althani A, Hoorfar M (2015) Microfluidics integrated biosensors: a leading technology towards lab-on-a-chip and sensing applications. Sensors 15(12):30011–30031. https://doi.org/10.3390/s151229783
42. Das J, Cederquist KB, Zaragoza AA, Lee PE, Sargent EH, Kelley SO (2012) An ultrasensitive universal detector based on neutralizer displacement. Nat Chem 4(8):642–648. https://doi.org/10.1038/nchem.1367
43. Das J, Ivanov I, Montermini L, Rak J, Sargent EH, Kelley SO (2015) An electrochemical clamp assay for direct, rapid analysis of circulating nucleic acids in serum. Nat Chem 7(7):569–575. https://doi.org/10.1038/nchem.2270
44. Cabuzu D, Cirja A, Puiu R, Grumezescu AM (2015) Biomedical applications of gold nanoparticles. Curr Top Med Chem 15(16):1605–1613. https://doi.org/10.2174/1568026615666150414144750
45. Ribeiro HS, Chu BS, Ichikawa S, Nakajima M (2008) Preparation of nano dispersions containing β-carotene by solvent displacement method. Food Hydrocoll 22(1):12–17. https://doi.org/10.1016/j.foodhyd.2007.04.009
46. Meng X, Kim S, Puligundla P, Ko S (2014) Carbon dioxide and oxygen gas sensors-possible application for monitoring quality, freshness, and safety of agricultural and food products with emphasis on importance of analytical signals and their transformation. J Korean Soc Appl Biol Chem 57(6):723–733. https://doi.org/10.1007/s13765-014-4180-3
47. Sangerlaub S, Muller K (2017) Long-time performance of bottles made of PET blended with various concentrations of oxygen scavenger additive stored at different temperatures. Packag Technol Sci 30(1–2):45–58. https://doi.org/10.1002/pts.2283
48. Pihlasalo S, Perez IM, Hollo N, Hokkanen E, Pahikkala T, Härmä H (2016) Luminometric label array for quantification and identification of metal ions. Anal Chem 88(10):5271–5280. https://doi.org/10.1021/acs.analchem.6b00453
49. Taitt CR, Shriver-Lake LC, Ngundi MM, Ligler FS (2008) Array biosensor for toxin detection: continued advances. Sensors 8(12):8361–8377. https://doi.org/10.3390/s8128361
50. Zhou L, Lin Y, Huang Z, Ren J, Qu X (2012) Carbon nanodots as fluorescence probes for rapid, sensitive, and label-free detection of Hg^{2+} and biothiols in complex matrices. Chem Commun 48(8):1147–1149. https://doi.org/10.1039/C2CC16791C
51. Fang Z, Hou G, Chen C, Hu L (2019) Nanocellulose-based films and their emerging applications. Curr Opin Solid State Mater Sci 23(4):100764. https://doi.org/10.1016/j.cossms.2019.07.003
52. Ravichandran R (2010) Nanotechnology applications in food and food processing: innovative green approaches, opportunities and uncertainties for global market. Int J Green Nanotechnol Phys Chem 1(2):72–96. https://doi.org/10.1080/19430871003684440
53. Dong Y, Wang R, Li G, Chen C, Chi Y, Chen G (2012) Polyamine-functionalized carbon quantum dots as fluorescent probes for selective and sensitive detection of copper ions. Anal Chem 84(14):6220–6224. https://doi.org/10.1021/ac3012126
54. Varshney M, Li Y (2007) Interdigitated array microelectrode based impedance biosensor coupled with magnetic nanoparticle–antibody conjugates for detection of Escherichia coli O157:H7 in food samples. Biosens Bioelectr 22(11):2408–2414. https://doi.org/10.1016/j.bios.2006.08.030
55. Arshak K, Adley C, Moore E, Cunniffe C, Campion M, Harris J (2007) Characterisation of polymer nanocomposite sensors for quantification of bacterial cultures. Sens Actuat B Chem 126(1):226–231. https://doi.org/10.1016/j.snb.2006.12.006
56. Bonilla JC, Bozkurt F, Ansari S, Sozer N, Kokini JL (2016) Applications of quantum dots in food science and biology. Trends Food Sci Technol 53:75–89. https://doi.org/10.1016/j.tifs.2016.04.006

57. Mayhew TM, Muhlfeld C, Vanhecke D, Ochs M (2009) A review of recent methods for efficiently quantifying immunogold and other nanoparticles using TEM sections through cells, tissues and organs. Ann Anat Anatomischer Anzeiger 191(2):153–170. https://doi.org/10.1016/j.aanat.2008.11.001
58. Blazkova M, Koets M, Rauch P, van Amerongen A (2009) Development of a nucleic acid lateral flow immunoassay for simultaneous detection of *Listeria* spp. and *Listeria monocytogenes* in food. Eur Food Res Technol 229(6):867–874
59. Duffy GF, Moore EJ (2017) Electrochemical immunosensors for food analysis: a review of recent developments. Anal Lett 50(1):1–32. https://doi.org/10.1080/00032719.2016.1167900
60. Ricci F, Volpe G, Micheli L, Palleschi G (2007) A review on novel developments and applications of immunosensors in food analysis. Anal Chim Acta 605(2):111–129. https://doi.org/10.1016/j.aca.2007.10.046
61. Yilmaz B, Yilmaz F (2018) Lab-on-a-chip technology and its applications. Omics Technol Bio-Eng Towards Improv Qual Life 1:145–153. https://doi.org/10.1016/B978-0-12-804659-3.00008-7
62. Duffy GF, Moore EJ (2017) Electrochemical immunosensors for food analysis: a review of recent developments. Analyt Lett 50(1):1–32. https://doi.org/10.1080/00032719.2016.1167900
63. Rho S (2014) Analysis of RFID standard patent data for RFID technology trends. J Adv Navigat Technol 18(2):185–190
64. Lin X, Ni Y, Kokot S (2013) Glassy carbon electrodes modified with gold nanoparticles for the simultaneous determination of three food antioxidants. Anal Chim Acta 765:54–62. https://doi.org/10.1016/j.aca.2012.12.036
65. Wu MYC, Hsu MY, Chen SJ, Hwang DK, Yen TH, Cheng CM (2017) Point-of-care detection devices for food safety monitoring: proactive disease prevention. Trends Biotechnol 35(4):288–300. https://doi.org/10.1016/j.tibtech.2016.12.005
66. Burgess R (1998) Avoiding supply chain management failure: lessons from business process re-engineering. Int J Logist Manag 9:15–23. https://doi.org/10.1108/09574099810805717
67. Finch BJ (2006) Best operations now: supply chain profitability and performance. Mcgraw-Hill Irwin Series Operations Management, Irwin/McGraw-Hill
68. Marra M, Ho W, Edwards JS (2012) Supply chain knowledge management: a literature review. Exp Syst Appl 39(5):6103–6110. https://doi.org/10.1016/j.eswa.2011.11.035
69. Evoy S, DiLello N, Deshpande V, Narayanan A, Liu H, Riegelman M, Martin BR, Hailer B, Bradley JC, Weiss W, Mayer TS, Gogotsi Y, Bau HH, Mallouk TE, Raman S (2004) Dielectrophoretic assembly and integration of nanowire devices with functional CMOS operating circuitry. Microelectron Eng 75(1):31–42. https://doi.org/10.1016/j.mee.2003.09.010
70. Neethirajan S, Jayas DS (2011) Nanotechnology for the food and bioprocessing industries. Food Bioprocess Technol 4:39–47. https://doi.org/10.1007/s11947-010-0328-2
71. Li J, Lu YJ (2004) Nanostructure engineered chemical sensors for hazardous gas and vapor detection. In: Islam MSD (ed) Nanosensing: materials and devices, pp 222–231
72. Liu Y, Li X, Dokmeci MR, Wang ML (2011) Carbon nanotube sensors integrated inside a microfluidic channel for water quality monitoring. In: Tomizuka MYC (ed) Sensors and smart structures technologies for civil, mechanical, and aerospace systems, p 7981. https://doi.org/10.1117/12.880142
73. Iliadis AA, Ali HA (2011) Properties of fast response room temperature ZnO–Si heterojunction gas nanosensors. IEEE Trans Nanotechnol 10:652–656. https://doi.org/10.1109/TNANO.2010.2089637
74. Kuo CY, Chan CL, Gau C, Liu CW, Shiau SH, Ting JH (2007) Nano temperature sensor using selective lateral growth of carbon nanotube between electrodes. IEEE Trans Nanotechnol 6:63–69. https://doi.org/10.1109/NANO.2005.1500668
75. Li Z, Zhang H, Zheng W, Wang W, Huang H, Wang C, MacDiarmid AG, Wei Y (2008) Highly sensitive and stable humidity nanosensors based on LiCl doped TiO_2 electrospun nanofibers. J Am Chem Soc 130(15):5036–5037. https://doi.org/10.1021/ja800176s

76. Hongnan Z, Zhenyu L, Wei W, Ce W, Li L (2010) Na^{+}-doped zinc oxide nanofiber membrane for high speed humidity sensor. J Am Ceram Soc 93(1):142–146. https://doi.org/10.1111/j.1551-2916.2009.03341.x
77. Cobianu C, Serban B, Georgescu I, Costea S, Bostan C (2010) Emerging all-differential chemical sensing. Rom J Inf Sci Technol 13:342–349
78. Elliott R, Ong TJ (2002) Science, medicine, and the future: nutritional genomics. BMJ 324(7351):1438–1442. https://doi.org/10.1136/bmj.324.7351.1438
79. Cushen M, Kerry J, Morris M, Cruz-Romero M, Cummins E (2012) Nanotechnologies in the food industry: recent developments, risks and regulation. Trends Food Sci Tech 24(1):30–46. https://doi.org/10.1016/j.tifs.2011.10.006
80. Xu X, Spencer HG (1997) Transport of electrolytes through a weak acid nanofiltration membrane: effects of flux and crossflow velocity interpreted using a fine-porous membrane model. Desalination 113(1):85–93. https://doi.org/10.1016/S0011-9164(97)00117-3
81. Arora A, Padua GW (2010) Nanocomposites in food packaging. J Food Sci 75(1):R43–R49. https://doi.org/10.1111/j.1750-3841.2009.01456.x
82. Azizi SMAS, Alloin F, Dufresne A (2005) Review of recent research into cellulosic whiskers, their properties and their application in nanocomposite field. Biomacromol 6(2):612–626. https://doi.org/10.1021/bm0493685
83. Gu HW, Ho PL, Tong E, Wang L, Xu B (2003) Presenting vancomycin on nanoparticles to enhance antimicrobial activities. Nano Lett 3(9):1261–1263. https://doi.org/10.1021/nl034396z
84. Kumar R, Munstedt H (2005) Silver ion release from antimicrobial polyamide/silver composites. Biomaterials 26(14):2081–2088. https://doi.org/10.1016/j.biomaterials.2004.05.030
85. Xiao-e L, Green ANM, Haque SA, Mills A, Durrant JR (2004) Light-driven oxygen scavenging by titania/polymer nanocomposite films. J Photochem Photobiol A Chem 162:253–259. https://doi.org/10.1016/j.nainr.2003.08.010
86. Lopez-Rubio A, Gavara R, Lagaron JM (2006) Bioactive packaging: turning foods into healthier foods through biomaterials. Trends Food Sci Technol 17:567–575. https://doi.org/10.1016/j.tifs.2006.04.012
87. Qhobosheane M, Santra S, Zhang P, Tan WH (2001) Biochemically functionalized silica nanoparticles. Analyst 126:1274–1278. https://doi.org/10.1039/B101489G
88. Wang LF, Shankar S, Rhim JW (2017) Properties of alginate-based films reinforced with cellulose fibers and cellulose nanowhiskers isolated from mulberry pulp. Food Hydrocoll 63:201–208. https://doi.org/10.1016/j.foodhyd.2016.08.041
89. Bowles M, Lu J (2014) Removing the blinders: a literature review on the potential of nanoscale technologies for the management of supply chains. Technol Forecast Soc Change 82:190–198. https://doi.org/10.1016/j.techfore.2013.10.017
90. Gutierrez-Tauste D, Domenech X et al (2007) Characterization of methylene blue/TiO_2 hybrid thin films prepared by the liquid phase deposition (LPD) method: application for fabrication of light-activated colorimetric oxygen indicators. J Photochem Photobiol A Chem 187:45–52. https://doi.org/10.1016/j.jphotochem.2006.09.011
91. Zhang C, Yin AX, Jiang R, Rong J, Dong L, Zhao T et al (2013) Time-temperature indicator for perishable products based on kinetically programmable Ag overgrowth on Au nanorods. ACS Nano 7(5):4561–4568. https://doi.org/10.1021/nn401266u
92. Kerry JP, O'Grady MN, Hogan SA (2006) Past, current and potential utilization of active and intelligent packaging systems for meat and muscle-based products: a review. Meat Sci 74(1):113–130. https://doi.org/10.1016/j.meatsci.2006.04.024
93. Wang L, Chen W, Xu D, Shim BS, Zhu Y, Sun F et al (2009) Simple, rapid, sensitive, and versatile SWNT paper sensor for environmental toxin detection competitive with ELISA. Nano Lett 9:4147–4152. https://doi.org/10.1021/nl902368r
94. Zamolo VA, Valenti G, Venturelli E, Chaloin O, Marcaccio M, Boscolo S et al (2012) Highly sensitive electrochemiluminescent nanobiosensor for the detection of palytoxin. ACS Nano 6(9):7989–7997. https://doi.org/10.1021/nn302573c

95. Fang Z, Zhao Y, Warner RD, Johnson SK (2017) Active and intelligent packaging in meat industry. Trends Food Sci Technol 61:60–71. https://doi.org/10.1016/j.tifs.2017.01.002
96. Ji X, Zheng J, Xu J, Rastogi VK, Cheng TC, DeFrank JJ et al (2005) (CdSe) ZnS quantum dots and organophosphorus hydrolase bioconjugate as biosensors for detection of paraoxon. J Phys Chem B 109(9):3793–3799. https://doi.org/10.1021/jp044928f
97. Rangan S, Schulze HG, Vardaki MZ, Blades MW, Piret JM, Turner RFB (2020) Applications of Raman spectroscopy in the development of cell therapies: state of the art and future perspectives. Analyst 145(6):2070–2105. https://doi.org/10.1039/C9AN01811E
98. Wang C, Irudayaraj J (2008) Gold nanorod probes for the detection of multiple pathogens. Small 4(12):2204–2208. https://doi.org/10.1002/smll.200800309
99. Karimi-Maleh H, Moazampour M, Yoosefian M, Sanati AL, Tahernejad-Javazmi F, Mahani M (2014) An electrochemical nanosensor for simultaneous voltammetric determination of ascorbic acid and Sudan I in food samples. Food Anal Methods 7:2169–2176. https://doi.org/10.1007/s12161-014-9867-x
100. Wei S, Zhao F, Xu Z, Zeng B (2006) Voltammetric determination of folic acid with a multi-walled carbon nanotube-modified gold electrode. Microchim Acta 152:285–290. https://doi.org/10.1007/s00604-005-0437-1
101. Bouwmeester H, Dekkers S, Noordam MY, Hagens WI, Bulder AS, de Heer C et al (2009) Review of health safety aspects of nanotechnologies in food production. Regulat Toxicol Pharmacol 53(1):52–62
102. Liao F, Chen C, Subramanian V (2005) Organic TFTs as gas sensors for electronic nose applications. Sens Actuat B 107:849–855. https://doi.org/10.1016/j.snb.2004.12.026
103. Singh RP (2000) Scientific principles of shelf-life evaluation. In: Man CMD, Jones A (eds) Shelf-life evaluation of foods. Springer, Berlin, pp 2–22. https://doi.org/10.1007/978-1-4615-2095-5_1
104. Robinson DKR, Morrison MJ (2010) Nanotechnologies for food packaging: reporting the science and technology research trends. Report Observ Nano 15:476–480
105. Arshak K, Adley C, Moore E, Cunniffe C, Campion M, Harris J (2007) Characterisation of polymer nanocomposite sensors for quantification of bacterial cultures. Sens Actuat B 126(1):226–231. https://doi.org/10.1016/j.snb.2006.12.006
106. Lee SK, Sheridan M, Mills A (2005) Novel UV-activated colorimetric oxygen indicator. Chem Mater 17:2744–2751. https://doi.org/10.1021/cm0403863
107. Gutierrez-Tauste D, Domenech X, Casan-Pastor N, Ayllon JA (2007) Characterization of methylene blue/TiO_2 hybrid thin films prepared by the liquid phase deposition (LPD) method: application for fabrication of light-activated colorimetric oxygen indicators. J Photochem Photobiol A Chem 187(1):45–52. https://doi.org/10.1016/j.jphotochem.2006.09.011
108. Mills A, Hazafy D (2009) Nanocrystalline SnO_2-based, UVB-activated, colourimetric oxygen indicator. Sens Actuat B Chem 136(2):344–349. https://doi.org/10.1016/j.snb.2008.12.048
109. Garcia M, Tamara F, Eric G (2010) Potential applications of nanotechnology in the agro-food sector. Ciencia e Tecnologia de Alimentos Campinas 30(3):573–581. https://doi.org/10.1590/S0101-20612010000300002

Techno-Economical Aspects and Nanobiosensors in the Food Industry

Sarika Kumari and Shubham Gudadhe

Abstract Nanobiosensor application in the food chain marks a positive development in the food industry about food safety quality management, and chain efficiency. This paper aims to identify the techno-economic feasibility of nanobiosensor technology for the food industry while introducing the possibility of disposing of conventional food testing systems. Potential benefits and costs, we discuss costs and benefits, the potential for further enlargement, and possible improvements of the supply chain. Furthermore, we review the underlying general properties of nanobiosensor platforms in terms of sensitivity, specificity, and fast response time. The use of these advanced sensors may result in better identification of the sources of contamination or disease-causing pathogens and minimize food-borne disease incidence among consumers. In addition, we highlight issues and constraints to the broad utilization of the concept such as regulatory concerns and technological impediments. Finally, this study reaffirms the call for cross-disciplinary fashion in realizing and deploying nanobiosensors in food systems across the value chain for better and safer food production.

Keywords Nanobiosensors · Food safety · Quality control · Economic analysis · Supply chain transparency · Contaminants detection · Regulatory challenges

1 Introduction

Today's food industry finds itself with an important question: Our industry faces different challenges that help us deliver good food effectively through advanced methods and keep both safety and client preference intact. Our current social system needs systems that help us control everything that affects food safety and quality while it travels its supply chain. Nanotechnology's development of nanobiosensors drives major changes throughout this sector according to research [1]. Integrated nanotechnology-based biosensing allows us to use nanomaterial features to spot

S. Kumari (✉) · S. Gudadhe
Department of Biotechnology, Central University of South Bihar, Gaya, Bihar, India
e-mail: 955simmisk1001@gmail.com

D. Sillu et al. (eds.), *Nanobiosensors for the Food Industry*, Smart Nanomaterials Technology, https://doi.org/10.1007/978-981-95-0136-6_13

chemical and biological substances faster and more effectively. These sensors will detect food problems promptly and effectively because they show strong results quickly. Standard food quality testing techniques often prove slow to respond and their test results tend to be less sensitive compared to modern alternative systems [2]. Standard tests often need multiple-step preparation processes that slow down detection time and affect product freshness testing results. Nanobiosensors lead as the best choice because they detect samples at the nano-level and deliver prompt test results.

Nanobiosensors in the food industry now combine breakthrough developments with fundamental transformations that alter our methods of ensuring food safety and quality [3]. These sensors help find low-level contaminants and pathogens including allergens at real-time detection rates [4]. These sensors help both preserve food quality and stop contamination while making companies follow regulations and trusting consumers. Biosensors that use nanotechnology dimensions improve both production methods and food business growth. Using these sensors throughout production lines plus distribution channels proves beneficial in two ways: they observe system changes caused by human error making the process more efficient [5]. These sensors will improve how companies make food and transport it because other businesses use digital methods and robots to run more efficiently.

Nonetheless, there are equally some drawbacks to nanobiosensors that are; high costs of developing and implementing, the policies and standards complicated, and there is need for rigorous identification [6]. However, there are enormous opportunities to apply nanobiosensors as the disadvantages do not swamp the benefits of nanobiosensors attracting the interest of researchers as well as investors in the food industry. An understanding of the principles and applications of nanobiosensors, current existing technology, and future technology, as well as its economic aspect, is provided in this chapter. Thus, in understanding the specifics of the activity of these sensors and their opportunities to change the sphere of food safety and quality management, the actors of the food industry could enhance their perception of such opportunities and threats. Our understanding is that a careful discussion of these issues will assist the reader in understanding how nanobiosensors will completely transform the food chain, making it safer and more efficient and, thus, possessing a greater value addition.

2 Fundamentals of Nanobiosensors

Nanobiosensors combine nanomaterials with biological sensing units to spot and measure specific chemical targets through nanoscale technologies. The nanoscale fundamental features of surface area quantum behavior and reactivity lead these sensors to detect targets more accurately and with lower concentration levels [7]. Nanobiosensors adopt nanotechnology to connect biosensing elements that precisely spot small biological signals.

Nanobiosensors operate at the nanoscale which measures between one billionth and one hundred billionths of a meter. Nanoscale dimensions produce substance properties that differ from regular-size materials. High edge count in a tight design lets many target readings interact better with the sensor thanks to its small form that consumes less space [8]. At the nanoscale, quantum effects show powerful changes in behavior. Because quantum dot fluorescence energy levels vary with size this particular feature makes quantum dots useful for optical detection systems [9]. Patterned interactions between quantum particles make it possible to measure and track light signals efficiently [10]. The best choice of material for biorecognition depends on its capacity to detect and engage the analyte preferentially and with high reliability. Sensor development typically employs various target analyte binders such as antibodies, enzymes, nucleic acids, and aptamers. The sensor components connect to the base platform for chemical interaction with one specific substance [11]. The biorecognition element within the sensor responds to target analytics through changes in its state. Our innovation lets us measure specific response levels from targeted molecules to obtain the needed analysis information.

3 Types of Nanobiosensors

Nanobiosensors exist in two forms depending on both how they detect targets and what materials they contain [12]. Nanobiosensors come in multiple forms and bring unique value to specific measurement tasks. Here's an overview of the primary types of nanobiosensors:

3.1 Optical Nanobiosensors

Nanobiosensors use optical methods to see how light reacts when sensing components contact target samples. The system detects changes when optical signals interact with specific materials according to pretest detection methods [13]. Sensors that use fluorescence rely on two types of fluorescent nanoparticles: quantum dots and organic dyes. The changes in fluorescent signals from these particles help us determine when they connect with target analytes. Our SPR sensor platform works with gold or silver metal nanoparticles to sense refractive index changes happening right at the sensor surface. SPR sensors respond quickly to minor changes in their environment and perform well for testing molecular connection discoveries. In metal-enhanced Raman scattering (MERS), particles detect small changes in vibrational molecular behavior. When the target analyte interacts with the sensing element, it changes the Raman signal which provides us with quantitative results.

3.2 Electrochemical Nanobiosensors

Nanobiosensors based on electrochemical detection track voltage and current changes because of how the target interact with biocomponents. The change in current output from an electrode surface during the electrochemical reaction with an analyte controls amperometric sensors. Carbon nanotubes and graphene provide electrical conductivity improvements while maintaining enhanced device sensitivity. When the target analysis contacts the biorecognition element, the Potentiometric Sensors demonstrate changes in the electrode potential reading. Ion-selective electrodes become more sensitive when connected to nanoscale materials [14]. The measurement of resistivity changes at electrode surfaces provides information about analyte binding. Sensor technology without labels works well for detection and works better when researchers add nanomaterials to the design.

3.3 Magnetic Nanobiosensors

By monitoring changes in magnetic nanoparticles' properties during target analysis interactions, magnetic nanobiosensors offer detection solutions. When magnetic nanoparticles relax their magnetization speed due to target analyte binding, the Magnetic Particle Relaxation method produces the results. The procedure reveals results with excellent precision while monitoring both complex and simple materials. Magnetic Resonance studies how changes in nanoparticles' magnetic characteristics affect their magnetic field response. This technique picks up trace amounts of analysis very well [15].

3.4 Mechanical Nanobiosensors

Mechanical nanobiosensors create mechanical measurements such as resonance frequency and mass alterations when the target analysis interacts with the biorecognition element. Cantilever-Based Sensors function by measuring how nanoscale cantilevers bend or change their resonance patterns when the target analyte connects to them. Scientists can record these modifications with great measurement accuracy. Surface Acoustic Wave (SAW) Sensors measure surface movements through acoustic energy transmitted along piezoelectric material surfaces. When a sample binds to our device its sound making elements shift in ways that scientists can track.

3.5 *Hybrid Nanobiosensors*

Nanobiosensors built with mixed sensing platforms achieve better results by identifying targets with higher precision. These sensors merge optical and electrochemical systems to measure target analysis from different perspectives simultaneously. The mixed sensing methods enable better analysis performance and wider working conditions [16]. Magnetic-Optical Sensors bring together magnetic and optical detection methods to make measurements more sensitive and readable especially when dealing with complicated analysis types. The system relies on magnetic nanoparticles for detection with optical measurement technology. These distinct nanobiosensors provide useful features that let them work well across many testing situations starting from food safety to environmental protection. The choice of nanobiosensor type matches the selected target analysis requirements and application conditions along with sensitivity perspectives (Table 1).

4 Application in the Food Industry

4.1 *Food Safety*

Salmonella and E. coli: Research shows that food-borne illnesses regularly happen when people consume Salmonella and E. coli pathogens [31]. The standard method to detect these pathogens needs sample culture which takes time (3–4 days) and only moderately determines their presence with accuracy. Nanobiosensors Approach: Nanostructures such as nanoparticles and nanowires enhance how well and precisely nanobiosensors detect pathogenic microorganisms. Each sensory unit includes molecular receptors that specifically attach to pathogen-specific features such as antibodies or DNA probes. The HPLC system delivers rapid analysis within 1–5 min while detecting tiny levels of substances and recognizing them precisely. By checking for contamination earlier in the process food hygiene standards improve progressively. Managed processing systems integrate these sensors as stationary sensors or mobile analytical tools. The technology can join computer systems that continuously examine food products during their processing run.

Investigators have proved that toxic compounds including pesticides alongside lead and cadmium display harmful effects at very minute concentrations [32]. Traditional methods used for contaminant measurement require extensive time and multiple sophisticated sample processing and examination procedures. The detection of small contaminant amounts occurs through nanobiosensors which use electronic and surface-oriented nanomaterials to achieve their detection task. The detection capability of functional carbon nanotubes and quantum dots enables them to target specific pesticide contaminants and heavy metals. The detection system provides high levels of performance because it identifies target substances at very small concentrations. A safety evaluation through necessary tests must pass ahead of food products

Table 1 Biosensor applications for detecting food contaminants

Application area	Type of biosensor	Contaminants detected	Methods used	Benefits	References
Meat products	Electrochemical biosensors	*Salmonella*, *E. coli*, *Listeria*	Current change detection	Rapid pathogen detection, enhanced safety	[17]
Dairy products	Optical biosensors	*Listeria*, antibiotics, spoilage bacteria	Light absorption	Improved quality control, shelf-life extension	[18, 19]
Fruits and vegetables	Magnetic biosensors	Pesticides, heavy metals	Magnetic field changes	Enhanced safety for consumers, better monitoring	[20, 21]
Grains and cereals	Enzymatic biosensors	Mycotoxins, fungal contamination	Enzymatic reactions	Early detection of spoilage, prevention of health risks	[22, 23]
Beverages	Impedimetric biosensors	Ethanol, pesticides, additives	Impedance measurement	Quality assurance, compliance with regulations	[24, 25]
Seafood	Surface plasma resonance	Histamine, pathogens	Surface plasmon resonance	Fast detection of toxins, improving food safety	[26]
Packaged foods	Colorimetric biosensors	Pathogens, chemical contaminants	Color change detection	User-friendly, easy to read results	[27, 28]
Food processing	Fiber-optic biosensors	Process indicators (pH, temperature)	Fiber-optic sensing	Real-time monitoring of production conditions	[29, 30]

reaching the consumer for use. Implementation: Sensor applications span from raw material quality assurance testing for pesticide residues found in farms to finished product safety examinations for heavy metal contamination after processing.

Scientists name the element within specific proteins which produces offending responses in sensitive bodily systems as allergens [33]. Different identification methods must be used to stop allergens from getting mixed while simultaneously protecting vulnerable food allergy sufferers. The detection of allergens through nanobiosensors relies on biomolecules that specifically bind to allergenic proteins at exact sites. The sensors combine signal-amplification approaches for detecting trace amounts of allergenic proteins. Advantages: The system enables fast precise identification of allergens which directly matter to manufacturers while also regarding the customers who use the commodities. The implementation of these detection methods helps lower expression of food allergies and improve food professionalism [34]. These sensors can be incorporated in food production companies to detect allergens in foods, in cleaning and sanitizing procedures to confirm that foods do not contain allergen remnants and in the labeling of foods to confirm claims made on the labels.

4.2 Quality Control

The freshness element remains extremely vital in most processed foods. Detecting signs of spoilage existence alongside changes in food chemical characteristics must inform product evaluations. Nanobiosensor Approach: Nanobiosensor platforms can actively identify biochemical markers which represent spoiling processes or component degradation. Many sensors detect off-gases created during spoilage which include ammonia and hydrogen sulfide [35]. These tools provide both precise and efficient assessment of perishable products which guide decisions about their use or disposal. By implementing this system, we can control food waste effectively while providing top-quality products to consumers. Freshness indicators serve two functions in perishable product monitoring since some operate directly within packaged products and other indicators fulfill their task by connecting with processing sites for performing regular quality assessments.

A physical reduction occurs in wholesome food because of both microbial growth and chemical changes. Even though identification methods may differ, this approach enhances detection efficiency for spoiled food items so we can properly address these cases and avoid unnecessary waste and safeguard consumers from potential risks [32]. Nanobiosensors serve as detectors which identify spoilage-related chemical tags including VOCs and altered pH levels. The sensors built to detect specific VOCs created by microorganisms demonstrate microbial activity information. Real-time spoilage monitoring through these subtle sensors helps investors make timely decisions based on the obtained data. Quality control of stockpiled market products benefits from these systems to ensure product excellence before commercial availability. The sensors enable high-quality food control throughout storage facilities and transportation systems alongside retail distribution networks where their use qualifies specific foods in each point of the chain.

All food products experience equivalent dietary value reduction owing to nutrient degradation. The processing of foods demands monitoring the nutrient levels to prevent damage to the nutritional value which will be preserved in final products. The focus level of vitamins and antioxidants can be measured by nanobiosensors for nutritional assessments. Sensors applied in this field enhance vitamin deterioration detection through heating and light exposure. The givers deliver precise immediate analyses of nutrients to help businesses modify storage processes and product formulations for maintaining nutritional property integrity. The same sensors can track food nutritional status during manufacturing production and later monitor stored foods to verify their declared nutritional information remains intact.

4.3 Process Monitoring and Control

Food processing and preservation through physical methods utilize pH changes and temperature and humidity control according to authors [36]. Deviation from these environmental standards leads to deteriorated product quality together with safety risks. Real-time environmental parameter monitoring happens through nanobiosensors that utilize nanoscale probes which incorporate biosensors. Nanobiosensors employ ion-sensitive materials to detect pH changes and thermoresponsive nanomaterials to read temperature through their sensing mechanisms [37]. These environmental parameters can be monitored using nanobiosensors through nanoscale probing that makes use of biosensors with real-time capabilities. For example, by using ion-sensitive materials, nanobiosensors can measure the change in pH, whereas by using thermoresponsive nanomaterials, they can measure temperature [37]. These instruments deliver both precision and sensitivity while offering direct real-time monitoring capabilities for continuous environment optimization. Nanobiosensors function by installation in processing equipment alongside product storage and transportation sequence to ensure constant environmental stability.

Production lines operated by automation systems demand continuous human oversight in order to maintain both high product standards and operational effectiveness. Real-time monitoring serves to shield systems from both critical failures and unproductive choices. Nanobiosensors that match production needs can continuously track contamination levels and component concentrations plus process environment conditions through their integration into manufacturing operations [13]. The same sensors help to measure both chemical reaction frequencies and contaminant presence levels in actual time. These devices enable continuous observation and accurate measurement without the involvement of human inspectors so they increase productivity. Manufacturing tools can integrate nanobiosensors, while in-line diagnostic tools with portability serve to monitor production processes which generate feedback information for optimization.

Food production and processing depend on sufficient compliance with both safety standards along with legal requirements. Nanobiosensors help meet quality control requirements through their ability to provide dependable data. Sensors can also be used in monitoring safety characteristics in products to include contaminants, allergens, and nutritional value as per the market standards [38]. Provide precise and timely information on, inter alia, compliance with food safety and quality measures, as well as measures for avoiding possible violations and, as a consequence, recalls.

5 Technological Advancements in Nanobiosensors

5.1 Material Innovations

Material advancement emerges as one of the primary trends in nanobiosensors development area according to the given overview. Modern innovations have led to developments in nanomaterials with groundbreaking performance features. Four notable nanomaterials—including gold nanoparticles, carbon nanotubes, graphene, and quantum dots—possess strong physical characteristics that boost device capability and operate at higher levels in nanobiosensors [39]. Tests show that graphene exhibits remarkable conductivity of electricity alongside enormous surface area properties thereby becoming optimal for developing efficient sensors with enhanced sensitivity. The quantum dot system demonstrates superior fluorescence capabilities when used to detect various biomolecules through its controlled fluorescent properties [40]. New discoveries in materials technology both enhance the superior characteristics of nanobiosensors by improving their sensitivity and specificity while enabling new clinical and environmental usage applications.

5.2 Advances in Nanomaterial Synthesis

Nanomaterials for synthesis have advanced significantly, and therefore, the formulation of nanobiosensors has been beneficial. Present technological tools facilitate a precise regulation of the dimensions, shape, and composition of nanoparticles; these parameters are relevant and influence the efficiency of sensors. Techniques such as chemical deposition, sol–gel process chemical vapor deposition, and electrochemical deposition of nanomaterials are practiced to meet requirement characteristics several times. For instance, the advancement in colloidal synthesis has enhanced the skills of synthesizing nanoparticles that possess specific sizes and shapes, hence increasing the sensor's reliability [41]. While it is crucial to advance in the functionalization of nanobiosensors, it also is necessary to fashion the development of more intricate forms of the nanoscale configuration that may enhance the elements of detection and integration of the sensors.

5.3 Enhanced Sensitivity and Specificity

The high sensitivity and specificity performance of skinny functions play some significant roles in the progress of nanobiosensor technology [42]. However, it is necessary to say that both the appearance of new, more advanced nanomaterials and the optimization of several specific geometries of sensors have improved the parameters mentioned above considerably. Now, they possess diagnostic capabilities for defining the prevalence of target analytes at the lowest concentrations possible but with reasonable accuracy. One of these improvements is signal amplification, whereby a single analyte molecule can produce a measurable signal change and the identification of selective ligands such as antibodies or aptamers where the sensor is only excited by the target analyte. These also raise the likelihood of diagnosis of monitoring applications but also allow for the detection of low-intensity biomarkers and contaminants that are otherwise undetectable [43].

5.4 Integration with Digital Technologies

Utilizing digitization combined with nanobiosensors brought about a spatial transformation for data collection and analysis processes. The modern nanobiosensors use digital interfaces to receive data that can connect to computer systems for processing. The integrated system enables real-time data transfer and allows users to operate devices remotely while additionally interacting with the interface. A user can establish Bluetooth or Wi-Fi connections between sensors and smartphones or online portals so centralized processing has global accessibility [44]. The trend demonstrates how nanobiosensors merge mathematically with other power generation technologies to increase both the ease of use and functional value of application software.

5.5 Data Processing and Analysis

These two have enhanced the functionality of the nanobiosensors: data processing and data analysis. Most sensors today produce substantial amounts of data that need analysis before processing. The monitoring process achieves meaningful sensor data analysis through real-time analysis enabled by high-performance computing technologies and the practice of mentioned algorithms [45]. Large data gets information extraction through the combination of three methods: Signal processing and conditioning, statistical, and data fusion and conditioning. The enhancements made to the sensor result in improved functionality through more precise readings and dependable data while ensuring dynamic system control and parallel parameter tracking capability [46].

5.6 Incorporation and Roles of Machine Learning and Artificial Intelligence

Creating and integrating machine learning (ML) and artificial intelligence (AI) with nanobiosensor technology is a huge step up in the advancement of sensor systems.

Many techniques of machine learning are employed for decoding the information present in the sensors or for categorizing patterns, and often for making convincing conclusions based on history. The various tasks that a nanobiosensor can handle are detecting anomalies, analyzing trends, and scheduling maintenance, all of which can be enhanced through the aid of artificial intelligence [47]. For instance, in the employment of supervised learning, algorithms can be presented for high accuracy and consistency for biomarkers detection or some constituents in opposition to noises and fluctuations [48]. Therefore, the integration of nanobiosensors with the help of the newly proposed architecture shall encourage the improvement of not only the effectiveness of the nanobiosensors but also the purposes they serve by affording a higher level of data control.

5.7 *Miniaturization*

From the above trends, there are two that have led to increased adoption of nanobiosensors. These include the following: This has led to the reduction of the size and, thus, the fabrication cost. Enhanced knowledge of fabrication and the characteristics of materials for the composite have enabled the miniaturization of the device in addition to increasing its performance as a sensor. This miniaturization makes it possible to culture sensors in more applications, such as portable identification kits and incorporated check instruments [49]. Also, the development of fabrication technologies and special materials played a role in reducing manufacturing costs, thus enhancing companies' sales of nanobiosensors. Further, improved and simple, low-cost fabrication, such as roll-to-roll processing of these complex sensors, has led to enhanced importance for these advanced sensors.

5.8 *Improvements in Sensor Design*

According to this kind of work, the fabrication of nanobiosensors has been enhanced by the evolution of the sensors' architecture. The current designs of sensors for enhancement of functionality, durability, and friendliness to the user include flexibility. There is the development of models of flexible and wearable sensors with prospects in health and environmental monitoring [50]. Other advances have also expanded to the design of multi-sensors that will be in a position to analyze not only one analyte or parameter but several of them simultaneously. Further, the enhancements of the coating and encapsulation of the sensor helped to prevent nanomaterials from degrading over time and remaining elastic. All these enhancements are relative to the general applicability and functionality of nanobiosensors over various disciplines.

5.9 Cost-Effective Manufacturing Techniques

This was a key factor in making nanobiosensors versatile in as many applications as possible and across many industries due to the quest for cheaper ways of production [51]. New trends in production technologies such as sophisticated production lines, automation, modular manufacturing, and other related production technologies have lowered the cost of fabricating advanced quality nanobiosensors. There are ink-jet methods, micro contact, and screen methods of manufacturing the nanoparticles are available, and they contribute to easy and cheap manufacturing of nanoparticles and large-scale manufacturing of the nanoparticles [52] (Fig. 1).

6 Economic Impact of Nanobiosensors

The economic impact of nanobiosensors can be attributed to the following factors: the cost related to the creation of the nanobiosensors and the process of putting them to use. The fabrication of nanobiosensors requires a lot of resources, and in most cases, there are no shortcuts. Therefore, it requires not only big money for research itself but also for the creation of a nontrivial amount of research equipment and personnel—or infrastructure—as any productive science requires [53]. That nearly always indicates a great deal of money spent on research and development but also implies that the sensors have to be incorporated into whatever systems they are supposed to gauge a part of. As to the question of what all of this may cost, one can only hope never mind that it is more than zero.

The application of nanobiosensors is costly but, at the same time has deep prospects in the sphere of health and environment. What has been learned from the actual uses of such highly sensitive devices will be worth the cost many times over when, where, and if these uses become standardized. Currently, nanobiosensors are applied in daily use very limited, while the potential gains of these gadgets are what make them so attractive [54]. A considerable part of the imagination of funding agencies is still driven by such notions as real time, in vivo health monitoring, better diagnostic capabilities, and the incredibly intimate association that one will bear with his/her computer—the intelligence system we are witnessing. The use of advanced technologies in the development of new products entails high costs, including those of the progressive physical and software, flexible and efficient electronic microscopes, special equipment for fabrication, employing of skillful scientists, engineers, and quite several trials before reaching the best nanobiosensors. The sensors in question might themselves entail innovative procedures and data analysis algorithms to get a handle on comprehensively [55].

Manufacturing and production costs are associated with a change from prototype construction of the nanobiosensors to small-scale manufacturing and production. Such expenses are the purchase of high-purity nanomaterials, the development of large-scale production methods, as well as measures taken to ensure the yield's quality

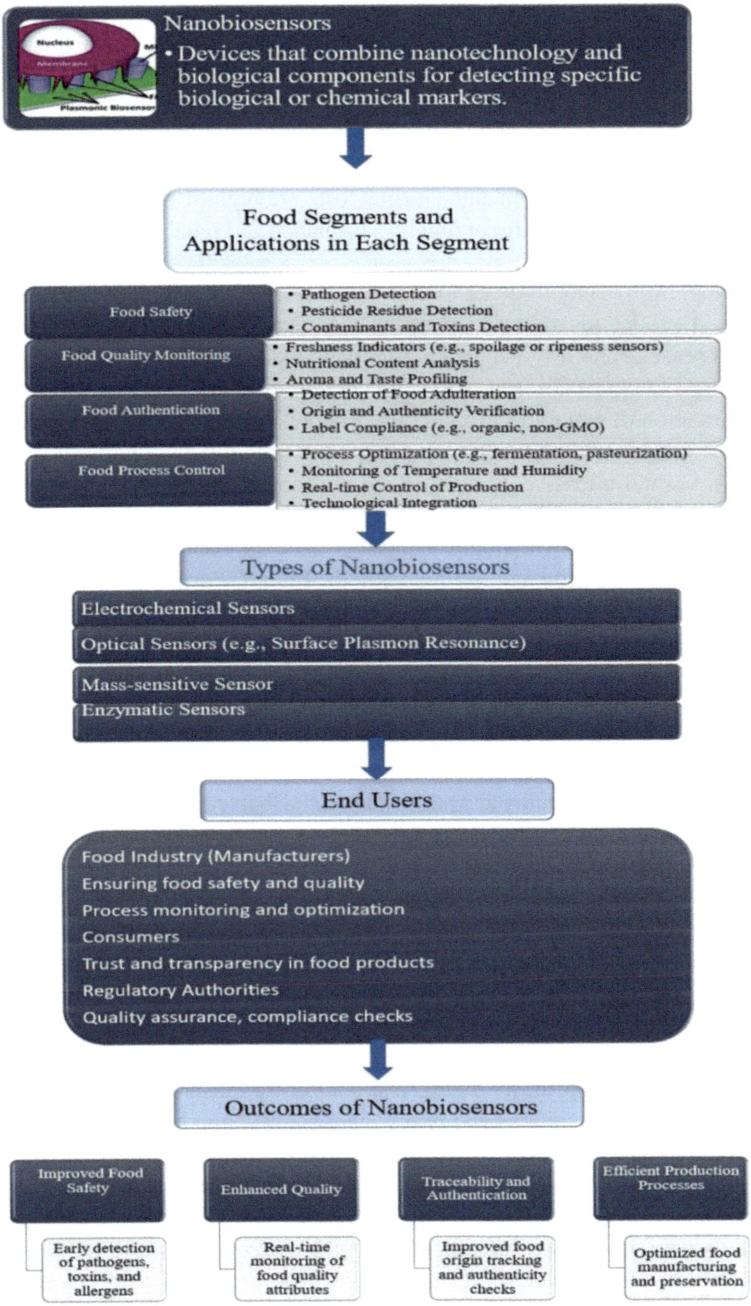

Fig. 1 Applications of nanobiosensors in different food segments

[56]. This has necessitated the use of sophisticated techniques in the fabrication of nanobiosensors, including lithography, chemical vapor deposition, and electrochemical deposition. Capital intensity is also another factor, in the sense that the production of goods involves specialized buildings as well as the integration of various systems to optimize manufacturing costs. Also, large production volumes require a standard quality system and performance, which in turn requires strict tests and validation processes. Although these manufacturing costs are high, it is compulsory to fabricate nanobiosensors that offer optimal performance and reliability for use across the market [57].

Moving from nanobiosensor prototype development to small-scale manufacturing and production requires investment in manufacturing expenses. Manufacturing expenses consist of acquiring pure nanomaterials together with establishing full-scale production platforms while implementing quality assurance protocols [56]. The fabrication of nanobiosensors requires sophisticated techniques which include lithography alongside chemical vapor deposition and electrochemical deposition methods. Capital-intensive production requires specialized buildings together with well-integrated systems to achieve optimized manufacturing costs. Larger production quantities remove variability in performance so standardized quality control demands strict validation and testing methods. Manufacturing nanobiosensors with optimal market performance requires high manufacturing costs but manufacturers must undertake this expense to deliver reliable devices across the market [57].

Return on investment serves as a key feature of nanobiosensors and is commonly recognized as ROI. Some nanobiosensors require notable investments for installation and system deployment although their long-term gains surpass the initial costs. The return on investment for Nanobiosensors emerges from two major sources: their ability to enhance data quality and performance and their capacity to generate iterative solutions across different human sectors such as health and environment and food sectors [58]. Healthcare diagnosis through nanobiosensors enables rapid and precise diagnostic tests while also minimizing healthcare expenses for improved patient health results. These benefits derived from nanobiosensors establish a pathway to achieve complete ROI optimization. The advantages of Nanobiosensors include I. Tangible Gains and II. Intangible Benefits which encompass a. Reputation and b. Competitive Advantage [59].

Among the many economic advantages attributed to nanobiosensors, some of the most impressive include the following: Nanobiosensors enable substantial cost reduction by optimizing operations and removing mistakes. The precision of diagnostics and measurements increases through nanobiosensors allowing for superior accuracy levels [60]. The decrease in error rates cuts down expenses related to workplace resubmission and diagnostic repetition and quality control actions. Nanobiosensors enable industrial processes to simplify operations reducing waste production and attaining maximum resource efficiency due to their efficiency level. Nanobiosensors employed in food production services quickly detect contaminants and nutrient-degrading agents to enable manufacturers implement remedies before contaminated products reach market consumers [61]. The combined effect of these efficiency improvements leads to unstable products with lower costs and better operational outcomes.

Strengthened compliance practices and product recalls might emerge from this limited financial boost which would improve operators' food business results. Self-operating nanoscale biosensors provide essential product recall capabilities that drive down expenses for compliance costs by swiftly detecting bacterial contaminants among other quality problems. Businesses specializing in food and pharmaceutical products face dramatic costs and severe social consequences when recalls are necessary so it becomes vital to identify product issues ahead of time [62]. The use of nanobiosensors enables businesses to produce higher quality products that fulfill both safety needs and regulatory standards [63]. Some examples of the new developments include the enhancement of the use of nanobiosensors in environmental monitoring, industrial automation and systems, and the area of health care, particularly in personalized medicine [64]. With advancements in technology, new prospects are emerging in numerous sectors, including smart farms, health monitoring gadgets, and real-time monitoring of the environment. An improved market environment is a result of the growing appreciation of the benefits accruing from nanobiosensors and through ongoing study and advancement. The same beneficial trends can be utilized to gain an advantage over the competition and to create new opportunities for earning. Market opportunities in nanobiosensors are starting to appear and boost the economy as the framework for revolutions in several fields is being created [65].

Nanotech continues its rapid evolution with expanding sensor market projections indicating future popularity for these devices. According to market analysts, there will be a very robust CAGR due to the combination of nanomaterial science development and increasing diagnostic and monitoring application demands. The nanobiosensor market continues to grow more rapidly according to estimates because market participants dedicate their resources to research and development for technological advancement. Nanobiosensors are projected to experience market expansion because they have broad applications throughout industrial marketplaces and health and environmental safety domains [66]. Revenues offer new prospects to businesses who wish to extract profits from advanced state sensing technology market expansion. The advancing popularity of nanobiosensor technology will drive market expansion through repeated usage in everyday operations.

Many big and small companies are active within the nanobiosensor market because it is a field where new entrants and traditional technology giants can equally act. Fluctuations in the market situation and the entrance of new participants, as well as the constant development of nanobiosensor technology, fuel competition. Industry leaders are bringing knowledge in nanomaterials and sensing technology into the competition to fend off start-ups that are bringing fresh ideas along with disruptive technologies. Currently, active cooperation with the industry, academicians, and technology innovators is creating a healthy nanobiotechnology environment for the development of nanobiosensor systems and the creation of novel applications for them [67].

7 Challenges and Barriers

7.1 Technical Challenges

Nanobiosensors require additional problem-solving to deliver accurate and affordable results when applied to food industry applications while resolving existing technical barriers. This is due to the following four special challenges: The food industry faces challenges to establish correct sensitization standards and specificity rates while maintaining method reliability between different scenarios and implementing emerging technologies into everyday routines [68]. A major technical challenge exists because nanobiosensors need to maintain high sensitivity and specificity for every kind of food product being tested. Food classifications change both the nanobiosensor's substances and the reactions induced in the target matter. Complex organic compounds from food products show misleading properties to human senses while producing incorrect test results because these compounds overlap between carbohydrate chemistry along with fat or protein content [69]. Finally, researchers together with developers must improve sensor fabrication while optimizing nanomaterial functionalization protocols. The sensors demand several calibration operations to achieve maximum chemical control measurements for food specimens.

7.2 Regulatory and Compliance Issues

The food industry uses nanobiosensors frequently as research shows multiple regulatory and compliance factors dominate their operation. Food safety's complex regulatory framework represents the key obstacle that researchers must solve while implementing these technologies according to given standards. Navigating Food Safety Regulations: Food safety regulations differ strongly between countries so a basic comprehension of national food governing laws becomes essential. Each potential use of nanobiosensors for food applications needs to complete stringent safety evaluations that require a complete testing process before gaining approval [37]. To gain approval, scientists must prove the sensors are harmless to sensor users and their touchable components showcase no adverse health effects [13]. The evaluation of new technologies requires compliance with rules issued by multiple regulatory bodies encompassing the U.S. Food and Drug Administration, European Food Safety Authority and additional agencies. The response to these regulations involves studying both the scientific principles of food safety together with legal codes.

The achievement of standardization and certification represents a fundamental regulatory requirement. The process of standardization requires developing shared assessment methods for nanobiosensor testing validation and calibration to ensure consistent measurement outcomes across different applications. The certification procedure demands extensive chapter work combined with many examinations to demonstrate that sensors fulfill necessary requirements [70].

7.3 Public and Ethical Considerations

The implementation of nanobiosensors in food industries brings public and ethical concerns about user receptiveness and proper utilization which needs evaluation. Ethical Concerns Regarding Nanomaterials: Food packaging and contact surfaces utilizing nanotechnology product urgent concerns related to nanocomponents' ethical acceptability. Health experts currently debate about potential risks which nanomaterials introduce both when consumed by humans and when evaluating their effects [71]. The effectiveness of nanomaterial uses on environmental protection remains a primary concern across production phases and disposal times. The deployment of nanomaterials demands both limited deployment scope and deep comprehension regarding their effects to ensure ethical implementation [72]. Risk management decision transparency needs to improve through more extensive dissemination of data about the safety measures used alongside nanobiosensors in environmental applications.

Consumer understanding and opinions about nano-based sensor products play an essential role when dealing with nanobiosensors. People resist new domains because the prolonged introduction of unfamiliar technologies results in either doubt or dismay from users mainly toward previously undiscovered materials and methods [73]. Consumer acceptance of nanobiosensors stands as a critical element which requires informed communication about nanobiosensors' benefits and safety characteristics and functionality features. Stakeholders' awareness campaign should aim at educating the public on the following aspects regarding nanobiosensors: Studies need to focus on explaining nanobiosensor functions and how their effectiveness helps achieve improved food security while covering nanobiosensor safety standards for certification purposes [74].

8 Future Directions and Emerging Trends

Nanobiosensors are ready for a positive evolution, which could cause radical changes in diagnostic, monitoring techniques and most analysis-related areas. With increased development from researchers and engineers working on nanotechnology-enhanced biosensors, the following main trends are expected to define the future of nanobiosensors. At the heart of these innovations are next-generation nanomaterials that augur well with the existing technologies [75]. They are synthesized at the atomic or molecular level, which presents several advantages in the operation of the sensors. On the other hand, two-dimensional materials like graphene and transition metal dichalcogenides deserve special attention [76]. Graphene, which exhibits high electrical conductivity, outstanding mechanical properties, and a large surface area, is prospective for ultra-sensitive detection. Transition metal dichalcogenides present tunable optoelectronic characteristics, which are suitable for different sensing applications. Also, the recent development of metamaterials or extricated materials that do not exist in nature but are crafted for special purposes is paving the way to the development of very sensitive and selective sensors through the control of electromagnetic waves. The increase in interest in synthetic polymers is complemented by bio-nanomaterials based on proteins and nucleic acids. These properties also make it possible to create sensors with little influence from other elements [77].

Microfluidics and lab on a chip are yet other major innovations that have emerged in the field. These innovations make it possible to perform a comprehensive and detailed sample processing and analysis in a single and small equipment while it is possible to analyze large numbers of samples using a minimal number of samples [78]. Moreover, the use of flexible and printable electronics broadens the possibility of using nanobiosensors in wearable technologies and implantable devices that can provide constant monitoring of the health state and smart clothing.

Nanobiosensors are still under development and optimizing; this technology is leading to various fields where innovative solutions are going to be applicable. Personalized Nutrition and Health Monitoring: Ever ever-increasing importance of human health care is going to be the role of nanobiosensors in the area of personalized nutrition and health. Such sensors may enable real-time, personal monitoring of the physiological and biochemical states in the human body and revolutionize personal health management. For instance, nanobiosensors can be incorporated into wearable technology or smart molecular structures that monitor metabolites, nutrients, and hydration status in real time [79].

That is like a wristband that will act as a biosensor that will read the patient's glucose level, electrolyte level, and water content. Such a device would provide feedback about the consumption of food and suggestions on a user's need for water or any other beverage to make healthy choices. Such an approach might prove to be quite helpful in dramatically improving people's health. Additionally, the coupling of these sensors with m-Health applications may open up possibilities for more objective distant supervision and teleconsultation, which would let healthcare managers offer relevant suggestions based on patients' biometric data [80]. Integration with Smart Packaging Solutions: The amalgamation of nanobiosensors via a smart packaging system shows great promise in the food industry, especially in developing safer foods. Their normal usage could involve smart packages containing nanobiosensors to check the status of the food products from production to consumption [81].

9 Conclusion

To conclude, nanobiosensors are one of the most promising areas in technology that can easily affect many fields, starting from the most vital one, namely food safety and quality. Such sensors that were enabled by innovations in nanomaterials such as graphene and transition metal dichalcogenides are incredibly sensitive and selective in identifying any number of substances. Its use in the food industry has the potential to improve safety by providing the right results of the presence of pathogens, toxins, and allergen foods and also sensing the freshness and spoilage of the food. Thus, compatibility with smart packaging systems additionally escalates their possible utilization for managers; they encompass monitoring possibilities and foods' proper utilization decrease. From an economic perspective, the use of nanobiosensors is considerably advantageous despite the high costs of nanostructure production and development. One of the benefits of outsourcing recurrence work is that it helps in trying to make it more efficient, accurate, and cheaper than recall costs, as shown by the market trends, which show that the market will demonstrate strong growth due to advancements in technology. However, some limitations include the capacity to retain the optimal performance of the sensor in different foods, conflict with regulatory policies on food packaging, and last but not least, the ethical issues associated with the usage of nanomaterials.

In the future, there should be positive trends in nanobiosensor technology development. The improvements in nanomaterials and sensors are expected to broaden their applications as well as boost the performance of microfluidics and wearable electronics. The prospect of wearable nanobiosensors and nanoscale packaging in particular revealed a promise of changing people's lives and food safety. Therefore, the application of nanobiosensors in a practical context appears to remain contingent on multidisciplinary engagements between scientists, industries, and regulatory agencies. Having built on existing obstacles and stimulated further advancement, nanobiosensors can open up rich possibilities of altering the foods safety and quality and can give a glimpse of the future where the application of technology and science to resolve sophisticated issues can contribute to healthcare in the world. In this chapter, we would like to conclude that nanobiosensors have accelerated and will continue to drive the development of future food safety and quality management systems. Further development of both will undeniably bring more solutions, advancements in terms of food production and distribution, and improved general stability of the food industry. The story of nanobiosensors is a promising one that has great expectations of changing the world in the future and is full of wonderful innovation in the years to come.

References

1. Raghu HV, Parkunan T, Kumar N (2020) Application of nanobiosensors for food safety monitoring. In: Dasgupta N, Ranjan S, Lichtfouse E (eds) Environmental nanotechnology, vol 4. Springer, Cham, pp 93–129. (Environmental chemistry for a sustainable world, vol 32). https://doi.org/10.1007/978-3-030-26668-4_3
2. Mathivanan S (2021) Perspectives of nano-materials and nanobiosensors in food safety and agriculture. Novel Nanomater. https://books.google.com/books?hl=en&lr=&id=Yb4zEAAAQBAJ&oi=fnd&pg=PA197&dq=Mathivanan,+S.+(2021).+Perspectives+of+nanomaterials+and+nanobiosensors+in+food+safety+and+agriculture.+Novel+nanomaterials,+197&ots=_16XdKUTO2&sig=qZ1q02N69_0nAvgU4QnNJBgh2Qo
3. Kannan P, Guo L (2020) Nanosensors for food safety. In: Nanosensors for smart cities. Elsevier, pp 339–54. https://www.sciencedirect.com/science/article/pii/B9780128198704000190
4. Chen J, Park B (2016) Recent advancements in nanobioassays and nanobiosensors for foodborne pathogenic bacteria detection. J Food Protect 79(6):1055–1069

5. Otles S, Yalcin B (2012) Review on the application of nanobiosensors in food analysis. Acta Scientiarum Polonorum Technologia Alimentaria 11(1):7–18
6. Attaallah R, Antonacci A, Arduini F, Amine A, Scognamiglio V (2020) Nanobiosensors for bioclinical applications: pros and cons. In: Patra JK, Fraceto LF, Das G, Campos EVR (eds) Green nanoparticles. Springer, Cham, pp 117–49. (Nanotechnology in the life sciences). https://doi.org/10.1007/978-3-030-39246-8_5
7. Bhattarai P, Hameed S (2020) Basics of biosensors and nanobiosensors. In: Wu A, Khan WS (eds) Nanobiosensors, 1st edn. Wiley, pp 1–22. https://doi.org/10.1002/9783527345137.ch1
8. Chopra N, Gavalas VG, Bachas LG, Hinds BJ, Bachas LG (2007) Functional one-dimensional nanomaterials: applications in nanoscale biosensors. Anal Lett 40(11):2067–2096
9. Stanisavljevic M, Krizkova S, Vaculovicova M, Kizek R, Adam V (2015) Quantum dots-fluorescence resonance energy transfer-based nanosensors and their application. Biosens Bioelectr 74:562–574
10. Tang B, Cao L, Xu K, Zhuo L, Ge J, Li Q et al (2008) A new nanobiosensor for glucose with high sensitivity and selectivity in serum based on fluorescence resonance energy transfer (FRET) between CdTe quantum dots and Au nanoparticles. Chem A Eur J 14(12):3637–3644
11. Willander M, Khun K, Ibupoto ZH (2014) Metal oxide nanosensors using polymeric membranes, enzymes and antibody receptors as ion and molecular recognition elements. Sensors 14(5):8605–8632
12. Malhotra S, Verma A, Tyagi N, Kumar V (2017) Biosensors: principle, types and applications. Int J Adv Res Innov Ideas Educ 3(2):3639–3644
13. Srivastava AK, Dev A, Karmakar S (2018) Nanosensors and nanobiosensors in food and agriculture. Environ Chem Lett 16:161–182
14. Pumera M, Sanchez S, Ichinose I, Tang J (2007) Electrochemical nanobiosensors. Sens Actuat B Chem 123(2):1195–1205
15. Varadan VK, Chen L, Xie J (2008) Nanomedicine: design and applications of magnetic nanomaterials, nanosensors and nanosystems. Wiley. https://books.google.com/books?hl=en&lr=&id=1sNpk-1ECIoC&oi=fnd&pg=PR5&dq=Varadan,+V.+K.,+Chen,+L.,+%26+Xie,+J.+(2008).+Nanomedicine:+design+and+applications+of+magnetic+nanomaterials,+nanosensors+and+nanosystems.+John+Wiley+%26+Sons&ots=yepgwy6NOg&sig=mkGWe2zmOYiZGe1R1RuZqT0OF_c. Accessed on 1 Sept 2024
16. Soto D, Orozco J (2022) Hybrid nanobioengineered nanomaterial-based electrochemical biosensors. Molecules 27(12):3841
17. Cossettini A, Vidic J, Maifreni M, Marino M, Pinamonti D, Manzano M (2022) Rapid detection of *Listeria monocytogenes*, *Salmonella*, *Campylobacter* spp., and *Escherichia coli* in food using biosensors. Food Control 137:108962
18. Singh AK, Bhunia AK (2018) Optical biosensors in foodborne pathogen detection. In: Smart biosensor technology. CRC Press, pp 443–68
19. Narsaiah K, Jha SN, Bhardwaj R, Sharma R, Kumar R (2012) Optical biosensors for food quality and safety assurance—a review. J Food Sci Technol 49(4):383–406
20. Audrey S, Beatriz PS, Jean-Louis M (2012) Biosensors for pesticide detection: new trends. Am J Analyt Chem. https://www.scirp.org/html/5-2200315_18264.htm
21. Verma N, Bhardwaj A (2015) Biosensor technology for pesticides—a review. Appl Biochem Biotechnol 175(6):3093–3119
22. Oliveira IS, da Silva Junior AG, de Andrade CAS, Oliveira MDL (2019) Biosensors for early detection of fungi spoilage and toxigenic and mycotoxins in food. Curr Opin Food Sci 29:64–79
23. Shrivastava A, Sharma RK (2022) Biosensors for the detection of mycotoxins. Toxin Rev 41(2):618–638
24. Rapini R, Marrazza G (2016) Biosensor potential in pesticide monitoring. In: Comprehensive analytical chemistry. Elsevier, pp 3–31. https://www.sciencedirect.com/science/article/pii/S0166526X16300472. Accessed on 29 Nov 2024
25. Rodriguez MLL, Benimeli C, Madrid RE, Giacomelli CE (2015) A simple *Streptomyces spore*-based impedimetric biosensor to detect lindane pesticide. Sens Actuat B Chem 207:447–454

26. Jiang S, Peng Y, Ning B, Bai J, Liu Y, Zhang N et al (2015) Surface plasmon resonance sensor based on molecularly imprinted polymer film for detection of histamine. Sens Actuat B Chem 221:15–21
27. Kim DM, Yoo SM (2022) Colorimetric systems for the detection of bacterial contamination: strategy and applications. Biosensors 12(7):532
28. Nguyen QH, Kim MI (2020) Nanomaterial-mediated paper-based biosensors for colorimetric pathogen detection. TrAC Trends Analyt Chem 132:116038
29. Wolfbeis OS (2006) Fiber-optic chemical sensors and biosensors. Anal Chem 78(12):3859–3874
30. Wolfbeis OS (2000) Fiber-optic chemical sensors and biosensors. Anal Chem 72(12):81–90
31. Pashazadeh P, Mokhtarzadeh A, Hasanzadeh M, Hejazi M, Hashemi M, de la Guardia M (2017) Nano-materials for use in sensing of salmonella infections: recent advances. Biosens Bioelectr 87:1050–1064
32. Kundu M, Krishnan P, Kotnala RK, Sumana G (2019) Recent developments in biosensors to combat agricultural challenges and their future prospects. Trends Food Sci Technol 88:157–178
33. Neethirajan S, Weng X, Tah A, Cordero JO, Ragavan KV (2018) Nano-biosensor platforms for detecting food allergens: new trends. Sens Bio-sens Res 18:13–30
34. Aquino A, Conte-Junior CA (2020) A systematic review of food allergy: nanobiosensor and food allergen detection. Biosensors 10(12):194
35. Meng X, Kim S, Puligundla P, Ko S (2014) Carbon dioxide and oxygen gas sensors-possible application for monitoring quality, freshness, and safety of agricultural and food products with emphasis on importance of analytical signals and their transformation. J Korean Soc Appl Biol Chem 57(6):723–733
36. Thakur M, Wang B, Verma ML (2022) Development and applications of nanobiosensors for sustainable agricultural and food industries: recent developments, challenges and perspectives. Environ Technol Innov 26:102371
37. Mahmoud AED, Fawzy M (2021) Nanosensors and nanobiosensors for monitoring the environmental pollutants. In: Makhlouf ASH, Ali GAM (eds) Waste recycling technologies for nanomaterials manufacturing. Springer, Cham, pp 229–46. (Topics in mining, metallurgy and materials engineering). https://doi.org/10.1007/978-3-030-68031-2_9
38. Curulli A (2021) Electrochemical biosensors in food safety: challenges and perspectives. Molecules 26(10):2940
39. Peña-Bahamonde J, Nguyen HN, Fanourakis SK, Rodrigues DF (2018) Recent advances in graphene-based biosensor technology with applications in life sciences. J Nanobiotechnol 16(1):75
40. Cinteza LO (2010) Quantum dots in biomedical applications: advances and challenges. J Nanophoton 4(1):042503
41. Howes PD, Chandrawati R, Stevens MM (2014) Colloidal nanoparticles as advanced biological sensors. Science 346(6205):1247390
42. Swierczewska M, Liu G, Lee S, Chen X (2012) High-sensitivity nanosensors for biomarker detection. Chem Soc Rev 41(7):2641–2655
43. Welch EC, Powell JM, Clevinger TB, Fairman AE, Shukla A (2021) Advances in biosensors and diagnostic technologies using nanostructures and nanomaterials. Adv Funct Mater 31(44):2104126
44. Harun-Ur-Rashid M, Foyez T, Jahan I, Pal K, Imran AB (2022) Rapid diagnosis of COVID-19 via nano-biosensor-implemented biomedical utilization: a systematic review. RSC Adv 12(15):9445–9465
45. Yan R, Wang T, Jiang X, Zhong Q, Huang X, Wang L et al (2020) Design of high-performance plasmonic nanosensors by particle swarm optimization algorithm combined with machine learning. Nanotechnology 31(37):375202
46. Sagadevan S, Periasamy M (2014) Recent trends in nanobiosensors and their applications: a review. Rev Adv Mater Sci 2014(36):62–69
47. Banerjee A, Maity S, Mastrangelo CH (2021) Nanostructures for biosensing, with a brief overview on cancer detection, IoT, and the role of machine learning in smart biosensors. Sensors 21(4):1253

48. Amethiya Y, Pipariya P, Patel S, Shah M (2022) Comparative analysis of breast cancer detection using machine learning and biosensors. Intell Med 2(2):69–81
49. Soleymani L, Li F (2017) Mechanistic challenges and advantages of biosensor miniaturization into the nanoscale. ACS Sens 2(4):458–467
50. Verma D, Singh KR, Yadav AK, Nayak V, Singh J, Solanki PR et al (2022) Internet of things (IoT) in nano-integrated wearable biosensor devices for healthcare applications. Biosens Bioelectr X 11:100153
51. Noah NM, Ndangili PM (2019) Current trends of nanobiosensors for point-of-care diagnostics. J Analyt Methods Chem 2019:1–16
52. Ulloa AM, Glassmaker N, Oduncu MR, Xu P, Wei A, Cakmak M et al (2021) Roll-to-roll manufactured sensors for nitroaromatic organophosphorus pesticides detection. ACS Appl Mater Interf 13(30):35961–35971
53. De Araujo WR, Lukas H, Torres MDT, Gao W, De La Fuente-Nunez C (2024) Low-cost biosensor technologies for rapid detection of COVID-19 and future pandemics. ACS Nano 18(3):1757–1777
54. Kulkarni MB, Ayachit NH, Aminabhavi TM (2022) Recent advancements in nanobiosensors: current trends, challenges, applications, and future scope. Biosensors 12(10):892
55. Yang T, Duncan TV (2021) Challenges and potential solutions for nanosensors intended for use with foods. Nat Nanotechnol 16(3):251–265
56. Mabrouk M, Das DB, Salem ZA, Beherei HH (2021) Nanomaterials for biomedical applications: production, characterisations, recent trends and difficulties. Molecules 26(4):1077
57. Chauhan N, Saxena K, Tikadar M, Jain U (2021) Recent advances in the design of biosensors based on novel nanomaterials: an insight. Nanotechnol Precis Eng 4(4):843
58. Handford CE, Dean M, Spence M, Henchion M, Elliott CT, Campbell K (2015) Awareness and attitudes towards the emerging use of nanotechnology in the agri-food sector. Food Control 57:24–34
59. Juanola-Feliu E, Colomer-Farrarons J, Miribel-Català P, Samitier J, Valls-Pasola J (2012) Market challenges facing academic research in commercializing nano-enabled implantable devices for in-vivo biomedical analysis. Technovation 32(3–4):193–204
60. Pramanik PKD, Solanki A, Debnath A, Nayyar A, El-Sappagh S, Kwak KS (2020) Advancing modern healthcare with nanotechnology, nanobiosensors, and internet of nano things: taxonomies, applications, architecture, and challenges. IEEE Access 8:65230–65266
61. Sharma TK, Ramanathan R, Rakwal R, Agrawal GK, Bansal V (2015) Moving forward in plant food safety and security through NanoBioSensors: adopt or adapt biomedical technologies? Proteomics 15(10):1680–1692
62. Vashist SK, Venkatesh AG, Mitsakakis K, Czilwik G, Roth G, von Stetten F et al (2012) Nanotechnology-based biosensors and diagnostics: technology push versus industrial/healthcare requirements. Bionanoscience 2(3):115–126
63. Chelliah R, Wei S, Daliri EBM, Rubab M, Elahi F, Yeon SJ et al (2021) Development of nanosensors based intelligent packaging systems: food quality and medicine. Nanomaterials 11(6):1515
64. Jain KK (2005) Role of nanobiotechnology in developing personalized medicine for cancer. Technol Cancer Res Treat 4(6):645–650
65. Keçili R, Büyüktiryaki S, Hussain CM (2019) Advancement in bioanalytical science through nanotechnology: past, present and future. TrAC Trends Analyt Chem 110:259–276
66. Rao CNR, Cheetham AK (2001) Science and technology of nanomaterials: current status and future prospects. J Mater Chem 11(12):2887–2894
67. Kumar S, Singh H, Feder-Kubis J, Duc ND (2023) Recent advances in nanobiosensors for sustainable healthcare applications: a systematic literature review. Environ Res 16:117177
68. Weng X, Zhang C, Jiang H (2021) Advances in microfluidic nanobiosensors for the detection of foodborne pathogens. Lwt 151:112172
69. Joyner JJ, Kumar DV (2015) Nanosensors and their applications in food analysis: a review. Int J Sci Technol 3(4):80

70. Ramburrun P, Khan RA, Choonara YE (2022) Design, preparation, and functionalization of nanobiomaterials for enhanced efficacy in current and future biomedical applications. Nanotechnol Rev 11(1):1802–1826
71. Khan A (2015) Ethical and social implications of nanotechnology. In: Engineering leaders conference 2014 on engineering education. Hamad bin Khalifa University Press (HBKU Press), Doha
72. Coles D, Frewer LJ (2013) Nanotechnology applied to European food production: a review of ethical and regulatory issues. Trends Food Sci Technol 34(1):32–43
73. Frewer LJ, Gupta N, George S, Fischer ARH, Giles EL, Coles D (2014) Consumer attitudes towards nanotechnologies applied to food production. Trends Food Sci Technol 40(2):211–225
74. Grieger K, Merck A, Kuzma J (2022) Formulating best practices for responsible innovation of nano-agrifoods through stakeholder insights and reflection. J Respon Technol 10:100030
75. Dar FA, Qazi G, Pirzadah TB (2020) Nano-biosensors: nextgen diagnostic tools in agriculture. In: Hakeem KR, Pirzadah TB (eds) Nanobiotechnology in agriculture. Springer, Cham, pp 129–44. (Nanotechnology in the life sciences). https://doi.org/10.1007/978-3-030-39978-8_7
76. Wen W, Song Y, Yan X, Zhu C, Du D, Wang S et al (2018) Recent advances in emerging 2D nanomaterials for biosensing and bioimaging applications. Mater Today 21(2):164–177
77. García-Mayagoitia S, Pérez-Hernández H, Medina-Pérez G, Campos-Montiel RG, Fernández-Luqueño F (2020) Bio-nanomaterials in the air pollution treatment. In: Nanomaterials for air remediation. Elsevier, pp 227–48. https://www.sciencedirect.com/science/article/pii/B9780128188217000129. Accessed on 1 Sept 2024
78. Kant K, Shahbazi MA, Dave VP, Ngo TA, Chidambara VA, Than LQ et al (2018) Microfluidic devices for sample preparation and rapid detection of foodborne pathogens. Biotechnol Adv 36(4):1003–1024
79. Kim J, Campbell AS, de Ávila BEF, Wang J (2019) Wearable biosensors for healthcare monitoring. Nat Biotechnol 37(4):389–406
80. Jain K (2009) Role of nanobiotechnology in the development of personalized medicine. Nanomedicine 4(3):249–252
81. Sundramoorthy AK, Kumar THV, Gunasekaran S (2018) Graphene-based nanosensors and smart food packaging systems for food safety and quality monitoring. In: Graphene bioelectronics. Elsevier, pp 267–306. https://www.sciencedirect.com/science/article/pii/B9780128133491000123

Advancements in Nanobiosensors for Food Production to Support Circular Economy

Gurleen Kaur Sodhi, Himani Saini, Anjali Chauhan, Pranshu Saxena, and Devendra Sillu

Abstract Globally, over 1.3 billion tons of food is wasted annually, largely due to inadequate monitoring along the supply chain. Nanobiosensors have emerged as a transformative solution, offering highly sensitive, specific, and real-time monitoring capabilities to enhance food safety, quality, and sustainability. Advances in nanotechnology have enabled the miniaturization and functionalization of these sensors, making them more efficient, versatile, and cost-effective. Key innovations include label-free detection systems, integration with IoT platforms, and the application of advanced nanomaterials. The use of artificial intelligence and machine learning further optimizes these technologies, enabling automated data analysis and informed decision-making throughout food production. Nanobiosensors play a critical role in agriculture, food processing, packaging, and distribution. In agriculture, they monitor soil health, detect crop diseases, and assess pesticide residues, promoting sustainable practices. In food processing, they enable real-time detection of spoilage and contamination, supporting circular economy principles by reducing waste. Embedded in packaging, nanobiosensors transform containers into smart systems, while in distribution, they ensure the safety and quality of perishable goods. Despite their commercial potential, challenges such as stability, scalability, and regulatory compliance persist. This chapter explores the advancements, applications, and future opportunities of nanobiosensors, emphasizing their pivotal role in building a sustainable and circular food production system.

Keywords Nanobiosensors · Circular economy · Food safety and quality · Sustainable food production · Smart packaging

G. K. Sodhi
University Institute of Biotechnology, Chandigarh University, Mohali, India

H. Saini · A. Chauhan · P. Saxena
Department of Biotechnology, Thapar Institute of Engineering and Technology, Patiala, India

D. Sillu (✉)
Department of Environmental Science and Engineering, Guangdong-Technion Israel Institute of Technology, Shantou, China
e-mail: devsillu@gmail.com

D. Sillu et al. (eds.), *Nanobiosensors for the Food Industry*, Smart Nanomaterials Technology, https://doi.org/10.1007/978-981-95-0136-6_14

1 Introduction

Every year, over 1.3 billion tons of food is wasted globally, much of it due to inadequate monitoring across the supply chain [1]. Nanobiosensors are revolutionizing food production by delivering highly sensitive, specific, and real-time monitoring capabilities that enhance food safety, quality, and sustainability [2]. These sensors detect trace amounts of biological and chemical agents with exceptional precision, enabling rigorous quality control across the supply chain [3]. Advances in nanotechnology have fuelled this transformation, allowing for the miniaturization and functionalization of biosensors, which are now more efficient and versatile than ever before. Key trends include the development of label-free detection systems, integration with Internet of Things (IoT) platforms, and the use of advanced nanomaterials [4]. These innovations not only boost sensitivity, selectivity, and durability but also lower costs and reduce environmental impact. The integration of artificial intelligence and machine learning further optimizes nanobiosensor applications, enabling automated data analysis and improving decision-making across food production [5, 6].

Nanobiosensors are now integral to each step of the food supply chain, from agriculture and processing to packaging and distribution [7]. In agriculture, they monitor soil health, detect crop diseases, and assess pesticide residues, promoting sustainable practices and efficient resource use [8]. In food processing, nanobiosensors support circular economy principles by enabling real-time detection of spoilage and contamination, thus reducing food waste [9, 10]. In packaging, nanobiosensors embedded in materials transform containers into smart systems that communicate food conditions in real time [11]. At the distribution stage, nanobiosensors maintain the safety and quality of perishable goods, ensuring food remains fresh and high-quality from production to consumption [12]. The commercial potential of nanobiosensors is evident in the rising number of patents filed worldwide, covering areas like multi-analyte detection, nanostructured electrode materials, and portable, on-site devices [13]. This intellectual property landscape underscores the relevance of nanobiosensors for the food industry. Several have already transitioned to commercial markets, including handheld devices for pathogen detection, freshness sensors, and portable systems for pesticide analysis, providing practical, reliable solutions for producers, distributors, and consumers alike.

While promising, nanobiosensors face technical and regulatory challenges that impact their real-world adoption [14]. Issues like stability, reproducibility, and scalability, along with concerns about regulatory compliance and potential toxicity, call for rigorous safety assessments and interdisciplinary collaboration. By addressing these challenges, stakeholders can develop robust, cost-effective solutions that maximize the impact of nanobiosensors in food production. This chapter explores these themes, offering a detailed overview of emerging trends, applications, commercial advancements, challenges, and future opportunities for nanobiosensors in food production. It highlights their transformative potential within the circular economy framework, tracing the current state and future trajectory of this essential technology.

2 Emerging Trends in Nanobiosensors

Nanobiosensors integrate nanostructures into biosensors to enhance their performance. The incorporation of nanostructures significantly improves efficiency, sensitivity, and response time by providing an increased surface area that enhances biomolecule interaction. Their compact design facilitates the miniaturization of biochips, requiring only minimal sample sizes for analysis [15]. This feature is particularly advantageous for clinical diagnostics, especially in scenarios such as outpatient testing, where obtaining large samples may be challenging. By requiring only, a small sample, nanobiosensors streamline testing procedures, making them faster, more convenient, and less invasive, thereby improving the overall diagnostic experience.

Nanobiosensors leverage advanced nanostructures, such as nanoparticles, nanotubes, nanowires, nanopores, and nanocomposites, combined with biomolecules like proteins and DNA, to create highly specialized detection systems [16]. Each type offers distinct advantages tailored to specific applications:

- Nanoparticle Biosensors: Utilize materials such as gold, quantum dots, or magnetic nanoparticles as transducer elements. Biomolecular detectors, such as antibodies, enzymes, or DNA probes, are functionalized onto the nanoparticle surfaces to enable targeted and sensitive detection.
- Nanowire Biosensors: Employ nanoscale wires, typically made from silicon or carbon nanotubes, offering excellent conductivity and sensitivity for detecting biomolecular interactions.
- Nanopore Biosensors: Use nanoscale pores to analyse and detect individual molecules, offering high precision in applications like DNA sequencing and protein characterization.
- Nanocomposite Biosensors: Combine multiple nanomaterials within a bulk matrix to synergize their unique properties. By incorporating at least one material at the nanoscale, nanocomposites achieve enhanced sensitivity, robustness, and functionality, making them suitable for advanced diagnostic and analytical tasks.

These innovative nanostructures open up new possibilities for precision diagnostics, environmental monitoring, and industrial applications.

2.1 In Agricultural Practices

The global population is projected to reach 9.8 billion by 2025, driving an urgent need for increased food production and efficient land utilization [17]. Nanobiosensors are poised to play a crucial role in addressing these challenges. By leveraging nanomaterials like gold and carbon, these sensors can detect specific diseases in plants and crops at remarkably low concentrations, enabling early intervention and effective disease management [18, 19]. Nanobiosensors also enable real-time monitoring of essential agricultural parameters, including soil fertility, pH levels, moisture content, mineral concentrations, and the presence of pests [3, 20]. This targeted approach supports precision agriculture, optimizing crop yields and resource utilization. The economic impact of nanobiosensors is significant, with their global market experiencing rapid growth. Valued at USD 13.6 billion in 2023, it is expected to surge to USD 23.15 billion by 2030, reflecting their expanding applications across various industries [7]. In the food industry, nanobiosensors are transforming safety and quality control by enabling the rapid and highly sensitive detection of pathogens and pesticide residues. Compared to conventional methods like HPLC and GC–MS, nanobiosensors offer faster results and greater efficiency, ensuring enhanced food security and safety [12]. This combination of innovation and practicality underscores their potential to address critical global challenges.

The integration of nanobiosensors into smart agriculture and precision farming systems is transforming crop management by providing real time, detailed data to enhance decision-making processes [21]. Precision farming leverages data-driven techniques to optimize crop production and maximize yields. Nanobiosensors play a pivotal role by monitoring critical parameters such as soil conditions, moisture levels, and environmental factors [22]. This enables targeted application of fertilizers, water, and pesticides at specific locations, reducing costs, enhancing safety, minimizing waste, and significantly lowering the environmental impact through reduced chemical usage. Smart nanobiosensors, enhanced with information technology, offer substantial benefits for farmers in remote regions. By incorporating global positioning systems (GPS)—a network of satellites, computers, and receivers that determine precise latitude and longitude based on signal timings—nanobiosensors can help manage resources efficiently [21]. Farmers can use these integrated systems to monitor and optimize fertilizer application, pesticide use, water levels, chemical and physical stress, and pathogen detection with ease.

Nanobiosensors also excel in pathogen detection, delivering results much faster than conventional methods. The field of nanodiagnostics leverages nanotechnology to provide rapid, precise detection of pathogens and contamination. For example, gold nanoparticles combined with DNA as a biological recognition element can be used to detect plant pathogens [23]. These advanced sensors not only identify pathogens but also monitor pathogen population genetics and the gene transfer dynamics between pathogens and their hosts. By integrating nanotechnology, nanobiosensors revolutionize agriculture with innovative tools for efficient resource management, improved crop health, and enhanced food security. Nanobiosensors are making a significant impact in agriculture, with various types tailored to address specific challenges:

- Quantum Dot Nanobiosensor: Utilizes gold nanoparticles as the sensing material and is effective in detecting mycotoxins such as ZEA, DON, and FB1/FB2 in crops like corn, oats, and barley.
- QD Nanosensor: Employs quantum dots as the nanomaterial for detecting a wide range of pathogens, offering high sensitivity and precision.

- DNA Nanobiosensor: Incorporates carbon nanotubes as the nanomaterial to identify infections in crops such as palm oil trees, aiding in early disease management.
- Pesticide Detection Nanobiosensor: Leverages graphene combined with molecularly imprinted polymers to detect pesticides like chlorothalonil and chlorpyrifos-methyl, enabling precise monitoring of chemical residues in agricultural produce.

With each designed for a specific function, they contribute to improved crop health, reduced chemical usage, and enhanced food safety, reinforcing their critical role in modern farming practices.

2.2 *In Food Industry*

Nanobiosensors are emerging as a transformative trend in the food industry, with advanced sensors and technologies being developed for the early detection of pathogens and toxins. Among these innovations are aptamer-based biosensors, which use synthetic molecules called aptamers to bind specifically to target pathogens and toxins [24, 25]. Aptamers, typically with molecular weights under 25 kDa, are highly selective, making them ideal for sensitive and accurate detection. Another breakthrough is the integration of microfluidics with Loop-Mediated Isothermal Amplification (LAMP) technology [26]. LAMP allows DNA amplification at a constant temperature, and when combined with microfluidics in nanobiosensors, it enables the rapid and highly sensitive detection of pathogens. This combination is particularly suited for real-time diagnostics in the food industry. Silver nanoparticles are also widely used in food safety applications due to their biocompatibility and health safety compared to other metals. Their ability to detect minute concentration changes, enhanced by high-energy laser interactions, makes them a preferred choice for developing highly sensitive nanobiosensors.

The Internet of Things (IoT) is a network of interconnected devices that communicate via the internet, enabling seamless data transfer and real-time analysis. When combined with biosensors, IoT can revolutionize the food industry by automating processes, enhancing monitoring capabilities, and streamlining data analysis [27]. In food management, IoT-enabled biosensors are used to monitor critical parameters such as temperature, humidity, and other environmental factors, ensuring optimal storage and quality. These systems can operate on a large scale, monitoring multiple locations simultaneously. For instance, the E-nose system analyses volatile compounds released by food to assess freshness, while the E-tongue system evaluates taste profiles, providing fast, accurate, and real-time data [28]. The integration of cloud computing further amplifies the utility of IoT biosensors [29]. By eliminating the need for local storage and processing, data can be efficiently analysed and stored in the cloud at reduced costs. This cloud-based data infrastructure enables real-time insights, trend analysis, and predictive analytics, making food quality management more proactive and efficient. Additionally, IoT-enabled nanobiosensors can generate active reports and alerts, ensuring immediate action when required. This integration of IoT, and nanotechnology is transforming the food industry, offering robust solutions for quality assurance, safety, and operational efficiency.

The development of portable biosensors offers a promising solution, enabling rapid analysis and action to prevent disease outbreaks [30]. These tools can benefit a wide range of users, including individual growers, regulatory agencies, exporters, and researchers. For example, *Ralstonia solanacearum* is among the most common plant pathogens, responsible for devastating diseases in crops such as bananas and potatoes [31]. Protecting plants from these pathogens is critical but challenging, as infections are often detected too late for effective intervention. Modern advancements have enabled the integration of biosensors with smartphones, utilizing auxiliary reusable devices (ARDs) and auxiliary disposable devices (ADDs) [30]. For instance, the Samsung Galaxy SII ARD facilitates the adaptation of immunochromatographic assays onto smartphone platforms, allowing for portable and convenient diagnostic capabilities. Additionally, an ADD system equipped with a surface plasmon resonance (SPR)-based biosensor can study biomolecule interactions, such as detecting β2 microglobulin—a biomarker for cancer, inflammatory disorders, and kidney diseases. However, challenges remain in seamlessly integrating biosensors with smartphones, particularly in fostering collaboration between smartphone manufacturers and ARD/ADD developers. Overcoming these barriers will unlock the full potential of biosensor-smartphone systems, offering practical and accessible solutions for plant disease management and other applications.

3 Agricultural Applications of Nanobiosensors

3.1 Biotic Stress Management

With every year that goes by, the world's population grows exponentially. A report indicates that by the year 2050, the world's population is expected to reach around 9.6 billion people. A 70–100% increase in agricultural productivity is needed to meet this population's growing food needs [32]. However, despite a noticeable increase in population, agricultural areas are becoming smaller and crops are encountering several challenges brought on by both biotic and abiotic factors. These have a detrimental effect on agricultural productivity. Around 40% of crop yield which amounts to ~2000 billion dollars worth of annual crops worldwide is lost due to numerous biotic stresses which include diseases, and infections caused by fungi, viruses, bacteria, insects, nematodes, and parasites [33]. These losses when combined with post-harvest losses pose a threat to resources in developing countries like India. Although diagnosing and detecting plant stressors might be difficult, the chance of recovering from infections increases with early diagnosis. Farmers would greatly benefit from an early warning system that detects plant stress before symptoms become apparent. Thus, to stop the disease from spreading widely, minimize crop loss and enhance food quality, quick and precise disease identification is crucial [34]. Detecting and monitoring biotic stresses in plants can be made easier by the use of nanobiosensors. The real-time analytical data provided by these small devices allows for intervention, minimizing crop losses, aids in precision farming practices and optimizes resource usage.

Nanomaterial-based sensors offer solutions for swiftly recognizing and monitoring biotic stresses in precision agriculture. These sensors help identify plant stressors early on by enabling the real-time detection of pathogens, stress signalling molecules, and disease indicators. For example, high-sensitivity electrochemical sensors may detect bacterial diseases like *P. syringae*, which causes bacterial speck in *Solanum lycopersicum*, while HeAptDNA-SWCNT sensors have been created to detect H_2O_2, a crucial stress signalling molecule [35–37]. The gold nanoparticle-based platforms such as impedimetric biosensors and lateral flow assays have successfully identified fungal diseases like *P. infestans* and viral agents like the *Citrus tristexa* virus [19]. For quick analysis, reducing crop losses, and guaranteeing sustainable farming methods, these nanosensors are irreplaceable.

Gold nanoparticle-based lateral flow biosensors have been used extensively to detect fungal pathogens like *Phytophthora* that cause tomato and potato blight diseases and soil-borne pathogens such *R. solanacearum* (which causes bacterial wilt in potatoes) [38, 39]. The surface-enhanced Raman scattering (SERS) can detect multiple pathogens simultaneously, whereas to quickly and accurately detect bacteria, there are urease sensors supported by magnetic nanoparticles [40]. In addition to this, the use of very sensitive quantum dots and interdigitated microelectrode array biosensors have made the identification of bacterial and viral infections precise, way ahead of its time and also helped to lessen the environmental impact of agricultural operations [41].

3.2 *Abiotic Stress Management*

Plants face numerous abiotic stresses such as drought, salinity, temperature extremes, light intensity fluctuations, and heavy metal contamination, which, hence, severely impact agricultural productivity. Many physiological and biochemical changes in plants such as alterations in membrane composition, hormone levels, photosynthesis efficiency, and the production of reactive oxygen species (ROS) are induced by these stresses. Around 51–82% of global crop loss annually is caused by abiotic stressors, emphasizing the urgent need for precise detection and effective mitigation strategies [42, 43]. Nanobiosensors have emerged as powerful tools for detecting stress-induced biomarkers, such as phytohormones like abscisic acid (ABA), indole-3-acetic acid (IAA), and salicylic acid (SA), which play pivotal roles in stress response. For example, an electrochemical biosensor was created with remarkable detection limits of 1.99 and 3.30 μM for IAA and SA in soybean seedlings under saline stress [44]. The sensitivity of stress biomarker detection can also be improved by multiwall carbon nanotubes (MWCNTs) and carbon-black-based devices. Nanotechnology can also be applied to improving stress tolerance during germination and early plant development via seed priming techniques. This involves treating seeds with nanoparticles, hence enhancing water uptake, regulating aquaporin genes and bolstering ROS/antioxidant systems.

This approach provides resilience to harsh environments like salinity and drought while increasing seed germination rates and vigour. Zinc-chitosan nanoparticles, e.g. have been shown to exhibit antifungal properties and improve maize crop germination [45, 46]. Similarly, carbon nanoparticles have been demonstrated to improve lettuce seedling growth when exposed to extreme salinity stress [47]. Beyond seed treatments, nanoparticle-based sensors for heavy metals like cadmium and mercury are revolutionizing environmental monitoring. Smartphone-integrated platforms and lateral flow assays using gold nanoparticles and aptamers enable quick, on-site detection with excellent sensitivity, addressing the limitations of traditional spectroscopic approaches. These innovations highlight the potential for nanotechnology to transform abiotic stress management.

3.3 Soil Quality Assessment

The basis of crop yield and environmental equilibrium is a significant concern in sustainable agricultural practices. It is a dynamic balance of physical, chemical and biological properties. However, keeping such balance amidst modern agricultural challenges calls for innovative solutions. Real-time soil health assessment is essential for making informed decision-making along with soil management [48]. With the ability to accurately and instantly monitor soil conditions, nanobiosensors are paving the way for precision agriculture. Nanomaterials provide an ideal platform for the development of biosensors due to their high surface area which enhances the chemical reactivity and catalytic activity. Because of their excellent specificity, stability, and sensitivity, these nanosensors can detect soil characteristics fluctuations within seconds [49]. The sensors, which use carbon nanodots and quantum dots, can detect heavy metals that may leach into water supplies to bioaccumulate due to overuse of agrochemicals or industrial runoff [50]. Likewise, proof-of-concept results found for controlled-environment testing of nanobiosensors capable of measuring levels of urea are promising and reputable for practical applications. These technologies aid in reconciling agricultural practices with sustainability goals by making the maximum use of resources and minimizing waste while also increasing crop yield.

Effective crop management requires constant monitoring of variables including pH, temperature, moisture, and nutrient levels since soil health is dynamic. Nanotechnology provides portable instruments that deliver real time, actionable information. Lipid-based nanobiosensors, for instance, can identify tiny amounts of nitrites in soil samples, providing information about the availability of nitrogen [51]. Proactive management may be possible if elevated ammonia or nitrate levels indicate microbial imbalances or even insect activity. Nanosensors are a particularly useful development for arid farming areas vulnerable to water scarcity since they allow for the precise measurement of soil moisture, a crucial factor in plant health. Additionally, soil-borne illnesses brought on by dangerous microbes like bacteria, viruses, or fungi may be detected by nanobiosensors as already mentioned.

3.4 Toxic Chemical Detection

Chemicals like pesticides, fungicides, and insecticides are commonly used in agriculture to improve crop yields. However, these chemicals can linger in the environment for long periods, and when used excessively, they can have toxic effects on other organisms, including humans. It is better to monitor the amount of these toxic chemicals in soil considering everyone's best interest. Traditional methods such as using chromatographic ones for detecting pesticides, often have significant drawbacks. They tend to be time-consuming, require expensive equipment, and demand highly trained personnel and specialized laboratory settings [52]. These limitations make it difficult to apply these methods for real time, on-site monitoring. Here, the need for more efficient techniques comes into the picture.

Nanotechnology has been a major factor in tackling these issues in the recent years. Nanosensors have been developed to tackle these issues precisely and quickly. For example an aptasensor with a detection limit of 0.4 nM for optical detection of the organophosphorus pesticide diazinon was recently reported [53]. This sensor makes use of reduced graphene quantum dots and multi-walled carbon nanotubes to increase its sensitivity. Similar to this, an electrochemical immunoassay was developed with an ultra-low detection limit of 10 fM to detect chlorpyrifos, another organophosphate insecticide, utilizing gold nanoparticles and fluorine-doped tin oxide [54]. These sensors are a good substitute for conventional pesticide detection techniques because they are not only quicker but also more affordable.

Enzyme-based nanosensors are also being used for pesticide detection [55, 56]. For example, enzyme biosensors which use acetylcholinesterase (AChE), organophosphate hydrolase (OPH), and organophosphorus acid anhydrolase (OPAA) are already available in the market to detect organophosphate pesticides [57, 58]. Not only this, new techniques such as fluorescent peptide probe-based detection are coming forward. These sensors work by using peptides that aggregate in the presence of pesticides, triggering fluorescence. With such advancements in nanobiosensor technology, detecting pesticides and other toxic chemicals in agricultural products and environmental samples is becoming more efficient and accessible, ultimately contributing to safer and more sustainable agricultural practices.

3.5 Heavy Metal Detection

The uncontrolled spread of heavy metals in the environment driven by industrialization, mining, and indiscriminate application of agrochemicals has emerged as a global challenge. Agricultural is particularly susceptible to this type of contamination, due to the excessive agricultural chemical usage that leads to the leaching of metallic ions into the agricultural soil [59]. The adverse ions remain in the body without decomposing easily and cause serious health diseases such as cancer and chronic diseases [60]. The historical tragedies of Minamata and Itai–Itai serve as stark reminders of the catastrophic consequences of heavy metal contamination [61]. Plants face an irritative stress due to the metal toxicity which can be referred to as oxidative stress, which makes the challenge backs up more again, to be solved [62].

Traditional approaches for identifying heavy metals for instance, spectroscopic and chromatographic methods are still useful, but may not be suitable for field testing [63]. These methodologies require expensive, complex equipment, proficient users, as well as sophisticated sample preparations, rendering them unsuitable for fast, in-field analysis. Nanobiosensors turn this field into a game-changing tool with their portability, sensitivity, and operation simplicity [64]. Noble nanoparticles such as gold nanoparticles (AuNPs) and carbon nanotubes are integral parts of sensors in detecting poisons such as mercury, cadmium, palladium, and arsenic [65–67].

Nanobiosensors are revolutionizing detection techniques in innovative ways. For instance, a platform incorporated in a smartphone was developed that could sense mercury ions (Hg^{2+}) and copper ion (Cu^{2+}) [68]. Lateral flow assays also using UCNPs, AuNPs, aptamers, and DNA probes exhibit similarly excellent sensitivity, successfully detecting mercury and cadmium [69]. Other advances are near-infrared fluorescent nanosensors for real-time arsenic detection [70]. This progress demonstrates the power of nanobiosensors to address traditional challenges to provide a readily adaptable and scalable solution to the environmental and agronomic issues resulting from heavy metal pollution.

3.6 Plant Health Monitoring

Nanobiosensors are transforming how we monitor plant health, providing precise and real-time information that can improve agricultural practices [71]. One of the most exciting new developments in this area is wearables tiny sensors that can be attached directly to a plant's leaves or stems. These sensors track aspects such as growth trends, temperature, humidity and exposure to poisons. They are very affordable, lightweight, and suitable for plants. Stretchable sensor is also used to monitor both plant growth and specific environmental conditions, providing a simpler method for obtaining important data directly from the plant's surface [72–74]. Graphene sensors can provide real-time imaging of water movement within plants, adding another layer of information on their physiology [75].

Another clear advantage is the possibility of using wearable sensors to monitor chemical exposure. Laser-induced graphene sensor for the detection of pesticides has been designed in the past [76, 77]. Adding gold nanoparticles could result in increased sensitivity improved, and the results can be wirelessly transmitted to smartphones. Further emphasizes the capacity of nanobiosensors to render chemical monitoring an enormously accurate and accessible procedure. Beyond chemical detection, wearable sensors are proving invaluable for tracking plant stress responses. For instance, a graphene-based sensor was developed that detects volatile organic compounds (VOCs) released by plants under stress [78, 79]. This system can identify VOCs linked to biotic stress, such as infections like tomato late blight, within four days. It can also detect VOC emissions within just 1 h of physical damage, offering an early warning system for plant health issues. Similarly, a polymer-based "tattoo" sensor capable of detecting oxidative damage in fruits caused by excessive ozone exposure was developed, broadening the application of these sensors beyond general plant health to include post-harvest monitoring [80].

Aside from wearable sensors, sophisticated electronic systems such as electronic tongues (e-tongues), noses (e-noses), and eyes (e-eyes) are finding applications in plant health diagnostics [81, 82]. These devices are especially helpful for the detection of plant disease and environmental conditions monitoring which provides another tool to farmers and helps researchers to keep crops healthy. Nanobiosensors are also proving to be a fantastic tool in diagnosing problems even before those telltale signs, the visible symptoms, are evident. They can identify nutrient deficiencies, soil health issues and water shortages so farmers can act early. They send the data straight to electronic devices like laptops or cellular phones where it may be analysed in real time. Overall, nanobiosensors represent a significant step forward in plant health monitoring. They allow for early detection of problems, efficient management of resources, and a more sustainable approach to agriculture. By integrating wearable technology and advanced electronic systems, farmers can make smarter, data-driven decisions that benefit both their crops and the environment.

4 Patents on Nanobiosensors

The rapid advancement in research on nanobiosensors for the food industry has led to a significant increase in the number of patents filed and granted in the recent years. These developments are a testament to the growing recognition of nanobiosensors as transformative tools in ensuring food safety and quality. Table 1 provides an insightful summary of patents filed or granted in this domain, showcasing the innovative strides made in nanobiosensors and their related technologies. These patents underscore the critical role of nanobiosensors in addressing key challenges in the agri-food sector, such as pathogen detection, toxin identification, and quality monitoring.

Nanobiosensors designed for pathogen detection have become a leading application area, driving their industrial adoption. These sensors often rely on attaching pathogen-specific antibodies to the surface of nanoparticles, typically fluorescent nanoparticles [83]. When exposed to UV irradiation, changes in fluorescence signal the presence of specific pathogens. This mechanism provides a reliable and quantitative means of pathogen detection, making it highly suitable for practical applications in the food industry. The accuracy and efficiency of such sensors have made them indispensable in early detection systems, helping to mitigate risks associated

Table 1 Patents granted in last few years for nanobiosensors with applications focused on food industry

Nanomaterial	Biological component	Application	Patent number
Fluorescence nanomaterial	Antibodies	Quantitative estimation of pathogens	CN110018303A
Nanomaterial electrode	Enzyme and antibodies	Detecting viruses and bacterial pathogens	US20240345066A1
Nanocrystalline diamond layer	Pathogen receptors	Pathogen detector	US20240345089A1
Nanotubes	DNA-probe	Pathogen detection	WO2010039941A9
Fluorescence nanomaterial	Aptamer	Antibiotic detection	CN104502585B
Gold nanograin	Aptamer	Antibiotic detection	CN104502585A
Ag-Fe_3O_4 nanoparticles	Enzyme	Organophosphorus pesticide detection	CN110596218A
Microfluidic chip	Antibodies	Detection of pesticide dithiocarbamate	CN109946256A
Fluorescent carbon dots	Enzyme	Insecticide	CN118726535A
Nanoparticles	Bacterial indicator	Food spoilage	US20240310348A1

with contaminated food products. Another notable class of nanobiosensors employs antibody affinity-based detection mechanisms [84]. These sensors capitalize on the highly specific immunogenic response generated by the interaction of primary antibodies with secondary antibodies in the presence of a target molecule [85]. This specificity enhances the sensitivity of the nanobiosensors, making them capable of detecting minute concentrations of pathogens or toxins. The optical signals produced by these interactions can be easily captured using a smartphone equipped with a camera, offering a user-friendly and portable solution for on-site analysis [7]. This integration of nanotechnology with everyday devices represents a significant leap forward in making sophisticated diagnostic tools accessible to non-specialist users.

In addition to antibody-based nanobiosensors, advancements have been made in utilizing single-stranded DNA molecules anchored to nanomaterials. These DNA-functionalized nanobiosensors have been patented for their ability to detect pathogens with remarkable speed and accuracy [86]. By targeting specific DNA sequences of pathogens, these sensors can rapidly identify contamination, enabling timely interventions in food processing and distribution chains. The rapid response time and high precision of these systems make them invaluable in maintaining food safety standards.

Beyond pathogen detection, nanobiosensors have shown exponential growth in their application for detecting toxins and biotoxins. This includes the identification of harmful antibiotics and allergens, which are critical for ensuring the safety of food for human consumption [87]. Aptamer-functionalized nanoparticles, for instance, serve as molecular probes that can bind with biotoxins or antibiotics with exceptional specificity. These nanobiosensor chips are designed to provide rapid and sensitive detection, making them ideal for real-time monitoring of food contaminants. The advent of colorimetric nanobiosensors for antibiotic detection has further simplified this process, allowing for visual identification of contaminants without the need for advanced instrumentation.

The scope of nanobiosensors extends beyond safety concerns to include quality assessment in food products [12]. Patented technologies have been developed to track changes in nutritional and organoleptic properties, detect fungal pathogens, and identify pesticide residues and other chemical agents. These sensors often employ cutting-edge materials, such as fluorescent nanoparticles, DNA aptamers, or immobilized enzymes, to achieve high sensitivity and reliability. For instance, enzyme-immobilized nanobiosensors have been used to measure specific biochemical changes in food products, offering insights into their freshness and nutritional value.

The increasing number of patents in this field reflects the remarkable progress in nanobiosensor technology and its integration into the agri-food industry (Table 1). These innovations are not only enhancing food safety but also contributing to sustainable practices by reducing waste and improving resource efficiency. As the field continues to evolve, nanobiosensors are poised to play a pivotal role in shaping the future of food quality and safety monitoring, addressing challenges with innovative, efficient, and scalable solutions.

5 Commercially Available Nanobiosensors

Ensuring the safety, authenticity, and quality of food products is crucial for food processing sector. The food industry is constantly under pressure to meet stringent regulatory standards, cater to consumer demands for transparency, and minimize risks associated with contamination [88]. Traditional techniques, such as liquid chromatography-mass spectrometry, polymerase chain reactions, and enzyme-linked immunoassays, while accurate, are highly laborious, time-consuming, often requires expensive equipment and unsuitable for on-site testing. In response to these challenges, the bridging between biology and nanotechnology has led to the development of nanobiosensors, devices that integrate biological recognition elements with nanomaterials [89]. Nanobiosensors demonstrate exceptional performance and the capability to 'nano-tune' diverse features to get optimal sensitivity and detection limits [88]. These nanobiosensors play a crucial role in monitoring food additives, mycotoxins, toxins, microbial contamination, nutritional content, allergens, pesticide residues, plant diseases, environmental factors, and genetically modified organisms [90]. The working principle of biosensors is generally categorized in four main steps [91]:

- Recognition Element: Biosensors employ a biological molecule, such as an enzyme, antibody, or DNA/RNA strand, that specifically interacts with the target analyte.
- Transducer: Nanomaterials such as gold nanoparticles, carbon nanotubes, graphene, or quantum dots convert the biorecognition event into a quantifiable signal (electrical, optical, or mechanical).
- Signal Processor: The detected signal is amplified, processed, and displayed for interpretation.

5.1 Pathogen Detection

The Foodborne pathogens such as *Salmonella, E. coli, Staphylococcus aureus, Campylobacter, Clostridium botulinum, Clostridium perfringens,* and *Listeria* are leading causes of foodborne illnesses [92]. The Biofilm producing microbes get themselves attach to storage surfaces, pipelines and equipment vessels, highlighting concerns regarding food safety in a food processing unit. Hence, rapid detection is crucial to prevent outbreaks. Numerous cost-effective and efficient nanosensors are currently available that can detect microbial growth in minutes or even seconds if needed urgently [93].

For instance, the food processing sector has lately employed magnetic nanoparticles to identify foodborne diseases. By interacting with food pathogens, the magnetic nanopartilces develops nanobioluminescent spray, which creates a visual impact that makes packed meals identifiable. This method identified various food pathogen in industries by combining magnetic relaxation data with fluorescence readings [94]. Pathogenic *E. coli* O157:H7 was quickly identified by multiparametric magneto-fluorescent nanosensors (MfnS) and magneto-plasmonic nanosensors (MPnS) [95]. Similarly, NanoBiT® detection platform, developed by Promega Corporation, utilizes a bioluminescence-based approach for detecting bacterial contamination in food products with high sensitivity and specificity [96]. The isothermal DNA amplification platform named, Veriflow® from Invisible Sentinel, employs gold nanoparticles for pathogen detection in food matrices, providing results within hours [97]. Another kit titled, BioScout Pathogen Detection Kit integrates quantum dots for fluorescence-based detection of bacterial pathogens, enabling high sensitivity in real-time applications.

5.2 Mycotoxin Analysis

Mycotoxins like aflatoxins and ochratoxin A, produced by fungi, are potent carcinogens and immunosuppressants posing significant health and economic risks worldwide. Rapid and accurate detection of these toxins is critical in the food industry to ensure safety and compliance with regulatory standards [98]. Several commercial biosensors are being tailored to detect a wide range of mycotoxins, offering effective alternatives to traditional methods like chromatography or mass spectrometry. For example, aptamer-based nanobiosensors from Biorex Diagnostics utilized graphene oxide with aptamers for selective and sensitive detection of multiple mycotoxins in complex food matrices, including spices and nuts [99]. In a similar manner, the RIDASCREENN fast mycotoxins test employs gold nanoparticle-based immunoassays for detecting aflatoxins and other toxins in cereals and nuts [100].

5.3 *Pesticide Residue Detection*

Pesticides used in agriculture can leave harmful residues in food products. Advanced sensor systems have been designed to detect pests, pathogens, and pesticide levels, enabling precise adjustments to the application of insecticides for effective crop management [101]. Additionally, nanosensors incorporated into platforms can oversee the controlled release of fertilizers and pesticides using nanoscale carriers [102]. This approach minimizes overuse, enhances productivity, and reduces waste. NanoSPR sensor from SPRinX Technologies utilized a surface plasmon resonance-based biosensor using gold nanostructures for detecting organophosphates in fruits and vegetables [103]. Similarly, Sensagri portable sensor combines graphene electrodes with enzymatic assays to offer a portable, on-site pesticide monitoring. Their compact design, ease of use, and rapid analysis time make them ideal for routine screening in food processing and supply chains. Furthermore, the recent developments integrating nanomaterials and smartphone-based platforms have expanded their applicability, enabling on-site testing with enhanced accuracy and lower detection limits [104]. These advancements are driving the adoption of biosensors as indispensable tools in the global effort to ensure food safety and compliance with regulatory standards.

5.4 *Allergen Identification*

Commercially available biosensors are necessary for allergen identification and food labelling in the food industry to safeguard consumer safety and comply with the stringent food regulatory standards. Devices such as enzyme-linked immunosorbent assay (ELISA)-based kits, including RIDASCREEN® and Neogen's Veratox®, are extensively utilized for detecting common allergens like gluten, peanuts, and milk [105]. Allergen management is critical for food labelling and consumer safety. The Biosensia's Allergen Sense uses quantum dots for fluorescence detection of allergenic proteins in complex food matrices [106]. The recent advancements, such as portable and real-time biosensors, are gaining attraction due to their ability to offer on-site testing with rapid results. Innovations integrating nanotechnology and electrochemical platforms have enhanced detection limits and specificity, addressing the challenges of complex food matrices [88].

SPR-based biosensors are increasingly being employed for multi-allergen detection. For instance, devices incorporating microarrays with optical transducers can detect up to 12 allergens, including cashew proteins (Ana o 1, Ana o 2, Ana o 3), with high precision. Such systems can achieve detection limits as low as 0.4 mg/kg in cookies and 0.9 mg/kg in chocolate samples, offering a robust solution for food safety monitoring [107]. Portable allergen biosensors, such as the Nima gluten sensor, have gained popularity among consumers for on-the-go testing. These devices use antibody-based technology to provide results within minutes, ensuring real-time allergen identification. Their compact design and user-friendly interface make them a valuable tool for individuals with food allergies and small-scale producers [108]. The continued development and integration of biosensor in the food supply chain hold promise for enhancing public health and ensuring transparency in food labelling [88].

5.5 *Monitoring Food Freshness*

The biosensors are crucial to monitor freshness and quality of the food products in food industries and provide effective solution for real-time assessment of spoilage and shelf life. These biosensors detect biochemical changes such as volatile organic compounds (VOCs), pH variations, or specific metabolites, which are indicators of microbial growth or degradation [109]. The FreshTag™ sensors by BluWrap use nanostructured polymer films to detect VOCs associated with spoilage in meat and seafood [110]. Another sensor named SmartLabels by Insignia Technologies Incorporates colorimetric nanobiosensors, to indicate the freshness of food products [111]. Biosensors like graphene oxide-based platforms and fluorescence sensors are now available commercially by several companies to monitor storage environments and providing actionable insights on food freshness during transit and storage [112]. The ToxinGuard® biosensor is specifically designed for detecting microbial toxins and spoilage in meat and seafood [113]. It utilizes advanced electrochemical sensing to identify spoilage markers with high sensitivity and precision. Apart from these, the portable and low-cost, paper-based biosensors are also employed to detect spoilage in food products by detecting ammonia and sulphides. These are suitable for real-time monitoring in supply chains, reducing food waste and ensuring product quality.

6 Nanobiosensor Challenges for Food Industry Applications/Technical Challenges

The nanobiosensors have revolutionized several industries, including the food sector by providing promising solutions for food safety, freshness, and quality control [88]. These nanosensors combine biological recognition elements (such as antibodies, aptamers, or enzymes) with nanotechnology to provide rapid, sensitive, and specific detection of foodborne pathogens, spoilage indicators, allergens, and contaminants [114]. Though, nanomaterial was proven to be the most efficient matrices for biosensor fabrications, there are a number of issues that make their commercialisation still a tough predicament [89]. The several technical challenges that need to be addressed for their widespread use in the food industries are:

- Attaining the required selectivity as well as sensitivity to detect target analytes even at very low concentration, especially in complex food products is one of the important technical challenged faced by the food industries. The high levels of interfering substances like fats, proteins, sugars, and salts make it difficult to ensure that the sensors can selectively target the analytes of interest without interference [115]. This issue requires the development of highly specific recognition elements and advanced materials that can distinguish between the target and non-target components effectively [31].
- The long-term commercial viability as well as reproducibility of biosensor is another factor to keep in mind for the real-world applications, where environmental conditions fluctuate rapidly [14]. Nanomaterials are highly susceptible to degradation due to factors such as temperature, humidity, and pH fluctuations, which can affect the performance of biosensors. Therefore, there is a need to develop robust nanobiosensor platforms with enhanced stability under various food processing and storage conditions [116].
- The effective integration of biosensors into existing food packaging and storage systems is essential for their potential use in the food industries to provide real-time monitoring of food quality and safety during transportation and storage [109]. Also, the integration of biosensors should not alter the food's sensory attributes, such as appearance, taste, or texture. Additionally, the biocompatibility of nanobiosensors with food products and packaging materials must be thoroughly evaluated to avoid any adverse effects on food safety [117].
- The cost of manufacturing remains a significant barrier to the widespread use of biosensors in the food industry [91]. The synthesis of high-quality nanomaterials, functionalization with biological recognition elements, and fabrication of sensor devices often require expensive reagents and complex processes. Advancements in nanomaterial synthesis and sensor fabrication techniques are necessary to make nanobiosensors more affordable and accessible to the food industry [109].
- Another challenge for the significant application of nanobiosensors in the food industry is the interpretation of data. Compared with traditional methods, nanobiosensors often generate complex signals that require advanced algorithms or software to interpret for real-time monitoring [118]. The lack of standardization

in biosensor performance, calibration, and data interpretation presents a hurdle for their widespread adoption across the food industry. To address this, standardized protocols for sensor calibration and data analysis need to be developed, allowing for consistent and reliable results [88, 91].

Despite the significant advancements in the nanobiosensor technology, several technical challenges remain in making them suitable for large-scale application in the food industry. Overcoming these challenges will require interdisciplinary efforts involving materials science, biotechnology, engineering, and regulatory agencies [109].

7 Future Prospects and Conclusion

Nanobiosensors are poised to revolutionize food industry, by enhancing precision and sustainability, critical components of a circular economy. The application ranging from pathogen detection, chemical residue detection to real-time monitoring of freshness and nutrient levels during and after production, present them as the ultimate saviour. These nanobiosensors offer feasible solutions for reducing waste and optimizing resource utilization. A number of nanobiosensors are being utilized in the food industry with demonstrated remarkable sensitivity and specificity in detecting chemical residues and pathogens, paving the way for widespread industrial adoption. Future innovations are likely to focus on scaling production of eco-friendly nanomaterials, such as cellulose nanocrystals or bio-derived quantum dots, to reduce the environmental footprint of sensor fabrication. The integration of nanobiosensors with machine learning algorithms and IoT systems could enable predictive monitoring, enhancing their utility in precision agriculture and post-harvest management. Moreover, collaboration between regulatory agencies and industry stakeholders will be critical to establish standardized frameworks for widespread implementation. In conclusion, advancements in nanobiosensors hold transformative potential for promoting a circular economy in food systems. The increasing number of patents filed and granted is testament of their capability to drive global sustainability goals. Furthermore, for the futuristic growth, interdisciplinary research, policy integration, and public–private partnerships will be essential to harness the full potential of this ground-breaking technology.

References

1. Cederberg C, Sonesson U (2011) Global food losses and food waste: extent, causes and prevention; study conducted for the international congress save food! At interpack. Food and Agriculture Organization of the United Nations, Rome

2. Rana L et al (2024) Nexus between nanotechnology and agricultural production systems: challenges and future prospects. Discov Appl Sci 6(11):555. https://doi.org/10.1007/s42452-024-06265-7
3. Mondal R et al (2022) Potential of nanobiosensor in sustainable agriculture: the state-of-art. Heliyon 8(12):e12207. https://doi.org/10.1016/j.heliyon.2022.e12207
4. Mathivanan S (2021) Perspectives of nano-materials and nanobiosensors in food safety and agriculture. In: Novel nanomaterials. IntechOpen. https://doi.org/10.5772/intechopen.95345
5. Zhang Z, Liu X, Zhou H, Xu S, Lee C (2024) Advances in machine-learning enhanced nanosensors: from cloud artificial intelligence toward future edge computing at chip level. Small Struct 5(4):2300325. https://doi.org/10.1002/sstr.202300325
6. Arellano-Vidal CL, Govan JE (2024) Machine learning techniques for improving nanosensors in agroenvironmental applications. Agronomy 14(2):2. https://doi.org/10.3390/agronomy14020341
7. Thakur M, Wang B, Verma ML (2022) Development and applications of nanobiosensors for sustainable agricultural and food industries: recent developments, challenges and perspectives. Environ Technol Innov 26:102371. https://doi.org/10.1016/j.eti.2022.102371
8. Miguel-Rojas C, Pérez-de-Luque A (2023) Nanobiosensors and nanoformulations in agriculture: new advances and challenges for sustainable agriculture. Emerg Top Life Sci 7(2):229. https://doi.org/10.1042/ETLS20230070
9. Yang T, Huang H, Zhu F, Lin Q, Zhang L, Liu J (2016) Recent progresses in nanobiosensing for food safety analysis. Sensors 16(7):7. https://doi.org/10.3390/s16071118
10. Kaur G, Bhari R, Kumar K (2024) Nanobiosensors and their role in detection of adulterants and contaminants in food products. Crit Rev Biotechnol 44(4):547–561. https://doi.org/10.1080/07388551.2023.2175196
11. Chelliah R et al (2021) Development of nanosensors based intelligent packaging systems: food quality and medicine. Nanomaterials 11(6):1515. https://doi.org/10.3390/nano11061515
12. Guruprasath N, Sankarganesh P, Adeyeye SO, Babu AS, Parthasarathy V (2024) Review on emerging applications of nanobiosensor in food safety. J Food Sci 89(7):3950–3972. https://doi.org/10.1111/1750-3841.17149
13. Yata VK, Tiwari BC, Ahmad I (2017) Research trends and patents in nano-food and agriculture. In: Ranjan S, Dasgupta N, Lichtfouse E (eds) Nanoscience in food and agriculture, vol 5. Springer, Cham, pp 1–20. https://doi.org/10.1007/978-3-319-58496-6_1
14. Yang T, Duncan TV (2021) Challenges and potential solutions for nanosensors intended for use with foods. Nat Nanotechnol 16(3):251–265. https://doi.org/10.1038/s41565-021-00867-7
15. Huang X, Zhu Y, Kianfar E (2021) Nano biosensors: properties, applications and electrochemical techniques. J Mater Res Technol 12:1649–1672. https://doi.org/10.1016/j.jmrt.2021.03.048
16. Darwish MA, Abd-Elaziem W, Elsheikh A, Zayed AA (2024) Advancements in nanomaterials for nanosensors: a comprehensive review. Nanoscale Adv 6(16):4015–4046. https://doi.org/10.1039/D4NA00214H
17. Kopittke PM, Menzies NW, Wang P, McKenna BA, Lombi E (2019) Soil and the intensification of agriculture for global food security. Environ Int 132:105078. https://doi.org/10.1016/j.envint.2019.105078
18. Shivashakarappa K, Reddy V, Tupakula VK, Farnian A, Vuppula A, Gunnaiah R (2022) Nanotechnology for the detection of plant pathogens. Plant Nano Biol 2:100018. https://doi.org/10.1016/j.plana.2022.100018
19. Kulabhusan PK, Tripathi A, Kant K (2022) Gold nanoparticles and plant pathogens: an overview and prospective for biosensing in forestry. Sensors 22(3):1259. https://doi.org/10.3390/s22031259
20. Bharti A, Jain U, Chauhan N (2024) From lab to field: nano-biosensors for real-time plant nutrient tracking. Plant Nano Biol 9:100079. https://doi.org/10.1016/j.plana.2024.100079
21. Garg S, Rumjit NP, Roy S (2024) Smart agriculture and nanotechnology: technology, challenges, and new perspective. Adv Agrochem 3(2):115–125. https://doi.org/10.1016/j.aac.2023.11.001

22. Yadav A, Yadav K, Ahmad R, Abd-Elsalam KA (2023) Emerging frontiers in nanotechnology for precision agriculture: advancements, hurdles and prospects. Agrochemicals 2(2):2. https://doi.org/10.3390/agrochemicals2020016
23. Hegde M, Pai P, Shetty MG, Babitha KS (2022) Gold nanoparticle based biosensors for rapid pathogen detection: a review. Environ Nanotechnol Monit Manag 18:100756. https://doi.org/10.1016/j.enmm.2022.100756
24. Kadam US, Hong JC (2022) Advances in aptameric biosensors designed to detect toxic contaminants from food, water, human fluids, and the environment. Trends Environ Analyt Chem 36:e00184. https://doi.org/10.1016/j.teac.2022.e00184
25. Léguillier V, Heddi B, Vidic J (2024) Recent advances in aptamer-based biosensors for bacterial detection. Biosensors 14(5):5. https://doi.org/10.3390/bios14050210
26. Garg N, Ahmad FJ, Kar S (2022) Recent advances in loop-mediated isothermal amplification (LAMP) for rapid and efficient detection of pathogens. Curr Res Microb Sci 3:100120. https://doi.org/10.1016/j.crmicr.2022.100120
27. Nath S (2024) Advancements in food quality monitoring: integrating biosensors for precision detection. Sustain Food Technol 2(4):976–992. https://doi.org/10.1039/d4fb00094c
28. Mahanti NK et al (2024) Enhancing food authentication through E-nose and E-tongue technologies: current trends and future directions. Trends Food Sci Technol 150:104574. https://doi.org/10.1016/j.tifs.2024.104574
29. Botta A, de Donato W, Persico V, Pescapé A (2016) Integration of cloud computing and internet of things: a survey. Fut Gener Comput Syst 56:684–700. https://doi.org/10.1016/j.future.2015.09.021
30. Srinivasan B, Tung S (2015) Development and applications of portable biosensors. SLAS Technol 20(4):365–389. https://doi.org/10.1177/2211068215581349
31. Wang Z, Luo W, Cheng S, Zhang H, Zong J, Zhang Z (2023) Ralstonia solanacearum: a soil borne hidden enemy of plants—research development in management strategies, their action mechanism and challenges. Front Plant Sci 14:1141902. https://doi.org/10.3389/fpls.2023.1141902
32. van Dijk M, Morley T, Rau ML, Saghai Y (2021) A meta-analysis of projected global food demand and population at risk of hunger for the period 2010–2050. Nat Food 2(7):494–501. https://doi.org/10.1038/s43016-021-00322-9
33. Saha B et al (2024) Sustainable Nano solutions for global food security and biotic stress management. Plant Nano Biol 9:100090. https://doi.org/10.1016/j.plana.2024.100090
34. Jafar A, Bibi N, Naqvi RA, Sadeghi-Niaraki A, Jeong D (2024) Revolutionizing agriculture with artificial intelligence: plant disease detection methods, applications, and their limitations. Front Plant Sci 15:260. https://doi.org/10.3389/fpls.2024.1356260
35. Jiang L et al (2024) Rapid detection of *Pseudomonas syringae* pv. actinidiae by electrochemical surface-enhanced Raman spectroscopy. Talanta 268:125336. https://doi.org/10.1016/j.talanta.2023.125336
36. Roper JM, Garcia JF, Tsutsui H (2021) Emerging technologies for monitoring plant health in vivo. ACS Omega 6(8):5101–5107. https://doi.org/10.1021/acsomega.0c05850
37. Ackermann J, Metternich JT, Herbertz S, Kruss S (2022) Biosensing with fluorescent carbon nanotubes. Angew Chem Int Edn 61(18):e202112372. https://doi.org/10.1002/anie.202112372
38. Zhan F, Wang T, Iradukunda L, Zhan J (2018) A gold nanoparticle-based lateral flow biosensor for sensitive visual detection of the potato late blight pathogen, *Phytophthora infestans*. Analyt Chim Acta 1036:153–161. https://doi.org/10.1016/j.aca.2018.06.083
39. Khaledian S, Nikkhah M, Shams-bakhsh M, Hoseinzadeh S (2017) A sensitive biosensor based on gold nanoparticles to detect *Ralstonia solanacearum* in soil. J Gen Plant Pathol 83(4):231–239. https://doi.org/10.1007/s10327-017-0721-z
40. Sun Y, Fang L, Wan Y, Gu Z (2018) Pathogenic detection and phenotype using magnetic nanoparticle-urease nanosensor. Sens Actuat B Chem 259:428–432. https://doi.org/10.1016/j.snb.2017.12.095

41. Varshney M, Li Y (2009) Interdigitated array microelectrodes based impedance biosensors for detection of bacterial cells. Biosens Bioelectr 24(10):2951–2960. https://doi.org/10.1016/j.bios.2008.10.001
42. Oshunsanya SO, Nwosu NJ, Li Y (2019) Abiotic stress in agricultural crops under climatic conditions. In: Jhariya MK, Banerjee A, Meena RS, Yadav DK (eds) Sustainable agriculture, forest and environmental management. Springer, Singapore, pp 71–100. https://doi.org/10.1007/978-981-13-6830-1_3
43. Teshome DT, Zharare GE, Naidoo S (2020) The threat of the combined effect of biotic and abiotic stress factors in forestry under a changing climate. Front Plant Sci 11:1009. https://doi.org/10.3389/fpls.2020.601009
44. Hu Y et al (2020) A multifunctional ratiometric electrochemical sensor for combined determination of indole-3-acetic acid and salicylic acid. RSC Adv 10(6):3115–3121. https://doi.org/10.1039/C9RA09951D
45. Choudhary RC et al (2019) Zinc encapsulated chitosan nanoparticle to promote maize crop yield. Int J Biol Macromol 127:126–135. https://doi.org/10.1016/j.ijbiomac.2018.12.274
46. Zungu B, Kamdem-Paumo H, Gaorongwe JL, Tsuene GN, Ruzvidzo O, Katata-Seru L (2023) Zn nutrients-loaded chitosan nanocomposites and their efficacy as nanopriming agents for maize (*Zea mays*) seeds. Front Chem 11:884. https://doi.org/10.3389/fchem.2023.1243884
47. Baz H, Creech M, Chen J, Gong H, Bradford K, Huo H (2020) Water-soluble carbon nanoparticles improve seed germination and post-germination growth of lettuce under salinity stress. Agronomy 10(8):8. https://doi.org/10.3390/agronomy10081192
48. Guo M (2021) Soil health assessment and management: recent development in science and practices. Soil Syst 5(4):4. https://doi.org/10.3390/soilsystems5040061
49. Thangadurai D, Shettar AK, Sangeetha J, Adetunji CO, Islam S, Al-Tawaha ARMS (2021) Chapter 10: nanosensors for detection and evaluation of organic compounds in soil. In: Amrane A, Mohan D, Nguyen TA, Assadi AA, Yasin G (eds) Nanomaterials for soil remediation. Elsevier, pp 205–219. https://doi.org/10.1016/B978-0-12-822891-3.00010-4
50. Torrinha Á, Oliveira TMBF, Islam S, Morais S (2023) Green carbon (nano)materials-based sensors for analysis of hazardous metal ions. In: Green carbon materials for environmental analysis: emerging research and future opportunities, ACS symposium series, vol 1441. American Chemical Society, pp 91–138. https://doi.org/10.1021/bk-2023-1441.ch005
51. Siontorou CG, Georgopoulos KN (2016) A biosensor platform for soil management: the case of nitrites. J Clean Product 111:133–142. https://doi.org/10.1016/j.jclepro.2015.07.038
52. Xu L et al (2022) Recent advances in rapid detection techniques for pesticide residue: a review. J Agric Food Chem 70(41):13093–13117. https://doi.org/10.1021/acs.jafc.2c05284
53. Talari FF, Bozorg A, Faridbod F, Vossoughi M (2021) A novel sensitive aptamer-based nanosensor using rGQDs and MWCNTs for rapid detection of diazinon pesticide. J Environ Chem Eng 9(1):104878. https://doi.org/10.1016/j.jece.2020.104878
54. Talan A, Mishra A, Eremin SA, Narang J, Kumar A, Gandhi S (2018) Ultrasensitive electrochemical immuno-sensing platform based on gold nanoparticles triggering chlorpyrifos detection in fruits and vegetables. Biosens Bioelectr 105:14–21. https://doi.org/10.1016/j.bios.2018.01.013
55. Naveen-Prasad S, Bansal V, Ramanathan R (2021) Detection of pesticides using nanozymes: trends, challenges and outlook. TrAC Trends Analyt Chem 144:116429. https://doi.org/10.1016/j.trac.2021.116429
56. Bucur B, Munteanu F-D, Marty J-L, Vasilescu A (2018) Advances in enzyme-based biosensors for pesticide detection. Biosensors 8(2):2. https://doi.org/10.3390/bios8020027
57. Simonian AL et al (2001) Enzyme-based biosensor for the direct detection of fluorine-containing organophosphates. Analyt Chim Acta 442(1):15–23. https://doi.org/10.1016/S0003-2670(01)01131-X
58. Songa EA, Okonkwo JO (2016) Recent approaches to improving selectivity and sensitivity of enzyme-based biosensors for organophosphorus pesticides: a review. Talanta 155:289–304. https://doi.org/10.1016/j.talanta.2016.04.046

59. Alengebawy A, Abdelkhalek ST, Qureshi SR, Wang M-Q (2021) Heavy metals and pesticides toxicity in agricultural soil and plants: ecological risks and human health implications. Toxics 9(3):3. https://doi.org/10.3390/toxics9030042
60. Afzal A, Mahreen N (2024) Emerging insights into the impacts of heavy metals exposure on health, reproductive and productive performance of livestock. Front Pharmacol 15:137. https://doi.org/10.3389/fphar.2024.1375137
61. Biswas HS, Poddar S (2024) Heavy metals in foodstuffs: presence, bioaccumulation, origins, health risk, and remediation. In: Biosorption processes for heavy metal removal. IGI Global Scientific Publishing, pp 85–99. https://doi.org/10.4018/979-8-3693-1618-4.ch004
62. Angulo-Bejarano PI, Puente-Rivera J, Cruz-Ortega R (2021) Metal and metalloid toxicity in plants: an overview on molecular aspects. Plants 10(4):4. https://doi.org/10.3390/plants10040635
63. Dong B et al (2019) Multiple methods for the identification of heavy metal sources in cropland soils from a resource-based region. Sci Total Environ 651:3127–3138. https://doi.org/10.1016/j.scitotenv.2018.10.130
64. Xu X, Yang S, Wang Y, Qian K (2022) Nanomaterial-based sensors and strategies for heavy metal ion detection. Green Analyt Chem 2:100020. https://doi.org/10.1016/j.greeac.2022.100020
65. Nandhakumar M, Thangaian DT, Kasi N (2023) Topical progress of gold nanoparticles towards diverse metal ion sensing through optical spectrometry and electrochemical techniques: a short review. J Mater Res Technol 22:1185–1209. https://doi.org/10.1016/j.jmrt.2022.11.182
66. Gezahegn TF, Ambaye AD, Mekoyete TM, Malefane ME, Oyedotun KO, Mokrani T (2024) Breakthroughs in nanostructured-based chemical sensors for the detection of toxic metals. Talanta Open 10:100354. https://doi.org/10.1016/j.talo.2024.100354
67. Sreekanth SP et al (2021) Multi-walled carbon nanotube-based nanobiosensor for the detection of cadmium in water. Environ Res 197:111148. https://doi.org/10.1016/j.envres.2021.111148
68. Hao Y, Ji F, Li T, Tian M, Han X, Chai F (2024) Portable smartphone platform utilizing AIE-featured carbon dots for multivariate visual detection for Cu^{2+}, Hg^{2+} and BSA in real samples. Food Chem 446:138843. https://doi.org/10.1016/j.foodchem.2024.138843
69. Jin B et al (2018) Lateral flow aptamer assay integrated smartphone-based portable device for simultaneous detection of multiple targets using upconversion nanoparticles. Sens Actuat B Chem 276:48–56. https://doi.org/10.1016/j.snb.2018.08.074
70. Lew TTS, Park M, Cui J, Strano MS (2021) Plant nanobionic sensors for arsenic detection. Adv Mater 33(1):2005683. https://doi.org/10.1002/adma.202005683
71. Zhang Q, Ying Y, Ping J (2022) Recent advances in plant nanoscience. Adv Sci 9(2):2103414. https://doi.org/10.1002/advs.202103414
72. Yan B, Zhang F, Wang M, Zhang Y, Fu S (2024) Flexible wearable sensors for crop monitoring: a review. Front Plant Sci 15:6074. https://doi.org/10.3389/fpls.2024.1406074
73. Phan TTH, Ngo TMV, Phan H-P (2024) Flexible mechanical sensors for plant growth monitoring: an emerging area for smart agriculture. Sensors 24(24):24. https://doi.org/10.3390/s24247995
74. Wang S et al (2024) Highly stretchable, robust, and resilient wearable electronics for remote, autonomous plant growth monitoring. Device 2(4):100322. https://doi.org/10.1016/j.device.2024.100322
75. Li D, Li G, Li J, Xu S (2022) Wearable crop sensor based on nano-graphene oxide for noninvasive real-time monitoring of plant water. Membranes 12(4):4. https://doi.org/10.3390/membranes12040358
76. Zhang W, Sun Q, Zhang X, Yuan W, Wu J (2023) A laser-induced graphene-based sensor modified with CeO_2 for determination of organophosphorus pesticides with improved performance. Sensors 23(23):23. https://doi.org/10.3390/s23239605
77. Wang L, Li M, Li B, Wang M, Zhao H, Zhao F (2023) Electrochemical sensor based on laser-induced graphene for carbendazim detection in water. Foods 12(12):12. https://doi.org/10.3390/foods12122277

78. Li Z et al (2021) Real-time monitoring of plant stresses via chemiresistive profiling of leaf volatiles by a wearable sensor. Matter 4(7):2553–2570. https://doi.org/10.1016/j.matt.2021.06.009
79. Lee S et al (2024) Wearable volatile organic compound sensors for plant health monitoring. Adv Sustain Syst 8(9):2300634. https://doi.org/10.1002/adsu.202300634
80. Kim JJ, Fan R, Allison LK, Andrew TL (2020) On-site identification of ozone damage in fruiting plants using vapor-deposited conducting polymer tattoos. Sci Adv 6(36):eabc3296. https://doi.org/10.1126/sciadv.abc3296
81. Wesoły M, Przewodowski W, Ciosek-Skibińska P (2023) Electronic noses and electronic tongues for the agricultural purposes. TrAC Trends Analyt Chem 164:117082. https://doi.org/10.1016/j.trac.2023.117082
82. Singh V, Sharma N, Singh S (2020) A review of imaging techniques for plant disease detection. Artif Intell Agricult 4:229–242. https://doi.org/10.1016/j.aiia.2020.10.002
83. Valenzuela-Amaro HM et al (2023) Emerging applications of nanobiosensors in pathogen detection in water and food. Biosensors 13(10):10. https://doi.org/10.3390/bios13100922
84. Nangare SN, Patil PO (2021) Affinity-based nanoarchitectured biotransducer for sensitivity enhancement of surface plasmon resonance sensors for in vitro diagnosis: a review. ACS Biomater Sci Eng 7(1):2–30. https://doi.org/10.1021/acsbiomaterials.0c01203
85. Guliy OI, Evstigneeva SS, Khanadeev VA, Dykman LA (2023) Antibody phage display technology for sensor-based virus detection: current status and future prospects. Biosensors 13(6):6. https://doi.org/10.3390/bios13060640
86. Tessaro L, Aquino A, de Almeida-Rodrigues P, Joshi N, Ferrari RG, Conte-Junior CA (2022) Nucleic acid-based nanobiosensor (NAB) used for salmonella detection in foods: a systematic review. Nanomaterials 12(5):5. https://doi.org/10.3390/nano12050821
87. Jigyasa S, Rajput JK (2022) Nanomaterial-based sensors as potential remedy for detection of biotoxins. Food Control 135:108686. https://doi.org/10.1016/j.foodcont.2021.108686
88. Neethirajan S, Weng X, Tah A, Cordero JO, Ragavan KV (2018) Nano-biosensor platforms for detecting food allergens: new trends. Sens Bio-Sens Res 18:13–30. https://doi.org/10.1016/j.sbsr.2018.02.005
89. Kundu M, Krishnan P, Kotnala RK, Sumana G (2019) Recent developments in biosensors to combat agricultural challenges and their future prospects. Trends Food Sci Technol 88:157–178. https://doi.org/10.1016/j.tifs.2019.03.024
90. Arora K (2018) Advances in nano based biosensors for food and agriculture. In: Nanotechnology, food security and water treatment. Springer, Cham, pp 1–52. https://doi.org/10.1007/978-3-319-70166-0_1
91. Chadha U et al (2022) Recent progress and growth in biosensors technology: a critical review. J Ind Eng Chem 109:21–51. https://doi.org/10.1016/j.jiec.2022.02.010
92. Nehra M, Kumar V, Kumar R, Dilbaghi N, Kumar S (2022) Current scenario of pathogen detection techniques in agro-food sector. Biosensors 12(7):7. https://doi.org/10.3390/bios12070489
93. Xu L, Bai X, Bhunia AK (2021) Current state of development of biosensors and their application in foodborne pathogen detection. J Food Protect 84(7):1213–1227. https://doi.org/10.4315/JFP-20-464
94. Xiao F, Li W, Xu H (2022) Advances in magnetic nanoparticles for the separation of foodborne pathogens: recognition, separation strategy, and application. Compr Rev Food Sci Food Saf 21:4478–4504. https://doi.org/10.1111/1541-4337.13023
95. Banerjee T et al (2016) Multiparametric magneto-fluorescent nanosensors for the ultrasensitive detection of *Escherichia coli* O157:H7. ACS Infect Dis 2(10):667–673. https://doi.org/10.1021/acsinfecdis.6b00108
96. Pradanas-González F et al (2024) Biosensing strategies using recombinant luminescent proteins and their use for food and environmental analysis. Anal Bioanal Chem 416(30):7205–7224. https://doi.org/10.1007/s00216-024-05552-x
97. Ripolles-Avila C, Martínez-Garcia M, Capellas M, Yuste J, Fung DYC, Rodríguez-Jerez J-J (2020) From hazard analysis to risk control using rapid methods in microbiology: a practical

approach for the food industry. Compreh Rev Food Sci Food Saf 19(4):1877–1907. https://doi.org/10.1111/1541-4337.12592
98. Rovina K et al (2019) Recent biosensors technologies for detection of mycotoxin in food products. In: Mycotoxins and food safety. IntechOpen. https://doi.org/10.5772/intechopen.89022
99. Jubeen F et al (2024) Mycotoxins detection in food using advanced, sensitive and robust electrochemical platform of sensors: a review. Sens Actuat A Phys 367:115045. https://doi.org/10.1016/j.sna.2024.115045
100. Tripathi P, Malik S, Nara S (2016) Immunoassay methods in food analysis. In: Food safety. Wiley, pp 435–453. https://doi.org/10.1002/9781119160588.ch12
101. Thakur MS, Ragavan KV (2013) Biosensors in food processing. J Food Sci Technol 50(4):625–641. https://doi.org/10.1007/s13197-012-0783-z
102. Verma N, Bhardwaj A (2015) Biosensor technology for pesticides: a review. Appl Biochem Biotechnol 175(6):3093–3119. https://doi.org/10.1007/s12010-015-1489-2
103. NanoSPR Devices (2025) NanoSPR devices. https://nanospr.com/. Accessed on 11 Jan 2025
104. Lu Y, Shi Z, Liu Q (2019) Smartphone-based biosensors for portable food evaluation. Curr Opin Food Sci 28:74–81. https://doi.org/10.1016/j.cofs.2019.09.003
105. Park DL et al (2005) Performance tested method multiple laboratory validation study of ELISA-based assays for the detection of peanuts in food. J AOAC Int 88(1):156–160
106. Shahin H, Walczak R (2013) Mobile phone as a fluorescence reader. Opt Appl
107. Zhou J et al (2019) Surface plasmon resonance (SPR) biosensors for food allergen detection in food matrices. Biosens Bioelectr 142:111449. https://doi.org/10.1016/j.bios.2019.111449
108. Zhang J et al (2019) An integrated, accurate, rapid, and economical handheld consumer gluten detector. Food Chem 275:446–456. https://doi.org/10.1016/j.foodchem.2018.08.117
109. Sobhan A, Muthukumarappan K, Wei L (2021) Biosensors and biopolymer-based nanocomposites for smart food packaging: challenges and opportunities. Food Packag Shelf Life 30:100745. https://doi.org/10.1016/j.fpsl.2021.100745
110. BluWrap (2025) DigitalFoodLab. https://www.digitalfoodlab.com/en/foodtech-database/bluwrap. Accessed on 11 Jan 2025
111. Wang C, Yusufu D, Mills A (2019) A smart adhesive 'consume within' (CW) indicator for food packaging. Food Packag Shelf Life 22:100395. https://doi.org/10.1016/j.fpsl.2019.100395
112. Raghavan VS et al (2021) Emerging graphene-based sensors for the detection of food adulterants and toxicants: a review. Food Chem 355:129547. https://doi.org/10.1016/j.foodchem.2021.129547
113. Griesche C, Baeumner AJ (2020) Biosensors to support sustainable agriculture and food safety. TrAC Trends Analyt Chem 128:115906. https://doi.org/10.1016/j.trac.2020.115906
114. Raghu HV, Parkunan T, Kumar N (2020) Application of nanobiosensors for food safety monitoring. In: Dasgupta N, Ranjan S, Lichtfouse E (eds) Environmental nanotechnology. Springer, Cham, pp 93–129. https://doi.org/10.1007/978-3-030-26668-4_3
115. Biosensors for Sustainable Food (2016) New opportunities and technical challenges. https://shop.elsevier.com/books/biosensors-for-sustainable-food-new-opportunities-and-technical-challenges/scognamiglio/978-0-444-63579-2. Accessed on 11 Jan 2025
116. Neethirajan S, Ragavan V, Weng X, Chand R (2018) Biosensors for sustainable food engineering: challenges and perspectives. Biosensors 8(1):1. https://doi.org/10.3390/bios8010023
117. Osmólska E, Stoma M, Starek-Wójcicka A (2022) Application of biosensors, sensors, and tags in intelligent packaging used for food products—a review. Sensors 22(24):24. https://doi.org/10.3390/s22249956
118. Singh A et al (2021) Recent advances in electrochemical biosensors: applications, challenges, and future scope. Biosensors 11(9):9. https://doi.org/10.3390/bios11090336

Challenges and Limitations of Implementing the Nanobiosensors in Food Sector

Baneesh Patial, Pankaj Chelak, Ajay Bansal, Renu Gupta, and Susheel K. Mittal

Abstract The chapter Challenges and Limitations in nanobiosensors Implementation in the Food Sector extends a comprehensive investigation of the complications in including nanobiosensor technology in the systems of food safety and quality monitoring. This chapter strives to acknowledge and illustrate the technical difficulties like stability, repeatability, and scalability of nanobiosensors inclusive of the financial and legal restrictions, although profuse nanobiosensors have great sensitivity, specificity, and fast detection of pollutants. Four primary sections define the chapter. The first part presents the elementary ideas of nanobiosensors and their benefits above established detection strategies. The second part explores the technological hardships, including the contextual influences on sensor performance, material compatibility, and signal interference. The third part inspects the economic issues by prioritizing the factor of cost-effectiveness and market-ready nature of nanobiosensor technologies. The last part of the chapter highlights the requirement for detailed protocols and procedures to assure dependability and safety, thus addressing the standardizing and regulating concerns. This chapter intends to notify the researchers, industry stakeholders, and legislators about the crucial domains that demand attention and improvement for the effective integration of these advanced technologies into the food sector by presenting a balanced view of the current status of nanobiosensor implementation.

Keywords Nanobiosensors · Challenges · Limitations · Food sector

B. Patial · A. Bansal (✉) · R. Gupta
Department of Chemical Engineering, Dr. B R Ambedkar National Institute of Technology, Jalandhar, Punjab, India
e-mail: bansala@nitj.ac.in

P. Chelak · A. Bansal · R. Gupta
Centre for Energy and Environment, Dr. B R Ambedkar National Institute of Technology, Jalandhar, Punjab, India

S. K. Mittal
I.K. Gujral Punjab Technical University, Jalandhar, Punjab, India

D. Sillu et al. (eds.), *Nanobiosensors for the Food Industry*, Smart Nanomaterials Technology, https://doi.org/10.1007/978-981-95-0136-6_15

1 Introduction

The incorporation of nanobiosensors in food safety and quality control embodies a critical innovation in the detection of pathogens with uniformity, specificity, and rapid response time [1, 2]. Technology in this era has the capability to transform the way the food industry secures the quality of its products and safety, offering emergency care outside the conventional methods [1, 3]. In spite of possessing potential characteristics like stability, reproducibility, and scalability of the sensors, the extensive use of nanobiosensors in the food industry is restricted due to multiple problems, including issues related to equipment, signal interference, and environmental effects [3–5]. Numerous technical, economic, and regulatory issues display the complexity of using nanobiosensors. The cost-effectiveness and market-ready nature of nanobiosensors have become a matter of concern, affecting their commercial potential among the stakeholders. In addition, the lack of a standard regulatory framework also complicates the matter hindering the reliability and safety concerns for using nanobiosensors in food. The constraints that are required to be addressed are the overall analysis. The chapter thus evaluates the core concepts of nanobiosensor technology, highlighting its advantages of being incorporated into the food industry above the use of traditional detection methods, establishing the foundation to understand the business process operations and management, which somehow prevents its wider use. The purpose of this article is to guide researchers, business professionals, and policymakers regarding solutions that enhance the use of nanobiosensors and thus help in creating a safe and reliable food system.

2 Nanobiosensors and Their Benefits Over Conventional Detection

For the identification of specific elements with high sensitivity and specificity, nanobiosensors act as an analytical tool that integrates biological recognition elements with a nanomaterial-based transducer. These devices leverage the unique properties of nanomaterials—such as high surface area, quantum effects, and tunable electrical, optical, and mechanical characteristics—to enhance the performance of biosensors. Nanobiosensors are particularly useful in various fields, including medical diagnostics, environmental monitoring, and food safety, where rapid, accurate, and sensitive detection is crucial.

2.1 *Types of Nanobiosensors*

The nanobiosensors are categorized as first is on the basis of transducer-based (mainly types of nanomaterial used in sensors) and second one is biological recognition element based. Electrochemical nanobiosensors utilize noble metallic nanoparticles of carbonaceous materials (carbon nanotubes and graphene) for amplifying electrochemical signals arising between the analyte of interest and biological recognition elements such as enzymes or antibodies or DNA [6, 7]. Optical Nanobiosensors detect the binding of analyte molecules; optical biosensors rely on changes in fluorescence, surface plasmon resonance, or light scattering that occur upon molecular interaction with a sensing element [8, 9]. Piezoelectric nanobiosensors incorporate nanomaterials with piezoelectric properties to detect mass changes on the sensor surface due to the binding of target molecules [10].

2.2 *Benefits Over Conventional Detection Techniques*

Nanobiosensors provide several advantages compared to conventional detection methods, including improved sensitivity, higher specificity, fast detection, mobility, and reduced sample volume [1–3, 11]. Thanks to their large surface area, nanomaterials can accommodate more binding sites, leading to the ability to detect very low concentrations of target molecules, but the fact that nanomaterials can be conjugated with selective bio-recognition elements, e.g., antibodies or aptamers, offers even better specificity and lower chances of false positives. The very short response time inherent in nanobiosensors due to the very active nanomaterials makes it possible to carry out detection activities almost in real time compared to prolonged procedures necessitated by classical methods such as ELISA or polymerase chain reaction (PCR) [12]. In addition, their relatively small size and combination with microfluidics make it possible for the nanobiosensors to achieve point-of-use testing, thus facilitating decentralization of testing and improving the speed of decision-making. Nanobiosensors are also useful in consuming smaller sample volumes, which is an advantage, especially when the sample volume is limited or hard to acquire [13]. They are highly adaptable, making it possible for them to be used for many purposes, for instance, from pathogens detection in foods to biomarkers for medical diagnosis, and it is also likely to be cheaper over time due to a decrease in the consumption of precursors. Moreover, nanobiosensors offer the potential for continuous, real-time monitoring, providing valuable insights for maintaining food safety and quality throughout the supply chain, something traditional methods typically cannot achieve [14].

2.3 *Applications in the Food Sector*

Food safety and quality control have been positively impacted by nanobiosensors in several key areas of application. They allow the quick and accurate identification of food contaminants such as *Salmonella*, *E. coli*, and *Listeria*, thus preventing epidemics and protecting populations [15, 16]. Moreover, nanobiosensors have also

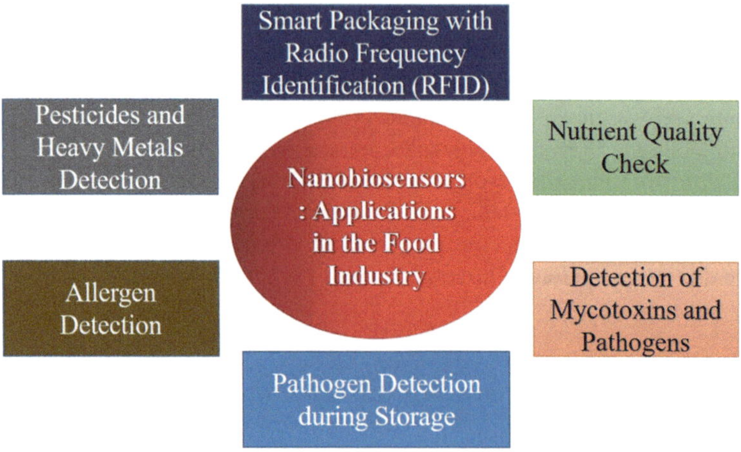

Fig. 1 Applications of nanobiosensors in the food sector

proven to be helpful in determining the presence of any harmful elements or certain food allergens and alerting people in the process [17]. In addition, the devices are also essential in the processes that are geared toward the quality management of the foods, through checking of freshness and conditions in which the food wow lee and vegetables, e.g., during storage, detection of volatile organic compounds that are produced as food is spoiling is very helpful in making sure the quality of the products is guaranteed throughout the consumer use period. Figure 1 demonstrates the various applications of nanobiosensors in the food sector.

3 Technological Difficulties in Nanobiosensor Implementation

Figure 2 illustrates the technological difficulties in nanobiosensor implementation.

3.1 Ambient Influences on Sensor Performance

Nanobiosensors, like all sensitive analytical devices, are exposed to various environmental factors that can significantly affect their performance. These environmental factors include temperature, humidity, pH, and the presence of interfering substances in the sample matrix. The performance of nanobiosensors is often highly temperature dependent. Changes in temperature can affect the binding affinity between the biological recognition element and the target analyte, leading to fluctuations in sensor performance [18]. For example, the reaction kinetics are quickened by a rise in the

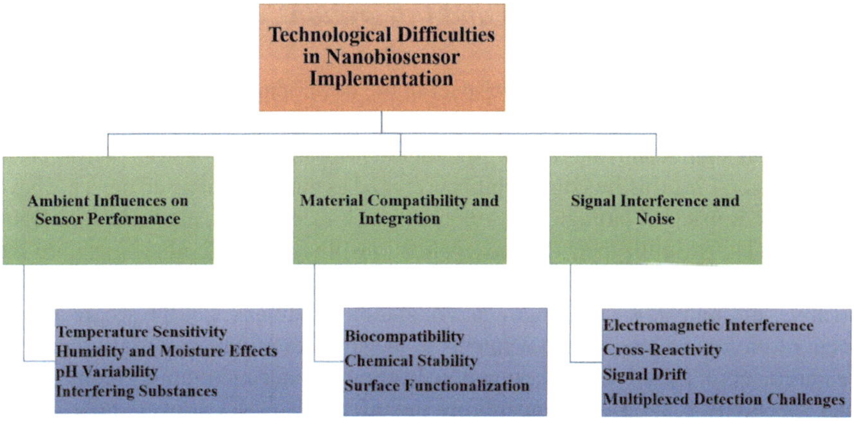

Fig. 2 Technological difficulties in nanobiosensor implementation

temperature, leading to false positive results and exaggerating the signal response vis a vis lower temperature, slowing down the reaction rate and reducing the sensor's sensitivity and response time. Particularly in field applications with less controlled environmental conditions, it is difficult to maintain temperature stability. The efficacy of the nanobiosensors, particularly those which are based on electrochemical transduction mechanisms, is influenced by humidity and moisture. High humidity can lead to condensation of water on the sensor surface, which in turn alters the electrical properties like conductivity or impedance of the nanobiosensors. Moisture affects the refractive index of the sensor environment in the case of optical sensors, which leads to faulty optical signals [19]. Thus, these two factors can deviate the sensor readings from the original, thus reducing the accuracy of detection. The sample's pH level has the greatest influence on the performance of nanobiosensors, even more so those that make use of enzymes or other pH-altering recognition elements [5, 20]. For example, enzymes have an ideal range of pH where maximum activity occurs. Any pH outside this range results in an inappropriate enzyme activity such that the sensor will have difficulties in detecting the target analyte. In the case of food safety applications where the samples will be acidic or alkaline depending on the food, it is very important to control the pH range, but this is frequently difficult. Some other substances present in the sample matrix can assist or hinder detection. For example, in complicated food samples, the sensor will be exposed to various organic solvents, salts, and even pollution that will not specifically sense on the sensor surface or change the output signal [1]. Even with a bit of interference, considerable errors are likely due to false alarms, poor fidelity, or even total loss. Hence, precise and dependable results become hard to obtain and, therefore, difficult to perform tasks successfully.

3.2 Material Compatibility and Integration

Material compatibility is a critical aspect in the development of nanobiosensors, as it affects both the biological detection elements and the nanomaterial-based transducers, thus significantly influencing the stability, durability, and overall performance of the sensor. Ensuring biocompatibility is crucial as the materials used must not have a negative impact on biological components such as enzymes, antibodies, or DNA probes [21]. To fulfill their desired applications, they must be able to endure the specific environmental conditions that often come across in food safety monitoring, such as varying pH, ionic strength, and oxidation conditions. For the secure attachment of biological recognition elements to the surfaces of nanomaterials, there is a requirement for effective surface functionalization without undermining their activity and specificity [22, 23]. Incompatibility in surface chemistry triggers poor immobilization and increased non-specific binding, giving rise to reduced sensitivity and producing inaccurate readings. Regarding the development of reliable and efficient nanobiosensors, one must consider factors such as careful selection and optimization of materials.

3.3 Signal Interference and Noise

In the practical execution of nanobiosensors in operational and realistic settings such as the food industry, signal interference poses one of the biggest challenges, as it can come from numerous sources, electronic devices, or power lines giving rise to noise in the sensor output, hindering the recognition of the actual signal generated by the target analyte [24]. Especially in those nanobiosensors which are based on electrical or electrochemical transduction mechanisms, thus affecting the accuracy and reliability of the detection method. This issue is problematic where multiple electronic devices are operating simultaneously, practically in industrial settings. Cross-reactivity can lead to false positive signals as when there is the availability of structurally similar compounds, the sensor binds not only to the target analyte but also to other non-target molecules [19], which in turn reduces the specificity and accuracy of the nanobiosensor. When the base signal changes, nanobiosensors may encounter signal drift. The gradual degradation of the detection elements, the accumulation of contaminants on the sensor surface, or changes in the environment surrounding the sensor [25] are some of the causative factors of signal drift. To handle this issue, frequent recalibration is required, which is time-consuming and impractical. Thus, drift complicates the interpretation of results. Nanobiosensors are often designed for the simultaneous detection of multiple analytes (multiplex detection). However, when detecting multiple targets in a single assay, signal interference can occur between different detection channels. This crosstalk can lead to overlapping signals, making accurate quantification of each target difficult [4]. Developing strategies to minimize or eliminate crosstalk is essential for effective multiplex detection.

4 Cost-Effectiveness and Market-Ready Nature of Nanobiosensor Technologies

In the present era of competition in the food industry, the widespread approval of any new technology remarkably depends upon the critical feature of its cost-effectiveness. The economic viability of nanobiosensors promotes their acceptance in the various sectors of the food industry even though it has advanced detection capabilities. This section of the chapter basically analyzes the economic considerations associated with nanobiosensors and compares them with conventional detection methods in the aspect of nanobiosensors being a cost-effective solution.

4.1 Production Costs of Nanobiosensors

The production costs of nanobiosensors are influenced by several factors, including the cost of raw materials, manufacturing processes, and the scalability of production techniques. Few specialized nanomaterials, such as gold nanoparticles, carbon nanotubes, or quantum dots, are generally utilized in nanobiosensors, but due to their complex synthesis processes and need for high purity, they come out to be expensive [26]. However, the synthesis techniques for these nanomaterials sustain improvement, and scale economy comes into play. The cost reduction factor can increase the affordability of nanobiosensors. Even though nanobiosensor fabrication can be costly and time-consuming as it involves complex techniques such as lithography, self-assembly, or chemical vapor deposition. However, due to the latest innovations in microfabrication and nanotechnology, efficient and cost-effective production of nanobiosensors has become possible. Currently, various other techniques, such as roll-to-roll printing and inkjet printing of nanomaterials, are being explored for large-scale production of nanobiosensors at reduced costs [27]. However, manufacturing can be costly for production rates to meet industrial demands. In comparison with conventional sensor technologies, the development of scalable manufacturing processes, such as batch processing and automated assembly lines, makes the production nanobiosensor superior to other techniques, thus dealing with the challenge of scalability in making nanobiosensors cost-effective [28].

4.2 Cost Comparison with Conventional Detection Methods

When evaluating the cost-effectiveness of nanobiosensors, it is important to compare them with conventional detection methods such as ELISA, PCR, or chromatography, which are commonly used in food safety and quality monitoring. The foundational expenses of developing and purchasing nanobiosensors may be higher, but reduced reagent consumption, faster analysis times, and lower labor costs are some of the savings in operational costs, which are long-term indeed. Nanobiosensors can be designed as portable, user-friendly devices that do not require extensive infrastructure or specialized training [29]. This reduces major initial investment in laboratory infrastructure, equipment, and skilled personnel, which is mandatory in traditional detection methods. The employment of nanobiosensors permits downscaling of sensors, which reduces the amount of reagent and sample volume which is required for evaluation. Also, the need for repeated testing is reduced significantly by the real-time monitoring feature enabled due to the fast response mechanism of nanobiosensors. These characteristics allow lower operational costs than traditional methods. The calibration costs of conventional methods are eventually higher as they require regular maintenance, calibration, and reagent replenishment. Contrary to all this, nanobiosensors require minimal maintenance and can be used multiple times, thus lower total cost of ownership. Further advancements in sensor design also focus on minimizing the need for maintenance and manual intervention by developing self-calibrating and self-cleaning nanobiosensors [30].

4.3 Market Readiness of Nanobiosensor Technologies

The marketability depends on their consistent performance under real-world conditions, which entirely focuses on the ability of nanobiosensors to deal with issues like stability, reproducibility, and scalability, even though it has outstanding performance in the laboratory. Figure 3 represents the market readiness of nanobiosensor technologies. The commercial readiness of nanobiosensors relies on various factors like technological maturity, regulatory approval, industry acceptance, and consumer perception. For the product to be market fit, regulatory approval is one of the most necessary steps to be undertaken; in the food industry, agencies such as the FDA and EFSA execute strict testing and validation to confirm in order to testify that nanobiosensors meet safety and performance standards. For a product that needs to be commercialized it must adhere to changing policies. Another crucial step is industry acceptance of the product, which relies on factors such as ease of use, reliability, cost-effectiveness, and smooth incorporation into the existing setup. Nanobiosensors that offer clear advantages over conventional methods, such as faster detection times, higher sensitivity, and lower costs, are more likely to gain acceptance in the food sector [31]. In food, consumer perception plays a major role in a product's acceptance, and some issues about the safety and ethical implications of nanotechnology in food applications can occur. We can gain consumer acceptance of nanotechnology-based food safety solutions by building trust, which can be achieved by educating consumers about the benefits of nanobiosensors and eradicating the misconceptions that might follow. There are different market strategies to enhance market readiness, as shown in Fig. 4.

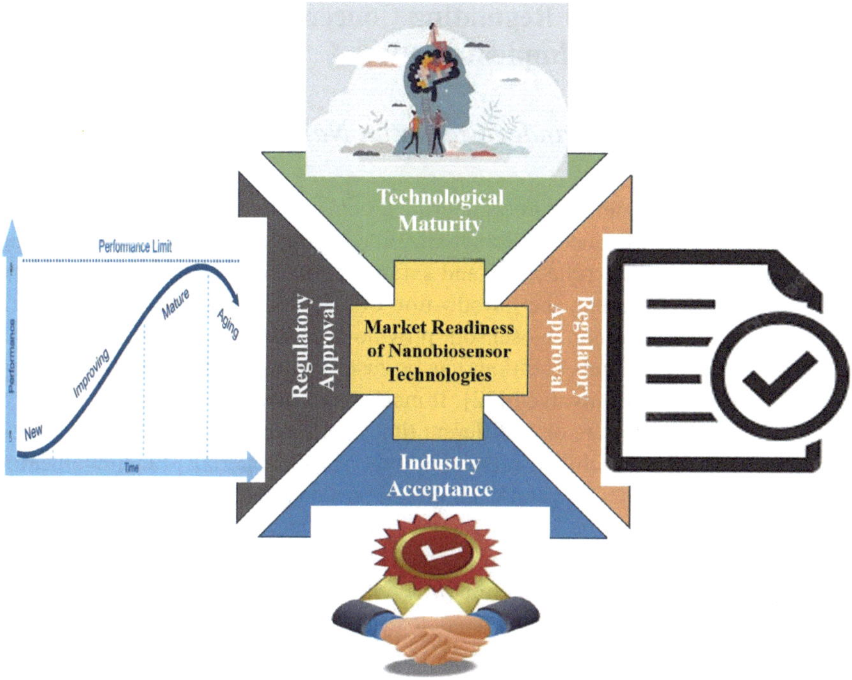

Fig. 3 Market readiness of nanobiosensor technologies

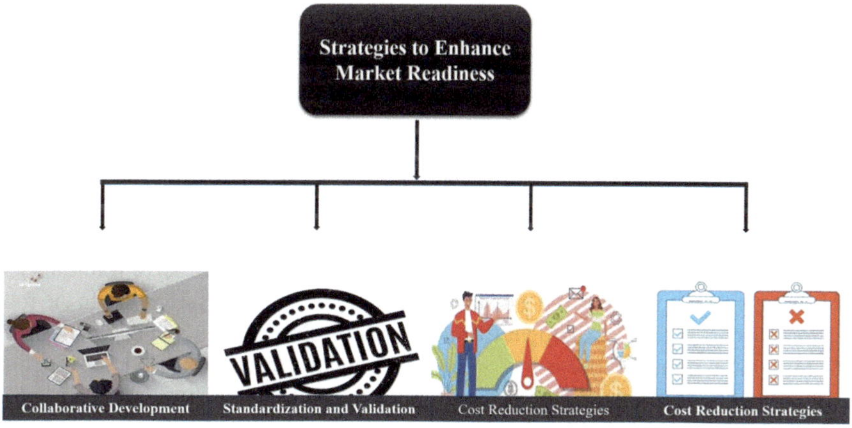

Fig. 4 Strategies to enhance readiness of nanobiosensor technologies

5 Standardizing and Regulating Concerns in Nanobiosensor Implementation

5.1 Importance of Standardization in Nanobiosensors

For the adoption of nanobiosensors in critical fields such as food safety and quality monitoring, the significance of Standardization becomes very important to ensure consistent performance, reliability, and safety. As the complexity and diversity in nanotechnology is present, the introduction of standardized protocols and benchmarks is necessary. Nanonobiosensors provide accurate and repeatable results in diverse environments, and conditions can be ensured only when standardized manufacturing processes are involved [32]. It makes sure that nanobiosensors produced by different manufacturers or at different times perform consistently as differences at checkpoints, like variability in the synthesis of nanomaterials, functionalization, and assembly of sensors, can cause a major impact on sensor performance. For each nanobiosensors to meet the same high quality and performance standards, standardized production protocols must be followed to reduce the variations that can arise. Furthermore, standardized production protocols, laboratories, and facilities are used to authenticate the performance of nanobiosensors under similar conditions to make sure that acceptance of nanobiosensors is trustworthy and comparable to conventional methods, especially in food safety, where any kind of contamination affects public health. By formulating standardized benchmarks for nanobiosensor performance, such as sensitivity, specificity, detection limits, and response time, along with standardized calibration procedures, the detection of target analytes at required conditions is also aided invariably [33].

5.2 Regulatory Concerns and the Need for Clear Guidelines

In the present era of advancement, the field of nanotechnology is also evolving day by day, and therefore, the need for the establishment of a comprehensive framework is also arising to keep a check that advanced technologies are also safe to use in the food sector and comply with existing laws and standards. Proper rules and instructions for the evaluation of nanobiosensors focusing on potential threats like toxicity, biocompatibility, and environmental impact must be developed by regulating bodies [34]. For approval, nanobiosensors must undergo strict assessment and certification processes exhibiting their efficiency, reliable performance posing no threat to consumers and safety. It is important to receive certification from recognized authorities such as the FDA or EFSA to guarantee stakeholders and the public consumers that nanobiosensors adhere to guidelines and meet quality standards considering the global nature of the food sector another crucial factor is the standardization of rules and instructions in different countries and regions. To eliminate any kind of discrepancies in the process of commercialization of nanobiosensors international collaboration can be a possible solution to achieve worldwide acceptance of nanobiosensors [27, 35].

5.3 Challenges in Developing Regulatory Frameworks for Nanobiosensors

Ongoing research and advancements in the field of nanotechnology also face difficulties in the development of comprehensive regulatory frameworks for nanobiosensors. There is a lack of adequate existing regulatory frameworks to tackle the unique challenges that are being presented. Based on the new research and experiences, guidelines and standards are sometimes created from scratch. Due to all this collaboration between industry experts, regulators, and scientists, it must be aligned, which is time-consuming [36]. Regulators must keep up with the constant need for alignment between the advancement in nanotechnology in terms of new materials, synthesis processes, applications, safety, and flexibility to allow progress and improvement. Regulators should have a holistic understanding and expertise in different fields comprising nanotechnology, biology, chemistry, materials science, and food safety. This requires collaboration with research institutions and industry experts, thus increasing the capability of regulators to analyze complex concepts [37].

5.4 The Role of Industry and Stakeholders in Standardization and Regulation

Industry experts, including manufacturers, stakeholders, and food producers, play a vital role in the standardization and regulation of nanobiosensors. By ensuring that frameworks that are established meet the industry demands and are practical and feasible in real-world conditions, Public–private partnerships play a crucial role by integrating the expertise and resources of both the public and private sectors [38]. Thus, the collective effort facilitates the standardization process, also validating the comprehensive nature of frameworks. Industry stakeholders should actively engage with regulators in the development of new standards and regulations. Their input and feedback are very important in certifying that regulations are practical and reflect the realities of nanobiosensor manufacturing and application. The involvement also helps to build trust and transparency between regulators and the industry.

6 Conclusion

In summary, we can conclude that nanobiosensors offer significant advantages over traditional methods like higher sensitivity, specificity, and fast detection. In addition to this, they are capable of miniaturization and adapted for various applications, and their portable nature makes them suitable for food safety and quality monitoring. Despite possessing all these properties, their successful incorporation into the food industry requires conquering various obstacles, such as technological difficulties related to environmental influences, material compatibility, and signal interference. Innovation in sensor design, material science, and signal processing is key to addressing these issues; furthermore, in order to achieve widespread acceptance, challenges related to production costs, scalability, and market readiness need to be taken care of technological advances, along with strategical efforts undertaken, can pave the way for nanobiosensors to become a standard in food safety management. By building partnerships between regulators, industry, and researchers, we can achieve the aim of higher-quality food for consumers with the help of nanobiosensors, which have the potential to revolutionize food safety.

References

1. Sharma R, Ragavan KV, Thakur MS, Raghavarao KSMS (2015) Recent advances in nanoparticle based aptasensors for food contaminants. Biosens Bioelectron 74:612–627. https://doi.org/10.1016/j.bios.2015.07.017
2. Guruprasath N, Sankarganesh P, Adeyeye SAO, Babu AS, Parthasarathy V (2024) Review on emerging applications of nanobiosensor in food safety. J Food Sci 89(7):3950–3972. https://doi.org/10.1111/1750-3841.17149
3. Naresh V, Lee N (2021) A review on biosensors and recent development of nanostructured materials-enabled biosensors. Sensors 21(4):1–35. https://doi.org/10.3390/s21041109
4. Huang X, Zhu Y, Kianfar E (2021) Nano biosensors: properties, applications and electrochemical techniques. J Mater Res Technol 12:1649–1672. https://doi.org/10.1016/j.jmrt.2021.03.048
5. Sabu C, Henna TK, Raphey VR, Nivitha KP, Pramod K (2019) Advanced biosensors for glucose and insulin. Biosens Bioelectron 141:111201. https://doi.org/10.1016/j.bios.2019.03.034
6. Estrela P, Hammond JL, Formisano N, Estrela P, Carrara S, Tkac J (2016) Electrochemical biosensors and nanobiosensors. Essays Biochem 60(1):69–80. https://doi.org/10.1042/EBC20150008
7. Sharifi M et al (2019) Cancer diagnosis using nanomaterials based electrochemical nanobiosensors. Biosens Bioelectron 126:773–784. https://doi.org/10.1016/j.bios.2018.11.026
8. Kneipp J, Kneipp H, Wittig B, Kneipp K (2010) Novel optical nanosensors for probing and imaging live cells. Nanomed Nanotechnol Biol Med 6(2):214–226. https://doi.org/10.1016/j.nano.2009.07.009
9. Cullum BM, Vo-Dinh T (2000) The development of optical nanosensors for biological measurements. Trends Biotechnol 18(9):388–393. https://doi.org/10.1016/S0167-7799(00)01477-3
10. Seo SE et al (2019) Smartphone with optical, physical, and electrochemical nanobiosensors. J Ind Eng Chem 77:1–11. https://doi.org/10.1016/j.jiec.2019.04.037
11. Banigo AT, Azeez TO, Ejeta KO, Lateef A, Ajuogu E (2020) Nanobiosensors: applications in biomedical technology. IOP Conf Ser Mater Sci Eng 805(1):1–28. https://doi.org/10.1088/1757-899X/805/1/012028

12. Swierczewska M, Liu G, Lee S, Chen X (2012) High-sensitivity nanosensors for biomarker detection. Chem Soc Rev 41(7):2641–2655. https://doi.org/10.1039/C1CS15238F
13. Altug H, Oh S-H, Maier SA, Homola J (2022) Advances and applications of nanophotonic biosensors. Nat Nanotechnol 17(1):5–16. https://doi.org/10.1038/s41565-021-01045-5
14. Chelliah R et al (2021) Development of nanosensors based intelligent packaging systems: food quality and medicine. Nanomaterials 11(6):515. https://doi.org/10.3390/nano11061515
15. Eleftheriadou M, Pyrgiotakis G, Demokritou P (2017) Nanotechnology to the rescue: using nano-enabled approaches in microbiological food safety and quality. Curr Opin Biotechnol 44:87–93. https://doi.org/10.1016/j.copbio.2016.11.012
16. Wang Y, Duncan TV (2017) Nanoscale sensors for assuring the safety of food products. Curr Opin Biotechnol 44:74–86. https://doi.org/10.1016/j.copbio.2016.10.005
17. Thakur M, Wang B, Verma ML (2022) Development and applications of nanobiosensors for sustainable agricultural and food industries: recent developments, challenges and perspectives. Environ Technol Innov 26:102371. https://doi.org/10.1016/j.eti.2022.102371
18. Wang C, Li X, Zhang F (2016) Bioapplications and biotechnologies of upconversion nanoparticle-based nanosensors. Analyst 141(12):3601–3620. https://doi.org/10.1039/C6AN00150E
19. Hassan MM, Zareef M, Xu Y, Li H, Chen Q (2021) SERS based sensor for mycotoxins detection: challenges and improvements. Food Chem 344:128652. https://doi.org/10.1016/j.foodchem.2020.128652
20. Pramod-Kumar EK, Um W, Park JH (2020) Recent developments in pathological pH-responsive polymeric nanobiosensors for cancer theranostics. Front Bioeng Biotechnol 8:586
21. Rowland CE, Brown CW, Delehanty JB, Medintz IL (2016) Nanomaterial-based sensors for the detection of biological threat agents. Mater Today 19(8):464–477. https://doi.org/10.1016/j.mattod.2016.02.018
22. Dhiman NK, Sidhu N, Agnihotri S, Mukherjee A, Reddy MS (2022) 18: role of nanomaterials in protecting building materials from degradation and deterioration. In: Iqbal HMN, Bilal M, Nguyen TA, Yasin G (eds) Biodegradation and biodeterioration at the nanoscale, in micro and nano technologies. Elsevier, pp 405–475. https://doi.org/10.1016/B978-0-12-823970-4.00024-5
23. Holzinger M, Le Goff A, Cosnier S (2014) Nanomaterials for biosensing applications: a review. Front Chem 2:63. https://doi.org/10.3389/fchem.2014.00063
24. Akyildiz IF, Jornet JM (2010) Electromagnetic wireless nanosensor networks. Nano Commun Netw 1(1):3–19. https://doi.org/10.1016/j.nancom.2010.04.001
25. Chen D, Xu T, Dou Y, Li T (2024) A calibration strategy for silicon nanowire field-effect transistor biosensors and its application in ultra-sensitive, label-free biosensing. ACS Nano 18(33):21873–21885. https://doi.org/10.1021/acsnano.4c01937
26. Chamorro-Garcia A, Merkoçi A (2016) Nanobiosensors in diagnostics. Nanobiomedicine 3:1849543516663574. https://doi.org/10.1177/1849543516663574
27. Kumari M, Gupta V, Kumar N, Arun RK (2024) Microfluidics-based nanobiosensors for healthcare monitoring. Mol Biotechnol 66(3):378–401. https://doi.org/10.1007/s12033-023-00760-9
28. Ulucan-Karnak F, Kuru Cİ, Akgöl S (2024) Commercial roadmap of nanobiosensor development. Front Nanotechnol 6:1–7. https://doi.org/10.3389/fnano.2024.1348308
29. Ghosh T, Raj GVSB, Dash KK (2022) A comprehensive review on nanotechnology based sensors for monitoring quality and shelf life of food products. Meas Food 7:100049. https://doi.org/10.1016/j.meafoo.2022.100049
30. Virk V, Deepak H, Taneja K, Srivastava R, Giri S (2024) Amelioration in nanobiosensors for the control of plant diseases: current status and future challenges. Front Nanotechnol 6:1–20. https://doi.org/10.3389/fnano.2024.1310165
31. Fröhlich P et al (2022) Effects of reliability indicators on usage, acceptance and preference of predictive process management decision support systems. Qual User Exp 7(1):1–23. https://doi.org/10.1007/s41233-022-00053-0

32. D'Souza AA, Kumari D, Banerjee R (2017) Nanocomposite biosensors for point-of-care—evaluation of food quality and safety. Nanobiosensors 63:629–676. https://doi.org/10.1016/B978-0-12-804301-1.00015-1
33. Kumar V, Guleria P, Mehta SK (2017) Nanosensors for food quality and safety assessment. Environ Chem Lett 15(2):165–177. https://doi.org/10.1007/s10311-017-0616-4
34. Mohammad ZH, Ahmad F, Ibrahim SA, Zaidi S (2022) Application of nanotechnology in different aspects of the food industry. Discov Food 2(1):13. https://doi.org/10.1007/s44187-022-00013-9
35. Gajdosova V, Lorencova L, Kasak P, Tkac J (2020) Electrochemical nanobiosensors for detection of breast cancer biomarkers. Sensors 20(14):4022. https://doi.org/10.3390/s20144022
36. Kumari R et al (2023) Regulation and safety measures for nanotechnology-based agri-products. Front Genome Ed 5:1–12. https://doi.org/10.3389/fgeed.2023.1200987
37. Baig N, Kammakakam I, Falath W, Kammakakam I (2021) Nanomaterials: a review of synthesis methods, properties, recent progress, and challenges. Mater Adv 2(6):1821–1871. https://doi.org/10.1039/d0ma00807a
38. Kulkarni MB, Ayachit NH, Aminabhavi TM (2022) Recent advancements in nanobiosensors: current trends, challenges, applications, and future scope. Biosensors 12(10):892. https://doi.org/10.3390/bios12100892